Series on Chemistry, Energy and the Environment

Bioinspired Chemistry

From Enzymes to Synthetic Models

Series on Chemistry, Energy and the Environment

ISSN: 2529-7716

The aim of this series on "Chemistry, Energy and the Environment" is to bring together authoritative contributions where the multidisciplinary character of Chemistry and its relationship to energy and the environment can be illustrated. In each volume, the latest advances in bio-, energy- and environmentally-related fields are described in a way that unambiguously shows the major contributions from chemists at the top of their field who describe new and improved technologies for the future.

Published

Vol. 5 *Bioinspired Chemistry: From Enzymes to Synthetic Models*
edited by Marius Réglier

Vol. 4 *Prospects for Li-ion Batteries and Emerging Energy Electrochemical Systems*
edited by Laure Monconduit and Laurence Croguennec

Vol. 3 *Advanced Green Chemistry*
Part 1: Greener Organic Reactions and Processes
edited by István T. Horváth and Max Malacria

Vol. 2 *Elaboration and Applications of Metal-Organic Frameworks*
edited by Shengqian Ma and Jason A. Perman

Vol. 1 *Perovskite Solar Cells: Principle, Materials and Devices*
edited by Eric Wei-Guang Diau and Peter Chao-Yu Chen

Volume 5

Series on Chemistry, Energy and the Environment

Bioinspired Chemistry

From Enzymes to Synthetic Models

Edited by

Marius Réglier

Centre National de la Recherche Scientifique and Aix Marseille Université, France

Series Editors

Karl M. Kadish

University of Houston, USA

Roger Guilard

Université de Bourgogne, France

World Scientific

NEW JERSEY • LONDON • SINGAPORE • BEIJING • SHANGHAI • HONG KONG • TAIPEI • CHENNAI • TOKYO

Published by

World Scientific Publishing Co. Pte. Ltd.

5 Toh Tuck Link, Singapore 596224

USA office: 27 Warren Street, Suite 401-402, Hackensack, NJ 07601

UK office: 57 Shelton Street, Covent Garden, London WC2H 9HE

Library of Congress Cataloging-in-Publication Data

Names: Réglier, Marius, 1952– editor.

Title: Bioinspired chemistry: from enzymes to synthetic models / edited by Marius Réglier (Centre National de la Recherche Scientifique and Aix Marseille Université, France).

Description: New Jersey : World Scientific, 2019. | Series: Series on chemistry, energy and the environment ; volume 5 | Includes bibliographical references.

Identifiers: LCCN 2018057444 | ISBN 9789813274433 (hardcover : alk. paper)

Subjects: LCSH: Green chemistry. | Bioorganic chemistry.

Classification: LCC TP155.2.E58 B567 2019 | DDC 660.6/3--dc23

LC record available at https://lccn.loc.gov/2018057444

British Library Cataloguing-in-Publication Data

A catalogue record for this book is available from the British Library.

For any available supplementary material, please visit
https://www.worldscientific.com/worldscibooks/10.1142/11103#t=suppl

Typeset by Stallion Press
Email: enquiries@stallionpress.com

Printed in Singapore

Dedication to Professor Silviu Balaban

This volume is dedicated to the memory of its initiator, Teodor Silviu Balaban, who passed away on November 16, 2016. We would like to express our sincere appreciation to Silviu for his pioneering contribution to the bioinspired solar energy conversion.

M. Réglier

To Prof. Silviu Balaban, who, with his heroic efforts in the lab, was one of the pioneers of responsive matrix research.

H. J. M de Groot

Dedication from the Series Editors

As Series Editors for the volumes entitled ***Chemistry of Energy and the Environment*** we would like to add our dedication of this work to Teodor Silviu Balaban who was one of the original authors in this series and had, at the time of his death, been working on two volumes related to the current work, one provisionally entitled *"Bioinspired Chemistry"* and the other *"Molecular Approaches to Photocatalytic Reactions."*

Silviu Balaban received his Ph.D. from the Polytechnic University, Bucharest in 1983 and his Habilitation from the University Louis Pasteur, Strasbourg, France, working under the supervision of Jean-Marie Lehn. He held major research positions in Strasbourg, Karslruhe, Germany and the University of Aix-Marseille, France where he held the position of Full Professor.

Dr. Balaban was one of the founding members of the Society of Porphyrins and Phthalocyanines in 2000 and a regular attendee and contributor to the society's biennial International Conference of Porphyrins and Phthalocyanines (ICPP). He will be remembered by his friends and colleagues, not only for his outstanding research contributions over the last three decades but also for his organizational ability, service to the scientific community of his peers and warm personality.

Karl M. Kadish, University of Houston
Roger Guilard, Université de Bourgogne Franche-Comté

Preface

The observation of Nature is an inexhaustible source of inspiration that can promote innovation in chemistry. Enzymes are smart molecules capable of catalyzing a reaction with high turnover rates and high stereoselectivity. These highly desirable properties have been acquired over billions of years through natural evolution of protein structure. In this context, mimicking the enzymes functioning represents an immense potential in the development of ecofriendly catalytic processes allowing for the production of many high value chemicals and clean fuels. The biomimetic and/or bioinspired approach can revolutionize our scientific paradigms as it is not based on what we can take from nature, but on what we can learn from it.

Solar energy conversion is a challenge that bioinspired chemistry is able to take-up. Nature has evolved organisms that are capable of converting sunlight into biochemical energy. The blueprint offered by natural photosynthesis, once understood in detail, has not to be reproduced as an exact copy in artificial devices. In a true biomimetic approach, we try to dissect the events of light-harvesting, charge separation, electron transport, and fuel production in order to mimic these with completely synthetic molecules. The basic principles of these elementary steps allow us to optimize their performance in order to produce solar fuels, which implies clean energy production combined with its storage (Chapters I–IV).

Due to the unceasing increase of energy need and the subsequent CO_2-induced global warming, the development of ecofriendly processes is another major scientific challenge of the 21st century. Hydrogen is a clean-burning fuel with high energy density. For its important ecologically-friendly potential, hydrogen utilization as a fuel has become an industrial and economical target for the future. However, the current processes for generating large quantities of hydrogen are far from the respect of the environment. Nowadays, high temperature and pressure are used to combust natural gas over rare-earth or precious-metal catalysts. Nature developed very efficient enzymatic systems, named hydrogenases that have a binuclear active site composed by a non-precious metal such as iron and nickel. Inspiration from the functioning of these enzymes (Chapter V) can allow for the development of biomimetic/bioinspired catalysts and definitively anchor our society to the hydrogen economical process.

Another critical problem is to reduce the CO_2-induced global warming by transformation of CO_2 into more valuable compounds. In this context, the chemical reduction of CO_2 is nowadays one of the most relevant and valuable reactions. However, CO_2 is a very stable molecule, chemically difficult to activate. Nonetheless, enzymes present in microorganisms are capable of effectively reducing CO_2 into formate. These enzymes, *i.e* formate dehydrogenases, have a Mo active site where the catalytic reaction takes place. Inspired by these enzymes, biomimetic/bioinspired Mo complexes (Chapter VI) have been developed in order to study and to decipher the molecular factors responsible for this remarkable reactivity and therefore for developing new, robust and efficient catalysts for the reduction of CO_2.

Dioxygen activation at metal centers (*e.g.* Fe, Cu, *etc.*) in enzymatic systems is a widespread strategy to catalyze oxidation reactions that are essential to living organisms. Bioinorganic chemistry, thanks to the development of small model complexes, has emerged as an important tool to get better insights on the structure and mechanism of metalloenzymes. In addition to the fundamental knowledge, model complexes achieving efficient dioxygen activation can lead to the development of efficient oxidation catalysts. As iron chemistry has been recently a subject of several reviews, we will focus our attention to the copper-containing monoxygenases either with dinuclear or mononuclear active sites and their different strategies for dioxygen activation. A comprehensive description of metalloenzymes being relevant to develop and efficient biomimetic/bioinspired models, the two last chapters (Chapters VII and VIII) devoted to copper-containing enzymes give an important place both to the enzymatic and biomimetic/bioinspired systems.

This volume aims to provide the interested readers with an introduction to enzyme mimics and a selected summary of the latest and advanced standings of their applications. The topic selection seeks to inspire researchers to perfect/improve the design and synthesis of enzyme mimics and to tailor their functionality for a much wider range of applications.

Marius Réglier, CNRS and Aix Marseille Université

Contents

1 *In vitro* and *In vivo* Synthesis of Bacteriochlorophyll Absorbing Near-Infrared Light

Yusuke Tsukatani,*,‡ and Hitoshi Tamiaki†

*Japan Agency for Marine-Earth Science and Technology (JAMSTEC), Yokosuka, Kanagawa 237-0061, Japan
†Graduate School of Life Sciences, Ritsumeikan University, Kusatsu, Shiga 525-8577, Japan
‡tsukatani@jamstec.go.jp

Table of Contents

List of Abbreviations

ALA	5-aminolevulinic acid
a-COR	*a*-type COR
b-COR	*b*-type COR
BChl	bacteriochlorophyll
Blc.	*Blastochloris*
Chl	chlorophyll
Chl.	*Chlorobium*
COR	chlorophyllide *a* oxidoreductase
DPOR	dark-operative POR
Hbt.	*Heliobacterium*
LPOR	light-dependent POR
POR	protochlorophyllide *a* oxidoreductase
Rba.	*Rhodobacter*

I. Introduction

Chlorophyll (Chl) and bacteriochlorophyll (BChl) are essential for harvesting light energy, migrating excited energy, and transferring electrons in photosynthetic organisms.[1] Oxygenic phototrophs such as plants, algae, and cyanobacteria produce Chl molecules, whereas anoxygenic phototrophic bacteria mainly have BChls. Chls are grouped into five molecular species (Chls *a*, *b*, *c*, *d*, and *f*), and naturally occurring BChls are grouped into six molecular species (BChls *a*, *b*, *c*, *d*, *e*, and *g*) based on their side chains and backbone structures.[2,3] In fact, BChls *c*, *d*, and *e*

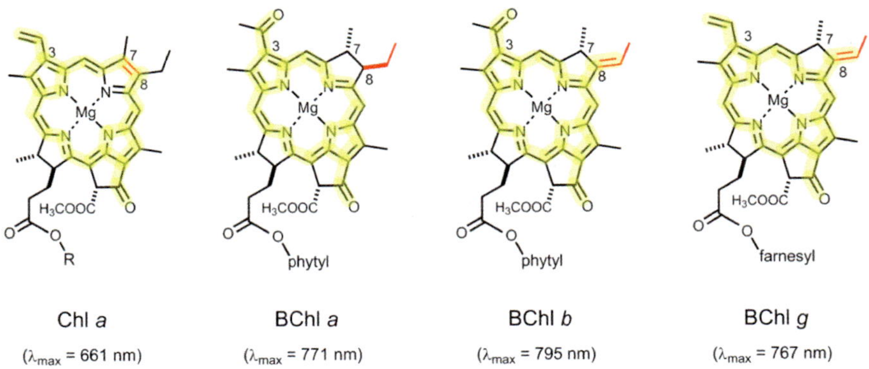

Figure 1. Molecular structures of Chl *a* and BChls *a, b,* and *g*. The presented λ_{max} peak positions of the Qy absorption bands are values obtained in diethyl ether. The π-conjugated systems are shaded by yellow. Conventional Chl *a* has phytyl ester (R = phytyl) and Chl *a* in green sulfur bacteria has 6,7-didehydrogenated phytyl ester (R = 2,6-phytadienyl).

are made of chlorin rings, dihydroporphyrins (not bacteriochloroins or tetrahydroporphyrins), but they have historically been called bacteriochlorophylls because they are found in bacteria. Hence, BChls *a*, *b*, and *g* are "true" BChls having bacteriochlorin rings as their tetrapyrrole backbone structures (Figure 1).

Photosynthetic bacteria in phyla *Chlorobi* and *Acidobacteria* produce BChl *c*, *d*, or *e* (depending on the species) as well as BChl *a* and Chl *a*.[4–7] For instance, in green sulfur bacteria (*Chlorobi*), BChl *c* harvests light energy in antenna systems, BChl *a* functions in light energy transfer and charge separation, and Chl *a* works as the primary electron acceptor in type-I photochemical reaction center complexes.[8,9] In addition to these three pigment species, *Chloracidobacterium thermophilum* (phylum *Acidobacteria*) was also revealed to produce Zn-BChl *a* bearing zinc instead of magnesium in the central position,[6] although it is still uncertain how and where this newly found pigment works. Filamentous anoxygenic phototrophs in the phylum *Chloroflexi* also possess BChl *c* and BChl *a*, but do not produce Chl *a* since they have type-II reaction center complexes.[10]

On the other hand, photosynthetic *Proteobacteria* or so-called "purple bacteria" produce only one pigment species, BChl *a* or *b*, depending on the bacterial species.[11] Two well-known model organisms in purple bacteria are *Rhodobacter* (*Rba.*) *capsulatus* and *Rba. sphaeroides*, both of which have BChl *a*.[12] A well-known bacterium containing BChl *b* is *Blastochloris* (*Blc.*) *viridis* (formerly called *Rhodopseudomonas viridis*) since the crystal structure of its photochemical reaction center was the subject of the Nobel Prize in Chemistry of 1988.[13] Phototrophic bacteria in the phylum *Firmicutes* are the only group that utilizes BChl *g* for light harvesting and light energy conversion.[14]

II. Characteristics of Bacteriochlorophylls *a, b,* and *g*

The absorption spectra of representative natural pigments are shown in Figure 2. Among these naturally occurring pigments, BChl *b* can absorb the longest wavelength light and is the sole pigment that has a Q_y absorption maximum at the near-infrared light region in the monomeric state. The peak positions of the Q_y absorption bands (λ_{max}) of BChls *a* and *b* locate at 771 nm and 795 nm, respectively: note that the near-infrared light region is

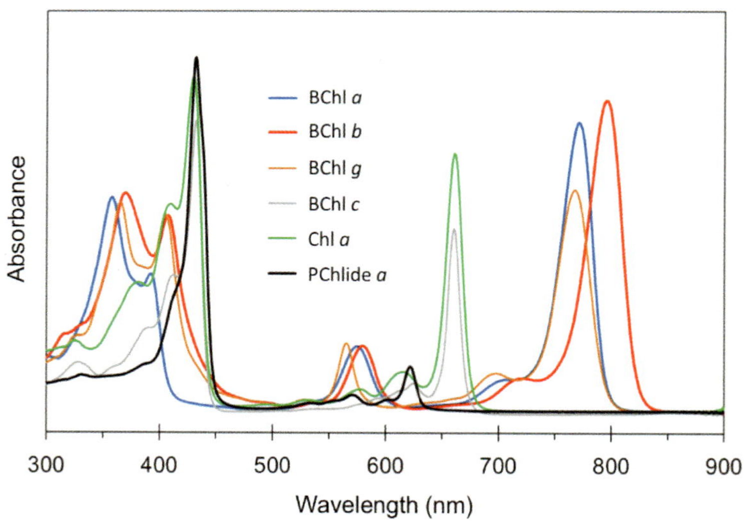

Figure 2. Absorption spectra of representative naturally occurring chlorophyllous pigments. BChl *a* (*blue line*), BChl *b* (*red*), BChl *g* (*orange*), BChl *c* (*gray*), Chl *a* (*green*), and protochlorophyllide *a* (PChlide *a*, *black*) were measured in diethyl ether (generously provided by Dr. Tadashi Mizoguchi of Ritsumeikan University).

typically considered above 780 nm. A characteristic of BChl *b* is an ethylidene group at the C8 position. Actually, the difference in chemical structures between BChl *a* and BChl *b* occurs only at the C8 position, i.e., an ethyl group and an ethylidene group, respectively (Figure 1). The C8-ethylidene group on BChl *b* extends the π-conjugated system on the tetrapyrrole ring and, therefore, BChl *b* absorbs longer wavelength light than BChl *a*. It should be noted that BChl *g* also has a C8-ethylidene group. However, the C3 position of BChl *g* is substituted with a vinyl group which is not converted into the acetyl group, in contrast to BChls *a* and *b* (see Figure 1). The C3-vinyl group on BChl *g* is less withdrawing than the C3-acetyl group on BChl *b*, and as a result, the Q_y absorption band of BChl *g* (767 nm in diethyl ether) is similar to that of BChl *a* (771 nm) (Figure 2).

III. BChl *a* Biosynthesis

A. Early Stages in Pigment Biosynthesis

Natural photosynthetic pigments are made from glutamate or glycine via 5-aminolevulinic acid (ALA).[15,16] Chl and BChl pigments share common biosynthetic pathways from ALA to chlorophyllide *a* (Figure 3). Eight ALA molecules are condensed into a tetrapyrrole compound, uroporphyrinogen III, which is also a common precursor for siroheme (Figure 3).[17,18] Uroporphyrinogen III is sequentially converted into protoporphyrin IX, which is based on a porphyrin backbone ring and a common precursor for heme.[18] Magnesium is inserted into protoporphyrin IX as the central atom by the three-subunit enzyme, Mg chelatase (named as ChlD/H/I or BchD/H/I),[19,20] and then divinyl protochlorophyllide *a* is formed through catalytic reactions by methyltransferase (ChlM or BchM) and oxidative cyclase (ChlE or BchE) (Figure 3).[19,21]

B. Divinyl Reductase

The C8-vinyl group of divinyl protochlorophyllide *a* is reduced by divinyl reductase to form protochlorophyllide *a* (Figure 3). There are two types of divinyl reductase: one is a plant-type divinyl reductase called BciA and another is a cyanobacterium-type divinyl reductase grouped as BciB.[22–24]

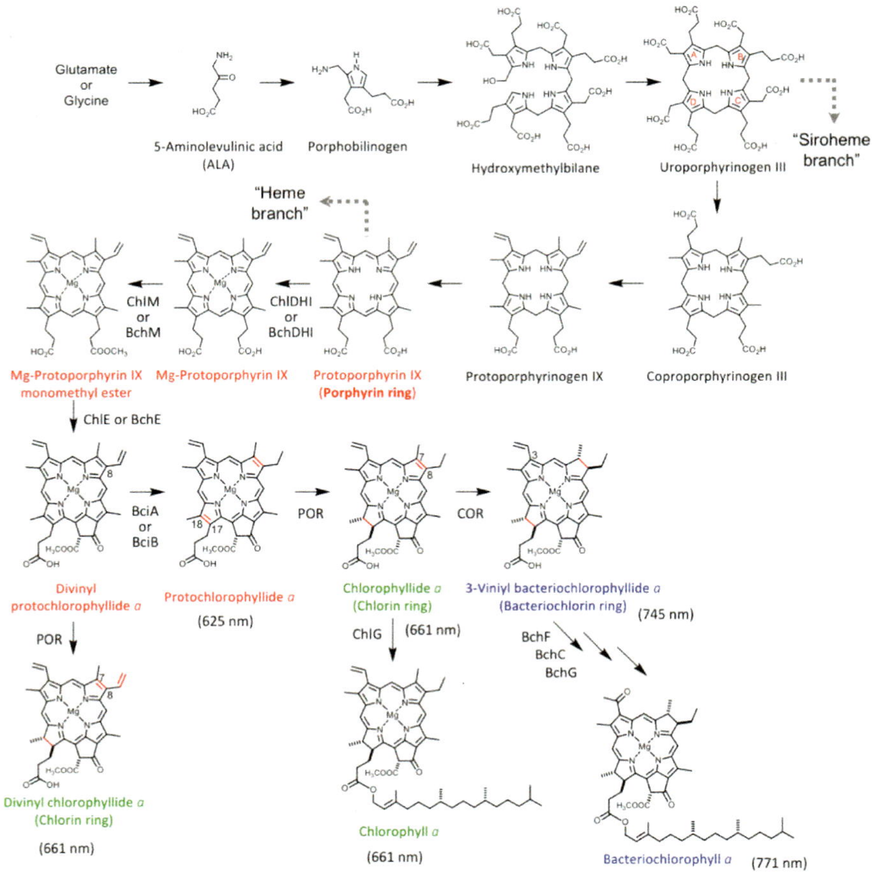

Figure 3. Schematic drawing of the biosynthetic pathways for natural chlorophyll pigments. Intermediates and final products of the pigment species presented in red, green, and blue have porphyrin, chlorin, bacteriochlorin ring backbone structures, respectively. The peak positions of the Qy absorption bands obtained in diethyl ether are presented in parentheses.

The primary structure and reaction mechanism of these two types of reductases are significantly different: BciA uses NADPH as electron sources, while BciB obtains electrons from ferredoxin for its reductase activities. Anoxygenic phototrophic bacteria have either BciA or BciB reductase, depending on the species.[23] Distribution of BciA/BciB among phototrophic bacteria is not related to phylogenetic grouping. For instance, even within green sulfur bacteria (phylum *Chlorobi*), *Chlorobaculum tepidum* has a gene for BciA, *Chlorobium* (*Chl.*) *limicola* has genes for BciB, and *Chl. clathratiforme* and *Prosthecochloris aestuarii* possess genes for both BciA and BciB.[25] In purple bacteria (phylum *Proteobacteria*), *Rba. sphaeroides* has BciA, but *Rhodopseudomonas palustris* possesses BciB.

Divinyl reductase is thought to recognize divinyl protochlorophyllide *a* as the substrate, but to some extent, the enzyme seems to work after divinyl protochlorophyllide *a* is converted into divinyl chlorophyllide *a*.[26–28] On the other hand, *in-vitro* enzymatic assays demonstrated that the C8-vinyl group of pigment intermediates before conversion to divinyl protochlorophyllide *a* (such as Mg-protoporphyrin IX) can be reduced by divinyl reductase.[29] The loose substrate specificity of enzymes may thus result in parallel routes to pigment biosynthetic pathways, which was also recently reported for 3-vinyl hydratase (BchF) and chlorophyllide *a* oxidoreductase (COR) in the late steps of pigment biosynthesis.[30]

C. Formation of a Chlorin Ring

The C17=C18 double bond of protochlorophyllide *a* is reduced by protochlorophyllide *a* oxidoreductase (POR), producing chlorophyllide *a*. The catalytic reaction is actually the conversion of a porphyrin ring (protochlorophyllide *a*) to a chlorin ring (chlorophyllide *a*). In oxygenic phototrophs such as plants and cyanobacteria, two distinct POR enzymes exist: light-dependent POR (LPOR) and dark-operative POR (DPOR), whereas anoxygenic phototrophic bacteria have only DPOR.[31] While LPOR is a single-subunit enzyme, DPOR consists of three subunits, BchL/N/B.[31–33] LPOR requires light for its catalytic reaction.[34] There are only two enzymes known for light-dependent reactions in chlorophyll pigment biosynthesis: LPOR and the recently found Chl *f* synthase.[35] The BchL subunit of DPOR forms a homodimer complex working as the electron-transfer component from ferredoxin to the BchN/B catalytic component.[31] The crystal structure of the BchN/B heterotetramer of DPOR was solved by two research groups independently, and is similar to that of the NifD/K catalytic component of nitrogenase.[36,37] The NifD/K complex of nitrogenase binds two kinds of catalytic clusters (P-cluster and Fe–Mo-cluster) to reduce the triple bond of the nitrogen molecules, but the BchN/B catalytic component instead binds the [4Fe–4S] cluster and the substrate, protochlorophyllide *a*.[36,37] The Fe–S cluster transfers electrons from the BchL component to protochlorophyllide *a*.[32] EPR measurements using wild-type and mutant DPOR and Chl *c* as the competitive inhibitor for protochlorophyllide *a* demonstrated that the reduction (electron addition) of the substrates precedes the addition of protons to the C17=C18 carbons.[38]

D. Formation of a Bacteriochlorin Ring

COR catalyzes the reduction of the C7=C8 double bond of chlorophyllide *a*, producing 3-vinyl bacteriochlorophyllide *a*: the conversion from a chlorin to a bacteriochlorin ring.[39] COR is also grouped in a nitrogenase-like enzyme family together with DPOR.[40,41] Although the crystal structure of COR has not yet been solved, the primary structures and subunit compositions of COR are relatively similar to those of DPOR and nitrogenase.[41] COR is encoded by *bchX*, *bchY*, and *bchZ* genes. BchX transfers electrons from ferredoxin to the $BchY_2BchZ_2$ catalytic component which binds the Fe–S clusters.[39,40] *In vitro* assays using both DPOR and COR showed that a porphyrin substrate was sequentially converted into 3-vinyl bacteriochlorophyllide *a* by the two enzymes.[28]

IV. BChl *b* Biosynthesis

A. Hypotheses on Possible Steps for C8-Ethylidene Formation

The biosynthetic pathways for BChl *b* and BChl *g* had long been unknown. Enzymes and/or the genes encoding enzymes to synthesize BChl *a* have all been identified so far.[15,16,42] The structural difference between BChl *a* and BChl *b* is found only to occur at the C8 position. BChl *a* has the C8-ethyl group, while BChl *b* has the C8-ethylidene group. BChl *g* also has the C8-ethylidene group, but the biosynthetic steps for BChl *g* may be much simpler than those for BChls *a* and *b*: the C3-vinyl group of BChl *g* is not transformed to an acetyl substituent as are BChls *a* and *b*. Excluding the formation of the C8-ethylidene group, the biosynthetic steps for BChl *g* can be covered by and are common to those for the known biosynthetic pathways of BChl *a*. Consequently, the enzyme for C8-ethylidene group formation may actually be the last missing step in BChls *b* and *g* biosynthesis. First, researchers have tried to find an enzyme for ethylidene formation by sequencing the genome of *Heliobacterium* (*Hbt.*) *modesticaldum* which produces BChl *g*.[43] They then found that *Hbt. modesticaldum* lacks the gene for divinyl reductase, *bciA* or *bciB*,[43] but excluding these, the organism has a complete gene set common to BChl *a* biosynthesis including the *bchX*/*Y*/*Z* genes for COR. That *Hbt. modesticaldum* lacks *bciA*/*bciB* but has DPOR and COR genes suggests that divinyl chlorophyllide is formed as the substrate for COR in the biosynthetic pathways. It was thus hypothesized that the C8-vinyl group on divinyl chlorophyllide *a* may be isomerized by an unknown isomerase (step A, Figure 4), forming bacteriochlorophyllide *g*, also called 3-vinyl bacteriochlorophyllide *b*.[43] Another pathway that could be hypothesized by the absence of *bciA*/*bciB* is the existence of an unknown third type of divinyl reductase. In this case, the C8-vinyl group and C7=C8 double bond is reduced by the third divinyl reductase and COR, respectively, then the C8–C8^1 single bond is oxidized by an unknown oxidase to synthesize the C8-ethylidene group (step B,

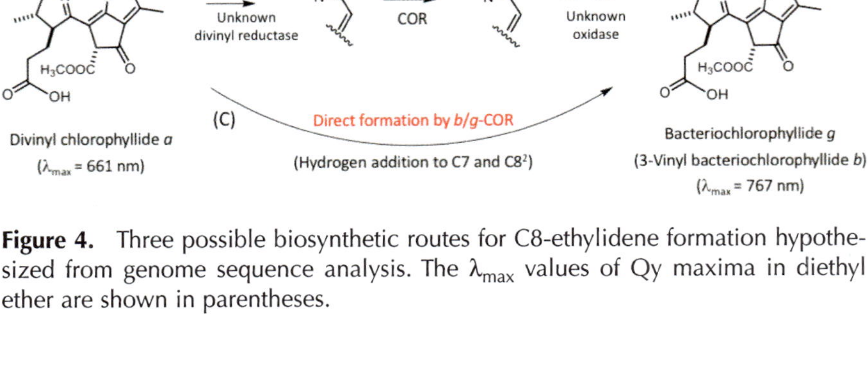

Figure 4. Three possible biosynthetic routes for C8-ethylidene formation hypothesized from genome sequence analysis. The λ_{max} values of Qy maxima in diethyl ether are shown in parentheses.

Figure 4). However, one oxidation step after two reduction steps for this hypothesis may be considered a waste of energy.

B. Conversion into Bacteriochlorophyllide *g* Possessing the C8-Ethylidene Group

To find other possibilities or strengthen the existing hypothesis on C8-ethylidene formation, we sequenced the genome of *Blc. viridis* that produces BChl *b*.[44] Interestingly, it was revealed that *Blc. viridis* also lacks the *bciA* nor *bciB* gene for divinyl reductase but still possesses the genes for COR. Upon these findings, we could conclude that if there is no third type divinyl reductase, the substrate for COR must be divinyl chlorophyllide *a* in BChl *b*- and *g*-producing photosynthetic bacteria, not C8-ethylated chlorophyllide *a* as in BChl *a*-producing bacteria. We then decided to test the enzymatic activity of COR of *Blc. viridis*. Initially, our intention was to test whether the COR could reduce the C7=C8 double bond, even when the C8 position was still vinylated. However, it came to our attention that the product of enzymatic assays using COR of BChl *b*-producing *Blc. viridis* and divinyl chlorophyllide *a* as the substrate possessed a C8-ethylidene group.[45] That is, COR of *Blc. viridis* reacts with divinyl chlorophyllide *a* and directly produces 3-vinyl bacteriochlorophyllide *b* (bacteriochlorophyllide *g*), which is the precursor to BChl *b*. It was also revealed that COR of *Blc. viridis* does not recognize chlorophyllide *a* which is the normal substrate for COR of BChl *a*-producing bacteria such as the *Rhodobacter* species.[45] In addition, *in vitro* assays using COR of BChl *g*-producing *H. modesticaldum* and divinyl chlorophyllide *a* provided the same product with the C8-ethylidene group, bacteriochlorophyllide *g*, the direct precursor to BChl *g*.[46] An unknown isomerase or oxidase was, therefore, unnecessary but COR of BChl *b*-producing *Blc. viridis* and BChl *g*-producing *Hbt. modesticaldum* (hereafter called *b*/*g*-type COR) was, in fact, ethylidene synthase (step C, Figure 4).

The two types of CORs show distinct substrate recognition and lead to the committed steps for BChl *a* and BChl *b*/*g* biosynthesis. Findings on new types of COR functions have meant that the biosynthetic pathways for BChls *b* and *g* are now completely identified. Also, since the enzymatic assays using *b*/*g*-type COR and chlorophyllide *a* did not exhibit any assay product,[42,45] an unknown divinyl reductase may probably not exist in *Blc. viridis* and *Hbt. modesticaldum*. If it did exist, 8-vinyl chlorophyllide *a* would be reduced to chlorophyllide *a* and no further biosynthetic process would proceed.

C. Possible Reaction Mechanism, Phylogeny, and Evolution of COR

To convert the C7=C8 double bond of the intermediates into a single bond, the COR of BChl *a*-producing bacteria (hereafter called *a*-type COR) initiates its reduction and adds two hydrogen atoms to the C7 and C8 positions; known in chemistry as the "1,2-addition" of hydrogen atoms (step A, Figure 4). However, the catalytic active center of *b*/*g*-type COR may surround the diene of the C7=C8 double bond and $C8^1$=$C8^2$ vinyl group, and hydrogen addition may occur at the C7 and $C8^2$ positions ("1,4-addition") to form the C8-ethylidene group (step C, Figure 4). On the one hand, the catalytic modes of *a*-type COR and *b*/*g*-type COR are the same: both carry out two-proton and two-electron additions. On the other hand, the two types of COR show distinct substrate recognition and different hydrogen addition fashion.

It is still unknown whether two hydrogen atoms for 1,2- and 1,4-additions are provided from the amino acid residues in the BchY/Z component of COR or other sources such as water molecules inside the component. For *a*-type COR to achieve the 1,2-addition of hydrogen atoms to the C7 and C8 positions, the distance between the two carbons of chlorophyllide *a* is about 1.37 Å.[45] In contrast, the distance between the C7 and $C8^2$ carbons of the cisoid C8–$C8^1$ rotamer in 8-vinyl chlorophyllide *a* is calculated to be 3.16 Å. If the two amino acid residues around the active center in *b*/*g*-type COR contribute to the 1,4-addition of hydrogen atoms, the relevant amino acid substitutions may extend the distance between the two reactive sites by about 1.8 Å for the *b*/*g*-type BchY/Z component. It will be interesting to solve and compare the crystal structures of *a*-type and *b*/*g*-type CORs, hopefully together with their substrate pigments.

Taking a closer look into the sequence alignments of *a*-type and *b*/*g*-type CORs did not provide a hint into the active center residues for the hydrogen addition.[45] Although it is known that the amino acid residue provides a hydrogen atom in DPOR from crystal structure analysis,[36] amino acid sequence alignments using DPOR and COR subunits showed that the amino acid of the hydrogen source in DPOR (Asp274 of BchB) is not conserved in BchY or BchZ. However, phylogenetic analysis based on *a*-type and *b*/*g*-type COR has revealed interesting results. The phylogenetic tree based on BchX, which is the electron-donating component, showed almost the same topology as that based on 16S rRNA, regardless of the host species having BChl *a* or *b*.[45] On the other hand, the phylogenetic trees based on BchY and BchZ, which consist of the catalytic component of COR, showed that BchY/Z of distantly related BChl *b*-producing bacteria, namely *Blc. viridis* in α-*Proteobacteria* and *Thioflavicoccus mobilis* in γ-*Proteobacteria* formed a coherent cluster.[45] This suggests that the BchY/Z component underwent selection pressure to achieve the 1,4-addition reaction mode in the course of evolution. Since only the genome sequences of two BChl *b*-producing species are available, the whole genome sequences of more species producing BChl *b* are necessary for further analysis.

Naturally occurring BChl *a* always has the C7-methyl group and C8-ethyl group on opposite sides of the tetrapyrrole backbone plane (Figure 1).[42] Therefore, proton addition by the catalytic reaction of *a*-type and *b*/*g*-type CORs may take place from opposite directions from the tetrapyrrole plane, although the proton-donating residues are yet unclear.

V. Another Function of *a*-Type COR

It was also tested whether *a*-type COR (*a*-COR) reacts with its abnormal substrate, divinyl chlorophyllide *a*, *in vitro*. Interestingly, the enzymatic assay demonstrated that *a*-type COR reduces both the C7=C8 double bond and C8-vinyl group of divinyl chlorophyllide *a*.[45] That is, *a*-COR has divinyl-reductase activity as well. The distances between the C7=C8 and $C8^1$=$C8^2$ carbons of divinyl chlorophyllide *a* are almost the same at approx. 1.37 Å. *a*-COR may probably reduce the C8-vinyl group of divinyl chlorophyllide *a* with the same reaction mode as the C7=C8 double bond reduction. We then investigated the C8-vinyl reduction activity (divinyl-reductase activity) of *a*-COR *in vivo*.[47] First, the *bciA* gene was disrupted in BChl *a*-producing *Rba. sphaeroides*, and the mutant was still found to produce normal BChl *a*. It is now easily understood that *a*-COR in the mutant works as an alternate divinyl reductase (note that *Rba. sphaeroides* does not have *bciB*). The *bchX/Y/Z* genes for *a*-COR were also disrupted in the *bciA* mutant of *Rba. sphaeroides*, accumulating the intermediate pigment, divinyl chlorophyllide *a*. Then, the COR genes of various phototrophic bacteria such as the BChl *a*-producing species in *Proteobacteria*, *Chlorobi*, and *Chloroflexi* were incorporated into the Δ*bciA*-Δ*bchX/Y/Z* mutant, and all the complementary strains were recovered to produce BChl *a*. This strongly indicates that the bifunctionality of *a*-COR is conserved in a wide range of photosynthetic bacteria.[47] Therefore, *a*-COR is a "hidden/forbidden third-type" divinyl reductase.

An interesting example is *Roseiflexus* (*Ros.*) *castenholzii* of filamentous anoxygenic phototrophs (phylum *Chloroflexi*) that naturally lacks both BciA and BciB but produces normal (C8-ethylated) BChl *a* (Y. Tsukatani and D. A. Bryant, unpublished results), so that a putative third-type divinyl reductase had been predicted in the bacterium. However, COR of *Ros. castenholzii* also compensated the double mutant of *Rba. sphaeroides* and was found to be bifunctional and exhibit divinyl-reductase activity.[47] This species may sustain the transition state of evolution for the pigment biosynthesis metabolism.

VI. Bioinspired Chemistry and Application Using COR

The reactivity of *b*/*g*-type COR provides us with clues to modify a well-known model organism of anoxygenic phototrophic bacteria, *Rba. sphaeroides*, to produce BChl *b*. As mentioned above, BChl *b* is the only naturally occurring pigment that efficiently absorbs near-infrared light in the monomer state. Materials absorbing long wavelength light at the near-infrared region are useful in the development of dye-sensitized solar cells and photodynamic therapies. As readers may realize, we first had to disrupt *bciA* in *Rba. sphaeroides* before incorporating *b*-type COR (*b*-COR), otherwise BciA reduces divinyl chlorophyllide *a* to chlorophyllide *a* which cannot be a substrate for *b*-COR. We then disrupted the genes for the intrinsic *a*-COR and incorporated the *b*-COR genes of *Blc. viridis* into the *Rba. sphaeroides* Δ*bciA*/Δ*a*-COR double mutant, and the resultant mutant successfully produced BChl *b*.[48]

The *b*-COR genes of *Blc. viridis* were also introduced into the *Rba. sphaeroides* mutant lacking only BciA, and the resultant mutant accumulated only BChl *a*. This means that the mutant sustains both intrinsic *a*-COR and extrinsic *b*-COR, and that *a*-COR overwhelmed *b*-COR activity in the cell. *b*-COR genes were then introduced into the *Rba. sphaeroides* mutant strain lacking BciA and BchJ, and the resulting strain produced both BChl *a* and BChl *b*. This is the first example showing that an anoxygenic photosynthetic bacterium can accumulate BChls *a* and *b* together in a cell. These results suggest that BchJ enhances the "1,2-addition" reductase activity of the intrinsic *a*-COR. In preliminary experiments, BchJ was found to stimulate the enzymatic activity of DPOR (K. Yamanashi and Y. Fujita, personal communications). BchJ may also interact with a wide range of pigment biosynthetic enzymes in addition to COR and DPOR (probably with divinyl reductase as well) to enhance their activities, leading to useful applications in the metabolic engineering of photosynthetic organisms to produce highly desired pigments.

VII. Conclusion and Perspectives

a-COR and *b*/*g*-COR have distinct substrate recognition and hydrogenation activities, and catalyze the committed steps for biosynthesis of BChl *a* and BChl *b*/*g*, respectively. Also, *a*-COR is revealed to have the divinyl-reductase activity in addition to its original function of the C7=C8 double bond reduction. DPOR is known to catalyze the C17=C18 double bond reduction. COR and DPOR are included in the nitrogenase-like enzyme family. Nitrogenase was originally found to reduce the multi-bonds of di-nitrogen, acetylene, etc, but also recently reported to catalyze the reductive catenation of carbon monoxide.[49] The flexibility of enzymatic activities is likely to be a common characteristic of nitrogenase and nitrogenase-like enzymes, and it would be interesting to solve enzymatic activities of their common ancestor. Bacteriochlorophyll *b* is the sole natural chlorophyllous pigment that shows its monomeric absorption maximum at the near-infrared light region and is useful for developing artificial light-driven systems such as dye-sensitized solar cells and photodynamic therapies. Based on the uncovered characteristics of *a*-type and *b*/*g*-

type CORs, the model phototrophic bacterium *Rba. sphaeroides* can be genetically engineered to produce a pigment able to absorb near-infrared light. Such research on the molecular basis and functional mechanisms of enzymes for pigment biosynthesis has great potential in aiding the development of systems for artificial photosynthesis and protein engineering.

VIII. Acknowledgments

We would like to express our sincere thanks to Dr. Tadashi Mizoguchi of Ritsumeikan University, Dr. Jiro Harada of Kurume University School of Medicine, and Drs. Yuichi Fujita and Haruki Yamamoto of Nagoya University for their support of our collaborative studies. Original research introduced in this chapter was partially supported by the Grants-in-Aid for Scientific Research on Innovative Areas "Artificial Photosynthesis (AnApple)" (No. 24107002, H.T.) and for Young Scientists (B) (No. 26840099, Y.T.) of the Japan Society for the Promotion of Science (JSPS). Y.T. additionally acknowledges support from the JST (Japan Science and Technology Agency) PRESTO (Precursory Research for Embryonic Science and Technology) research fellowship.

IX. References and Citations

1. Blankenship, R. E. *Molecular Mechanism of Photosynthesis*, **2014**, 2nd Ed. (Wiley Blackwell, West Sussex, UK).
2. Tamiaki, H.; Shibata, R.; Mizoguchi, T. *Photochem. Photobiol.* **2007**, *83*, 152.
3. Tamiaki, H.; Kunieda, M. Photochemistry of chlorophylls and their synthetic analogs In: Kadish, K. M.; Smith, K. M.; Guilard, R. (eds), *Handbook of Porphyrin Science*, **2011**, vol. 11, (World Scientific, Singapore) pp. 223–290.
4. Frigaard N. U.; Bryant, D. A. Chlorosomes: antenna organelles in photosynthetic green bacteria. In: Shively, J. M. (ed), *Complex Structures in Prokaryotes*, **2006**, vol. 2, (Springer, Berlin) pp. 79–114.
5. Costas, A. M. G.; Tsukatani, Y.; Rijpstra, W. I.; Schouten, S.; Welander, P. V.; Summons, R. E.; Bryant, D. A. *J. Bacteriol.* **2011**, *194*, 1158.
6. Tsukatani, Y.; Romberger, S. P.; Golbeck, J. H.; Bryant, D. A. *J. Biol. Chem.* **2012**, *287*, 5720.
7. Tsukatani, Y.; Harada, J.; Mizoguchi, T.; Tamiaki, H. *Photochem. Photobiol. Sci.* **2013**, *12*, 2195.

8. Hauska, G.; Schoedl, T.; Remigy, H.; Tsiotis, G. *Biochim. Biophys. Acta* **2001**, *1507*, 260.
9. Bryant, D. A.; Frigaard N. U. *Trends Microbiol.* **2006**, *14*, 488.
10. Hanada, S. *Microbes Environ.* **2003**, *18*, 51.
11. Stomp, M.; Huisman, J.; Stal, L. J.; Matthijs, H. C. P. *ISME J.* **2007**, *1*, 271.
12. Zubova, S. V.; Melzer, M.; Prokhorenko, I. R. *Biol. Bull.* **2005**, *32*, 168.
13. Deisenhofer, J.; Epp, O.; Miki, K.; Huber, R.; Michel, H. *Nature* **1985**, *318*, 618.
14. Brockmann Jr, H.; Lipinski, A. *Arch. Microbiol.* **1983**, *136*, 17.
15. Chew, A. G. M.; Bryant, D. A. *Annu. Rev. Microbiol.* **2007**, *61*, 113.
16. Masuda, T.; Fujita, Y. *Photochem. Photobiol. Sci.* **2008**, *7*, 1131.
17. Tripathy, B. C.; Sherameti, I.; Oelmüller, R. *Plant Signal Behav.* **2010**, *5*, 14.
18. Beale, S. I. *Trends Plant Sci.* **2005**, *10*, 309.
19. Gibson, L. C. D.; Willows, R. D.; Kannangara, C. G.; Vonwettstein, D.; Hunter, C. N. *Proc. Natl. Acad. Sci. USA* **1995**, *92*, 1941.
20. Walker, C. J.; Willows, R. D. *Biochem. J.* **1997**, *327*, 321.
21. Willows, R. D. *Nat. Prod. Rep.* **2003**, *20*, 327.
22. Chew, A. G. M.; Bryant, D. A. *J. Biol. Chem.* **2007**, *282*, 2967.
23. Ito, H.; Yokono, M.; Tanaka, R.; Tanaka, A. *J. Biol. Chem.* **2008**, *283*, 9002.
24. Saunders, A. H.; Golbeck, J. H.; Bryant, D. A. *Biochemistry* **2013**, *52*, 8442.
25. Liu, Z.; Bryant, D. A. *J. Bacteriol.* **2011**, *193*, 4996.
26. Nagata, N.; Tanaka, R.; Tanaka, A. *Plant Cell Physiol.* **2007**, *48*, 1803.
27. Ito, H.; Tanaka, A. *Plant Cell Physiol.* **2014**, *55*, 593.
28. Yamamoto, H.; Kato, M.; Yamanashi, K.; Fujita, Y. *Biochem. Biophys. Res. Commun.* **2014**, *448*, 200.
29. Kolossov, V. L.; Rebeiz, C. A. Evidence for various 4-vinyl reductase activities in higher plants. In: Rebeiz, C. A.; Benning, C.; Bohnert, H. J.; Daniell, H.; Hoober, J. K.; Lichtenhaler, H. K.; Portis, A. R.; Tripathy, B. C. (eds), *The Chloroplast: Basics and Applications. Advances*, **2010**, (Springer, Dordrecht, The Netherlands) pp. 25–38.
30. Lange, C.; Kiesel, S.; Peters, S.; Virus, S.; Scheer, H.; Jahn, D.; Moser, J. *J. Biol. Chem.* **2015**, *290*, 19697.
31. Reinbothe, C.; Bakkouri, M. E.I.; Buhr, F.; Muraki, N.; Nomata, J.; Kurisu, G.; Fujita, Y.; Reinbothe, S. *Trends Plant Sci.* **2010**, *15*, 614.
32. Nomata, J.; Kitashima, M.; Inoue, K.; Fujita, Y. *FEBS Lett.* **2006**, *580*, 6151.
33. Nomata, J.; Ogawa, T.; Kitashima, M.; Inoue, K.; Fujita, Y. *FEBS Lett.* **2008**, *582*, 1346.
34. Gabruk, M.; Mysliwa-Kurdziel, B. *Biochemistry* **2015**, *54*, 5255.
35. Ho, M. Y.; Shen, G.; Canniffe, D. P.; Zhao, C.; Bryant, D. A. *Science* **2016**, *353*, aaf9178.
36. Muraki, N.; Nomata, J.; Ebata, K.; Mizoguchi, T.; Shiba, T.; Tamiaki, H.; Kurisu, G.; Fujita, Y. *Nature* **2010**, *465*, 110.

37. Moser, J.; Lange, C.; Krausze, J.; Rebelein, J.; Schubert, W. D.; Ribbe, M. W.; Heinz, D.W.; Jahn, D. *Proc. Natl. Acad. Sci. USA* **2013**, *110*, 2094.
38. Kondo, T.; Nomata, J.; Fujita, Y.; Itoh, S. *FEBS Lett.* **2011**, *585*, 214.
39. Nomata, J.; Mizoguchi, T.; Tamiaki, H.; Fujita, Y. *J. Biol. Chem.* **2006**, *281*, 15021.
40. Fujita, Y.; Bauer, C. E. The Light-independent protochlorophyllide reductase: A nitrogenase-like enzyme catalyzing a key reaction for greening in the dark. In: Kadish, K. M.; Smith, K. M.; Guilard, R. (eds), *The Porphyrin Handbook. Chlorophylls and Bilins: Biosynthesis, Synthesis, and Degradation,* **2003**, vol. 13, (Elsevier Science Publishing, New York) pp. 109–156.
41. Raymond, J.; Siefert, J. L.; Staples, C. R.; Blankenship, R. E. *Mol. Biol. Evol.* **2004**, *21*, 541.
42. Tamiaki, H.; Teramura, M.; Tsukatani, Y. *Bull. Chem. Soc. Jpn.* **2016**, *89*, 161.
43. Sattley, W. M.; Madigan, M. T.; Swingley, W. D.; Cheung, P. C.; Clocksin, K. M.; Conrad, A. L.; Dejesa, L. C.; Honchak, B. M.; Jung, D. O.; Karbach, L. E.; Kurdoglu, A.; Lahiri, S.; Mastrian, S. D.; Page, L. E.; Taylor, H. L.; Wang, Z. T.; Raymond, J.; Chen, M.; Blankenship, R. E.; Touchman, J. W. *J. Bacteriol.* **2008**, *190*, 4687.
44. Tsukatani, Y.; Hirose, Y.; Harada, J.; Misawa, N.; Mori, K.; Inoue, K.; Tamiaki, H. *Genome Announc.* **2015**, *3*, e01006.
45. Tsukatani, Y.; Yamamoto, H.; Harada, J.; Yoshitomi, T.; Nomata, J.; Kasahara, M.; Mizoguchi, T.; Fujita, Y.; Tamiaki, H. *Sci. Rep.* **2013**, *3*, 1217.
46. Tsukatani, Y.; Yamamoto, H.; Mizoguchi, T.; Fujita, Y.; Tamiaki, H. *Biochim. Biophys. Acta* **2013**, *1827*, 1200.
47. Harada, J.; Mizoguchi, T.; Tsukatani, Y.; Yokono, M.; Tanaka, A.; Tamiaki, H. *J. Biol. Chem.* **2014**, *289*, 12716.
48. Tsukatani, Y.; Harada, J.; Nomata, J.; Yamamoto, H.; Fujita, Y.; Mizoguchi, T.; Tamiaki, H. *Sci. Rep.* **2015**, *5*, 9741.
49. Hu, Y.; Lee, C. C.; Ribbe, M. W. *Science* **2011**, *333*, 753.

2 Porphyrins in Molecular Solar Cells

Lydia Cabau,* John N. Clifford* and Emilio Palomares*,†,‡

*Institute of Chemical Research of Catalonia, The Barcelona Institute of Science and Technology (ICIQ-BIST). Avda. Països Catalans, 16. Tarragona. E-43007, Spain
†ICREA. Passeig Lluis Companys, 28. Barcelona E-08017, Spain
‡epalomares@iciq.es

Table of Contents

List of Abbreviations

BHJ	bulk heterojunction
BODIPY	boron-dipyrromethene
BTD	benzothiadiazole
CDCA	chenodeoxycholic acid
DPP	diketo-pyrrolopyrole
DSSC	dye sensitized solar cells
DTS	dithienosilole
ETM	electron transport material
FF	fill factor
FTO	fluorine-doped tin oxide
HOMO	highest occupied molecular orbital
HTM	hole transport material
ICT	intramolecular charge transfer
LUMO	lowest unoccupied molecular orbital
MLCT	metal to ligand charge transfer band
OSC	organic solar cell
POR	porphyrin
SQ	sqaraine
TPA	triphenylamine

I. Dye Sensitized Solar Cells

Dye sensitized solar cells (DSSC) are photoelectrochemical cells which use dyes to sensitize a wide band-gap semiconductor metal oxide (generally TiO_2) supported onto transparent conducting glass (fluorine-doped tin oxide, FTO) that works as a working electrode. The counter electrode consists of a layer of platinum coated on the FTO conducting glass. These two electrodes are sandwiched together and connected using a hole transporting material (HTM, i.e. a polymer or redox electrolyte) that serves to regenerate the dye ground state.

A. Principles of DSSC

A typical DSSC contains six components: a semiconductor photoanode or working electrode, a sensitizer, a hole transporting material (electrolyte or polymer), a spacer (usually Surilyn©) and a counter electrode (Figure 1).

First of all, the incoming light is absorbed by the sensitizer promoting an electron from the ground state to the excited state (Figure 1, (**1**)).

$$TiO_2|D + hv \rightarrow TiO_2|D^* \quad (1)$$

Thereafter, the electron is transferred to the semiconductor conduction band (**2**), and in an ideal case, the electron will flow through an external circuit to the counter electrode (Figure 1, (**3**)).

$$TiO_2|D^* \rightarrow TiO_2|D^+ + e_{cb} \quad (2)$$

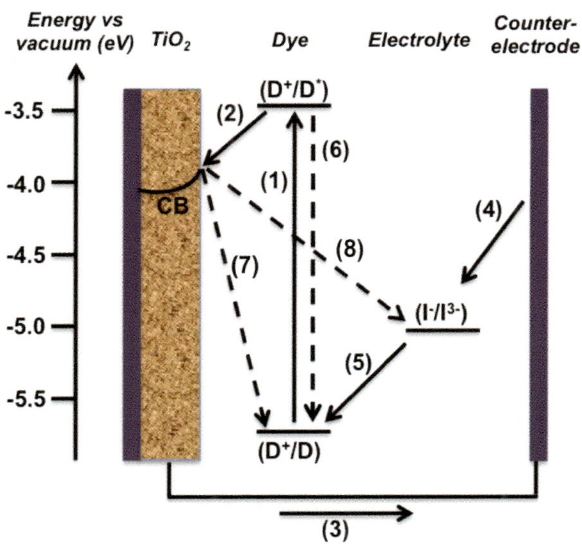

Figure 1. Scheme of a DSSC and the most relevant charge transfer events taking place upon illumination (reprinted from Dr. Lydia Cabau's Ph.D. thesis, 2014).

The HTM (**5**) regenerates the oxidized sensitizer, while the oxidized HTM is itself regenerated at the counter electrode (Figure 1, (**4**)).

$$2TiO_2 \mid D^+ + 3I^- \rightarrow 2TiO_2 \mid D + I_3^- \tag{3}$$

$$I_3^- + 2e^- Pt \rightarrow 3I^- \tag{4}$$

However, these devices have also undesirable charge recombination reactions (Figure 1), which are responsible for losses in the device efficiency. One of these reactions is the deactivation of the dye excited state (Figure 1, (**6**)). Nonetheless, this process occurs on the nanosecond time scale (10^{-9}s) while the electron injection from the excited state into the semiconductor conduction band occurs at least one order of magnitude faster, making electron injection more favorable than the deactivation of the dye excited state. Another undesired reaction is the recombination of photoinjected electrons at the semiconductor with the oxidized sensitizer (Figure 1, (**7**)). This process is slower (10^{-6}–10^{-3}s) than the regeneration of the sensitizer (Figure 1, (**5**)) (10^{-9}–10^{-6}s). In order to have slow recombination, we have to make sure that the regeneration of the dye occurs efficiently. To be sure of this, the HOMO (highest occupied molecular orbital) of the molecule has to be far from the surface of the semiconductor, and, moreover, its energy has to be lower in energy with respect to the redox electrolyte potential to provide sufficient driving force for regeneration to occur. Another recombination reaction can occur after the regeneration of the sensitizer, because the oxidized electrolyte is close to the surface of the semiconductor. Therefore, recombination of photo-injected electrons in the semiconductor with the oxidized electrolyte can occur with lifetimes in the range of 10^{-3}–10^{-1}s, making this mechanism one of the principal loss reactions in this type of device.[1]

B. Basic Solar Cell Parameters

When a solar cell is illuminated, a photocurrent and a voltage are generated producing a curve typically similar to that shown in Figure 2.

The overall energy conversion of a solar cell is defined as the ratio of the output power of the cell per incident irradiance (Equation 5)

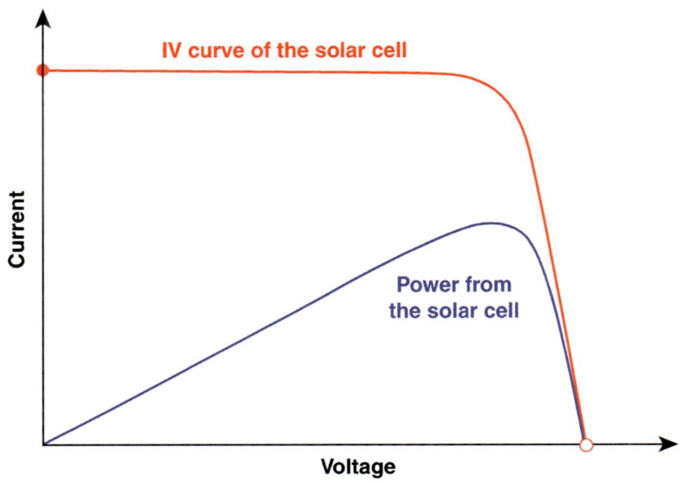

Figure 2. Typical IV-curve of a solar cell (reprinted from Dr. Lydia Cabau's Ph.D. thesis, 2014).

$$\eta = \frac{P_{max}}{P_{light}} = \frac{J_{sc} \cdot V_{oc} \cdot FF}{P_{light}} \tag{5}$$

where J_{sc} (mA/cm^2) is the photocurrent density at short circuit, V_{oc} (V) the Voltage at open circuit, FF is the Fill Factor that measures the how square the I–V curve is (the higher the Fill Factor the higher the efficiency at a given J_{sc} and V_{oc}), and P_{light} the power of the incident light.

As one could imagine, device efficiency, and thus, the shape of an IV curve is affected by the different charge transfer reactions detailed above.

C. Characteristics of Efficient Sensitizers in DSSC

The sensitizer in a DSSC plays a very important role in helping to achieve the maximum efficiency in devices. First of all, the sensitizer has to have the capability to capture the incident radiation, absorbing the photons from a wide range of wavelengths from the solar spectrum. Moreover, the sensitizer must have an anchoring group in order to bind strongly (covalently) to the semiconductor surface. Several anchoring groups have been described in the bibliography such as phosphonates and catechols etc.,[2] though the most common group used is the carboxylic acid.[3] The hydroxyl of this anchoring group reacts with the TiO_2 surface forming a covalent bond in the best cases.

The HOMO and LUMO (Lowest Unoccupied Molecular Orbital) energy of the sensitizer is key in achieving good efficiency by their control over unfavored electron transfer reactions. As already mentioned above, the dye HOMO level has to be located away from the semiconductor surface and has to be of lower energy than the oxidation potential of the redox active electrolyte. On the other hand, the LUMO of the sensitizer has to be close to the surface of the semiconductor in order to achieve efficient electron injection (so-called unidirectional electron transfer). Furthermore, it is paramount that the LUMO level is higher than the conduction band energy level of the TiO_2 to favor the electron injection from the dye excited state.

Secondly, dye solubility in organic solvents (preferably non-halogenated solvents) and the sensitizer's ability to minimize the formation of aggregates in solution and on the semiconductor surface after sensitization are also important. The latter can be partially solved with the addition of a co-sensitizer such as chenodeoxycholic acid (CDCA) (Figure 3).[4]

As an example of efficient dyes used in DSSC we will now detail the use of ruthenium complexes.

D. Ruthenium Complexes

The first efficient dyes used for DSSCs were Ru complexes (trinuclear ruthenium dye), giving a light-to-photoelectrical conversion efficiency between 7.1% and 7.9% ($RuL_2(\mu\text{-}(CN)Ru(CN)L'_2)_2$, where L is 2,2′bipyridine-4,4′-dicarboxylic acid and L′ is 2,2′-bipyridine).[5]

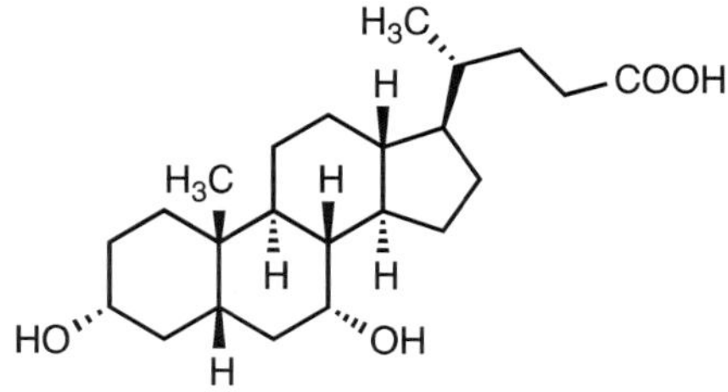

CDCA

Figure 3. Molecular structure of CDCA.

Ruthenium complexes were employed due to their broad absorbance from the visible region to the near-infrared.[6] The general structure of a ruthenium dye for DSSCs consists usually of a Ru(II) atom coordinated by polypyridyl ligands and thiocyanate moieties in an octahedral geometry. The carboxylic acids are used as anchoring groups and are attached to the bipyridyl moiety leading to efficient injection of electrons in the semiconductor from the excited state. These complexes show an absorption band in the visible region that can be tuned by manipulation of the MLCT (metal to ligand charge transfer band) transition.[7]

Since the seminal paper by Gratzel and O'Regan using ruthenium complexes, many studies involving modification of these complexes have been published. Only two years after their breakthrough work, Grätzel and coworkers increased the efficiency of a DSSC beyond 10% (J_{sc} = 18.2 mA/cm^2, V_{oc} = 720 mV, FF = 0.73) by employing cis-di(thiocyanato)bis(2,2′-bipyridyl-4,4′-dicarboxylate)ruthenium (II), most commonly known as **N3** dye (Figure 4).[8] In the following years, many ruthenium complexes were reported such as tristhiocyanato-4,4′,4″-tricarboxy-2,2′:6,2″-terpyridine ruthenium(II), also called the **Black dye** (Figure 4), with an initial efficiency

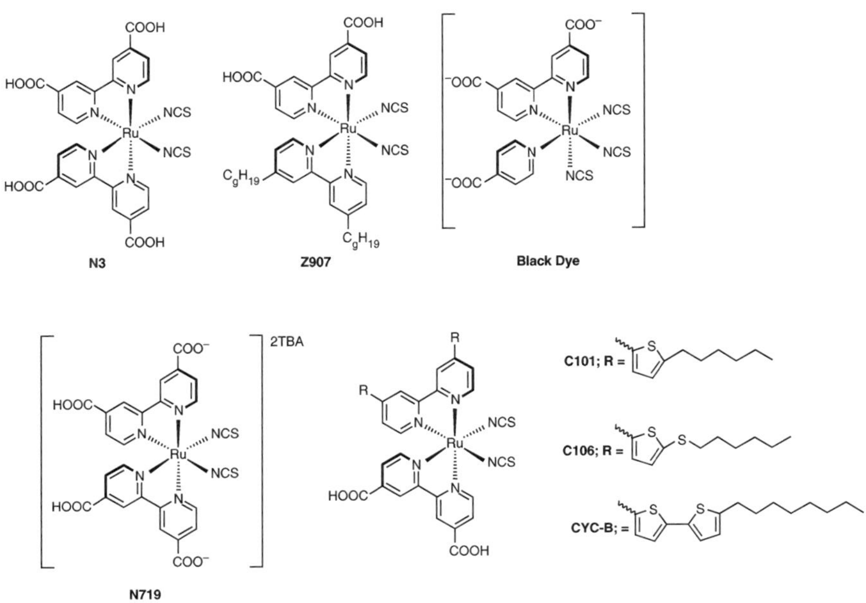

Figure 4. Efficient Ruthenium complexes used as sensitizers in DSSCs.

of 10.4%,[9] that was in later increased to 11.4% in 2012.[10] Moreover, other Ru-complexes have been published with efficiencies close to the paradigm dye **N719** (Figure 4)[11] such as **Z907** (Figure 4) which contains long alkyl chains to increase the solubility and slow the recombination between photo-injected electrons and oxidized electrolyte. However, a milestone was set in 2008 with the design of Ruthenium complexes bearing π-conjugated moieties containing thiophenes and other derivatives at the bipyridyl ligands. Such structures were designed to increase the absorption in the near-infrared region as well as to increase the molecular extinction coefficient of the dye. Many of these novel dyes have led to higher efficiencies such as the dye **C101**,[12] which has a 2-hexylthiophene in the bipyridyl ligand, and produces an efficiency of 11.0% (J_{sc} = 17.9 mA/cm^2, V_{oc} = 778 mV, FF = 0.78) (Figure 4). Similar efficiencies have been obtained with the dyes **CYC-B1**[13] and **CYC-B11**[14] (Figure 4). Indeed, the maximum performance achieved with a Ruthenium dye was done so with the dye **C106**[15] containing the moiety 2-(hexylthio)-5-methylthiophene, yielding an efficiency of 11.7% (J_{sc} = 19.8 mA/cm^2, V_{oc} = 758 mV, FF = 0.78) (Figure 4).

Ruthenium complexes have yielded high efficiencies and also broad absorptions in DSSCs; however, they also have many drawbacks. Most importantly perhaps is the issue of cost — ruthenium is a non-abundant earth metal. In addition, Ruthenium complexes only have moderate absorption coefficients. Such drawbacks have encouraged many researchers to work on the development of alternative sensitizers based on metal free organic dyes.

E. Metal Free Organic Dyes: Donor-π-Acceptor Dyes

Donor-π-acceptor dyes, also known as push–pull dyes, consist of an electron donor and electron acceptor molecular unit linked covalently with a π conjugated spacer (Figure 5). The photophysical properties associated with these dyes are directly related to the Intramolecular Charge Transfer (ICT) from the donor to the acceptor moiety. This ICT gives these dyes high molar extinction coefficients.

These dyes have absorption bands that are easy to tune and high molecular extinction coefficients. They are considered as feasible alternatives to ruthenium complexes in DSSC research.

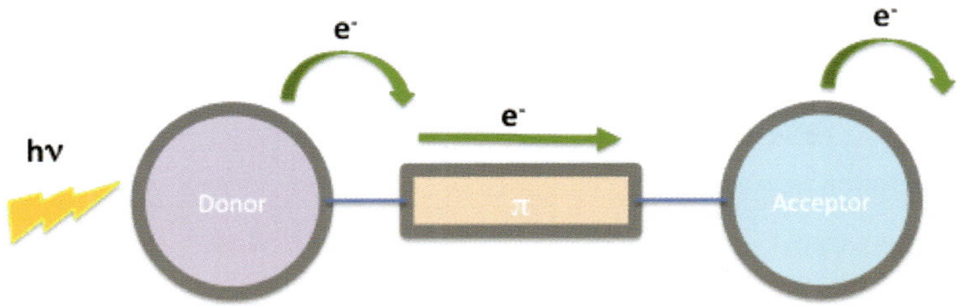

Figure 5. Structure of D-π-A dyes.

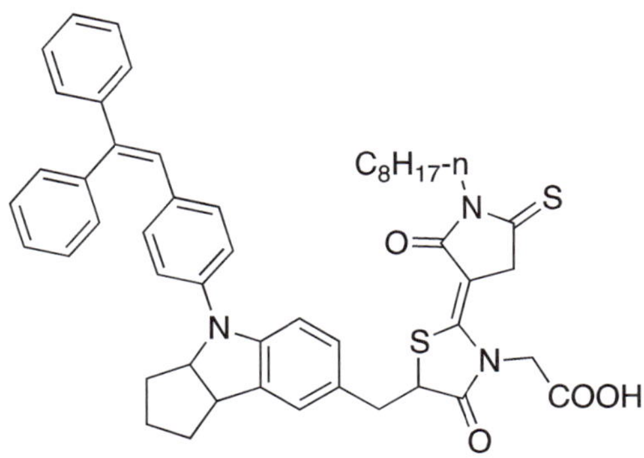

D205

Figure 6. Molecular structure of D205.

The design of these push–pull dyes is crucial to achieve high efficiencies in DSSCs. It is paramount to have a very good donor moiety, which remains stable when oxidized upon light irradiation. For example, Hara and co-workers compared thiophene and coumarin donor moieties in DSSCs and found efficiencies of 6% and 8.2%, respectively.[16]

In 2008, Grätzel and co-workers studied the **D205** dye (Figure 6) which achieved an efficiency of 9.52% (J_{sc} = 18.7 mA/cm^2, V_{oc} = 710 mV, FF = 0.71). The structure of this dye contains an indoline group with an *n*-octyl moiety on the rhodanine structure. The long alkyl chains were employed to decrease dye aggregation and to make the molecule more soluble in organic solvents. Control over the formation of aggregates is an important issue for organic sensitizers in DSSCs. In this particular work, they observed that the combination of a long alkyl chain and the use of CDCA led to outstanding increase in efficiency.[17]

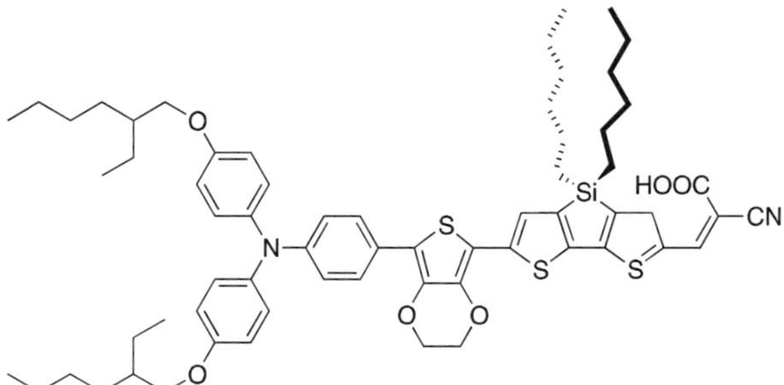

Figure 7. Molecular structures of C219.

More recently, Wang and co-workers reported that the dye **C219** (Figure 7), for the first time, could yield efficiencies close to or above 10%. **C219** was reported to deliver an efficiency of 10.1% (J_{sc} = 17.9 mA/cm^2, V_{oc} = 770 mV, FF = 0.73). This novel dye consists of a binary π spacer, a 3,4-ethylenedioxythiophene unit (EDOT) connected to the donor moiety (alkoxy-substituted triphenylamine) to lift the HOMO, and a dihexyl-substituted dithienosilole (DTS) attached to the acceptor to achieve the appropriate LUMO energy.[18]

All of the devices discussed so far were fabricated using iodide/triiodide electrolyte. However, in 2010, Hagfeldt, Sun and co-workers employed a Co^{2+}/Co^{3+} electrolyte in a DSSC device and produced a remarkable efficiency of 6.2% with the **D35** dye (Figure 8). In their work, they synthesized two new sensitizers, **D35** and **D29**, with the same π-bridge and identical acceptor moiety but with different donor groups. For **D29** dye the electron donating group was the group p-*N,N*-dimethylaminelinyl at the TPA (triphenylamine) moiety while for **D35** it was the *o,p*-dibutoxylpheny group at the TPA. These two dyes were investigated to compare the effect of bulky alkoxyl substituents in devices employing cobalt electrolyte. For this, a series of cobalt electrolytes was synthesized to determine the best structure and potential to achieve efficient dye regeneration and higher V_{oc}.[19]

This team also reported the same dye comparing device performance in the same conditions with iodine/iodide and cobalt electrolytes achieving 7.1% (J_{sc} = 13.6 mA/cm^2, V_{oc} = 720 mV, FF = 0.71) and 8.3% respectively (J_{sc} = 14.1 mA/cm^2, V_{oc} = 820 mV, FF = 0.73).[20] This study showed that using cobalt electrolyte increases the V_{oc} by 100 mV. Dye efficiency was

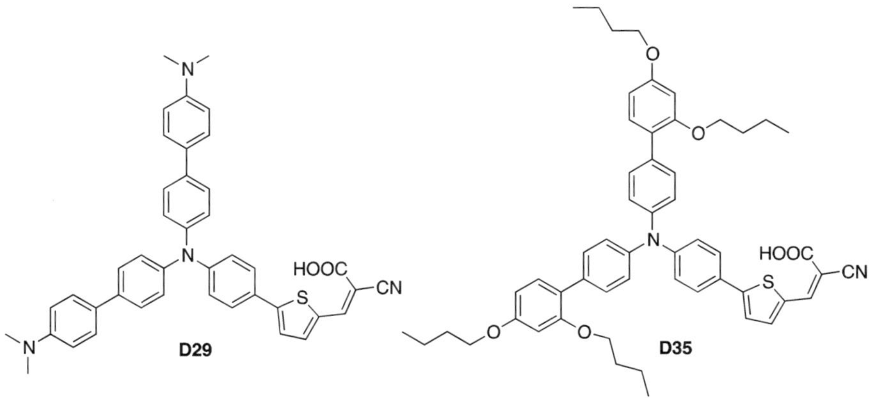

Figure 8. Molecular structures of **D29** and **D35**.

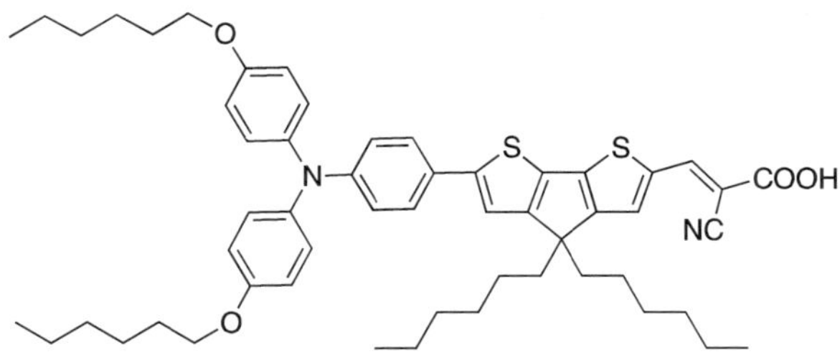

C218

Figure 9. Molecular structure of **C218**.

further improved to 9.4% (J_{sc} = 13.0 mA/cm^2, V_{oc} = 950 mV, FF = 0.76)[10] in a subsequent study.

Wang and his group focused on the synthesis of organic dyes for DSSC order to achieve record efficiencies by using cobalt electrolytes and, in 2011, reported a new dye, **C229** (Figure 10), with a similar structure to the C218 dye. However, in this work they enlarged the π spacer introducing two thiophene units and, thus, increasing the molar absorption coefficient owing to the improved delocalizability of π spacer. While, the efficiency achieved using iodine/iodide electrolyte was only 6.7% (J_{sc} = 15.20 mA/cm^2, V_{oc} = 680 mV, FF = 0.65), in the same conditions, employing the cobalt electrolyte yielded an efficiency of 9.4% (J_{sc} = 15.3 mA/cm^2, V_{oc} = 850 mV, FF = 0.73).[20]

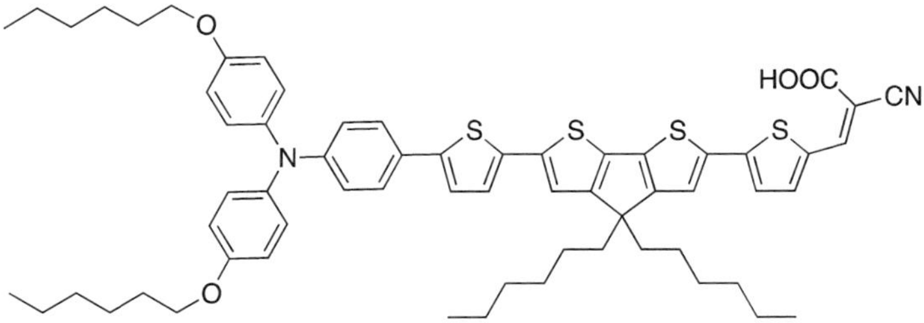

C229

Figure 10. Molecular structure of **C229**.

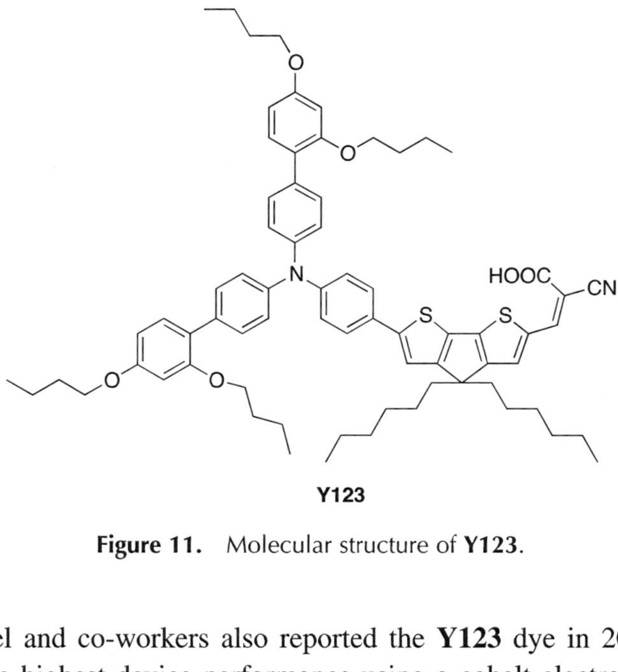

Y123

Figure 11. Molecular structure of **Y123**.

Grätzel and co-workers also reported the **Y123** dye in 2011 which yielded the highest device performance using a cobalt electrolyte at the time with 9.6% in their champion cell (Figure 11).[26] A further study with this dye reported an impressive V_{oc} of 1 V using a Cobalt electrolyte.[27]

F. Porphyrins

Porphyrins consist of a tetra pyrrole macrocycle incorporating four modified pyrroles connected at the α carbon by methine (Figure 12). They follow Huckel's rule of aromaticity (possessing 4n + 2π electrons). This feature makes porphyrins outstanding dyes with a high molecular extinction

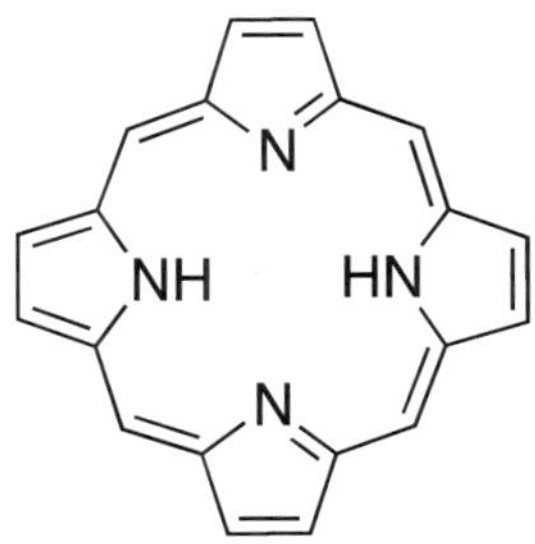

Figure 12. Basic porphyrin structure.

coefficient and also is responsible for the pleasant colours that porphyrin solutions often have.[28] This molecule is present in nature in many biological systems such as chlorophyll, hemoglobine, cytochromes, and many enzymes too. Due to their excellent optical properties, porphyrins are used in medicine[29–31] and electronic[32–34] devices. Due to their role in photosynthesis these molecules are very attractive and promising sensitizers for efficient organic photovoltaic devices.[35,36]

The typical absorption profile of a porphyrin consists of an intense absorption band close to 400 nm called the Soret band and moderate absorption bands between 500 and 700 nm.[28] In order to use these molecules for DSSC applications, it is important to take into account that the molecule requires an anchoring group to attach it to semiconductor surfaces, as is the case of the previously discussed organic dyes. Porphyrins can be functionalized at several different positions — the four *meso*-positions and the eight β-positions (Figure 13). In case of porphyrins, carboxylic acid is also considered the best anchoring group.[37,38]

During the last decades, many porphyrins have been synthesized for DSSC applications, with functionalization in the β and *meso*-positions. The first notable example was demonstrated by Grätzel and Kay in 1993 with a modest efficiency of 2.6%.[39] Thereafter, the first example of a *meso*-position substituted porphyrin was published by Cherian and Wamser in 2000 with an efficiency of 3.5%.[40]

During the last years, several works with different porphyrins have been published with the aims of increasing device efficiency, minimizing dye aggregation, and achieving good charge separation.[41,42] Nonetheless, it was not until Yeh and Diau added a donor group to a porphyrin structure, as in the case of **YD1** and **YD2** (Figure 14), that the efficiency of devices

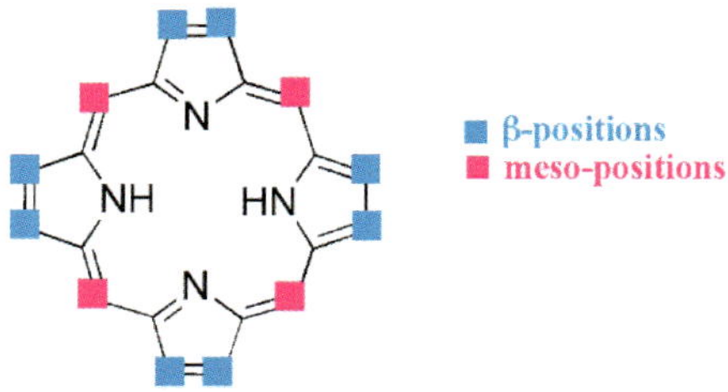

Figure 13. Available positions for the functionalization of porphyrins.

containing these sensitizers increased extended the absorption and improved the charge separation efficiency.[43]

Furthermore, aggregation problems were tackled by adding different concentrations of chenodeoxycholic acid (CDCA).[14] In this study, **YD1** and **YD2** were synthesized and the dye **YD0** was used as a reference. The dye **YD2** exhibited a cell performance close to 6.8% (J_{sc} = 13.7 mA/cm^2, V_{oc} = 711 mV, FF = 0.69). This efficiency is slightly smaller compared the Ruthenium paradigm, **N719** (J_{sc} = 13.8 mA/cm^2, V_{oc} = 760 mV, FF = 0.70, n = 7.3%). Higher efficiencies were obtained with **YD2** and **YD1** due to the slower recombination of the electrons with the oxidized electrolyte. In 2010, the device performance for **YD2** was improved by Grätzel and co-workers with an overall efficiency of 10.9% obtained (J_{sc} = 18.6 mA/cm^2, V_{oc} = 770 mV, FF = 0.76).[17]

Several studies were published with porphyrins though none indicated efficiencies greater than that of **YD2**.[25,44–49] It was not until the end of 2011 when Yella and co-workers[50] achieved this with **YD2-0-C8** (Figure 15).

The structure of this porphyrin is similar to that of **YD2** with the difference being that it incorporates two octyloxy groups in the *ortho* positions of each *meso*-phenyl ring increasing the electronic density on the porphyrin π-system compared to **YD2**. This increase in electronic density results in a higher LUMO energy level. Devices with this new dye showed very promising results using cobalt electrolyte, achieving a record in efficiency of 11.9% (J_{sc} = 17.3 mA/cm^2, V_{oc} = 965 mV, FF = 0.71). This new value is higher than the previous record achieved with organic dyes.[26] In the same work, aiming to increase the efficiency, a co-sensitizer was added (**Y123**). This "cocktail" achieved a remarkable efficiency of 12.3% (J_{sc} = 17.7 mA/cm^2, V_{oc} = 935 mV, FF = 0.74).

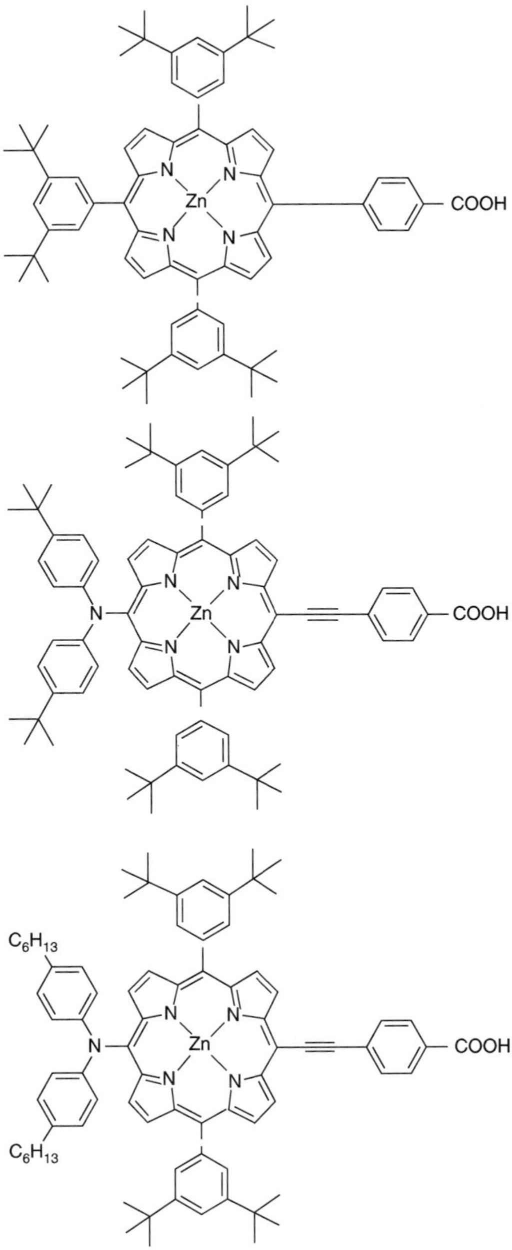

YD0

YD1

YD2

Figure 14. Dye molecular structures of **YD0**, **YD1** and **YD2**.

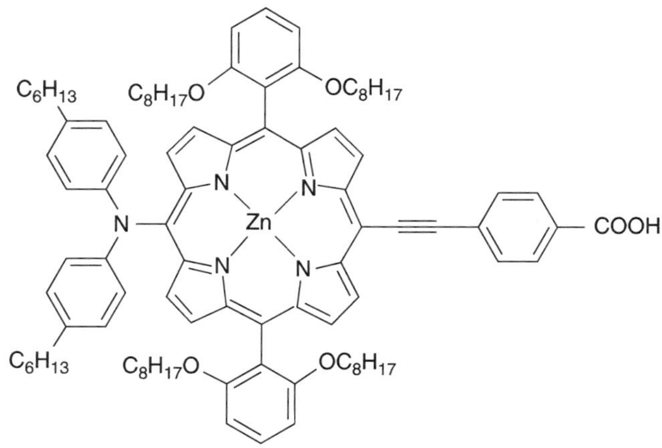

YD2-o-C8

Figure 15. Molecular structure of **YD2-o-C8**.

The same group carried out further efforts to increase the efficiency achieved by using **YD2-o-C8** and in 2014 published two new porphyrins, **GY21** and **GY50** (Figure 16). In this work, their strategy was the introduction of the benzothiadiazole (BTD) unit as the π-conjugated linker between the anchoring group and the porphyrin core to broaden the absorption spectrum. Moreover, a phenyl group was introduced into this structure a spacer between the BDT moiety and the carboxylic group. Devices made with **GY21** and **GY50** yielded efficiencies of 2.5% (J_{sc} = 5.03 mA/cm^2, V_{oc} = 615 mV, FF = 0.80) and 12.75 (J_{sc} = 18.53 mA/cm^2, V_{oc} = 885 mV, FF = 0.77), respectively.

On the one hand, the higher efficiency achieved with **GY50** is due to panchromatic absorption, which means a second dye is not needed as a co-sensitizer.

On the other hand, the lower conversion efficiency of **GY21** compared to **GY50** is due to the lack of directionality of the excited state and, thus, the observation of much less photocurrent.[52]

Subsequent to this, Chin-Li Wang and co-workers synthesized **LD31** and the **LD14**[51] by inserting an ethynyl-anthracenyl moiety between the donor unit and the porphyrin core to extend the π-conjugation in order to improve the light-harvesting efficiency (Figure 17).[44]

These new porphyrins were employed in devices both with and without the co-sensitizer **AN-4** and achieved 9.95% (J_{sc} = 20.02 mA/cm^2,

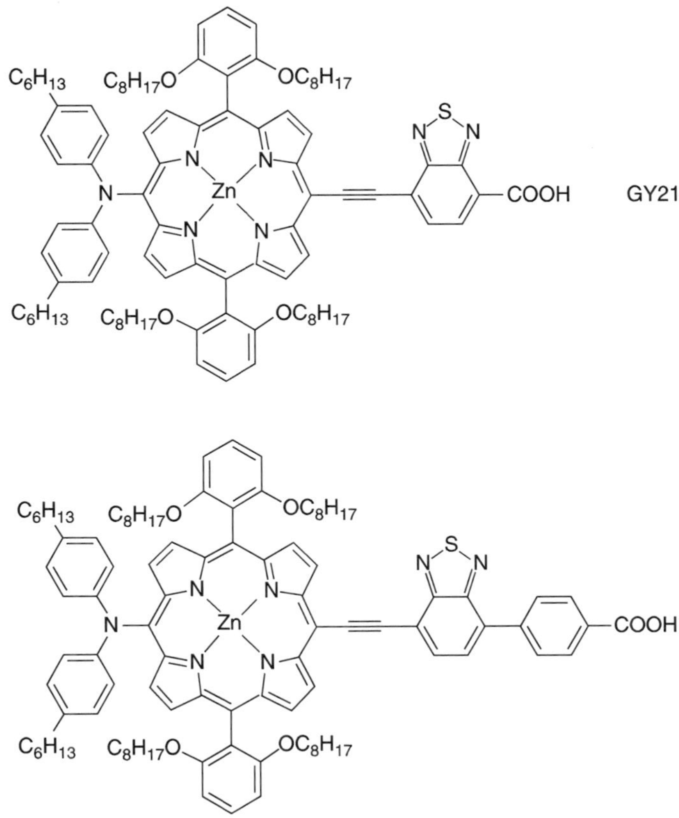

GY21

GY50

Figure 16. Molecular structures of **GY21** and **GY50**.

V_{oc} = 699 mV, FF = 0.71) and 10.3% (J_{sc} = 20.3 mA/cm^2, V_{oc} = 704 mV, FF = 0.72) respectively.

In 2014, the current highest efficiency DSSC was achieved with a power conversion of 13% by Grätzel and co-worker using two new porphyrins called **SM371** and **SM315** (Figure 18).

The structure of these porphyrins is similar to that previously reported by the same group (**GY21** and **GY50**). However for **GY21** and **GY50**, the donor moiety used was bis(2′,4′-bis(hexyloxy)-[1,1′-biphenyl]-4-yl)amine. This donor group has also been used in several studies of organic dyes which have reported good efficiencies in DSSCs based on the cobalt electrolyte.[23,26] The efficiencies achieved for **SM371** and **SM315** were 12% (J_{sc} = 15.9 mA/cm^2, Voc = 960 mV, FF = 0.79) and 13.0% (Jsc = 18.1 mA/cm^2, Voc = 910 mV,

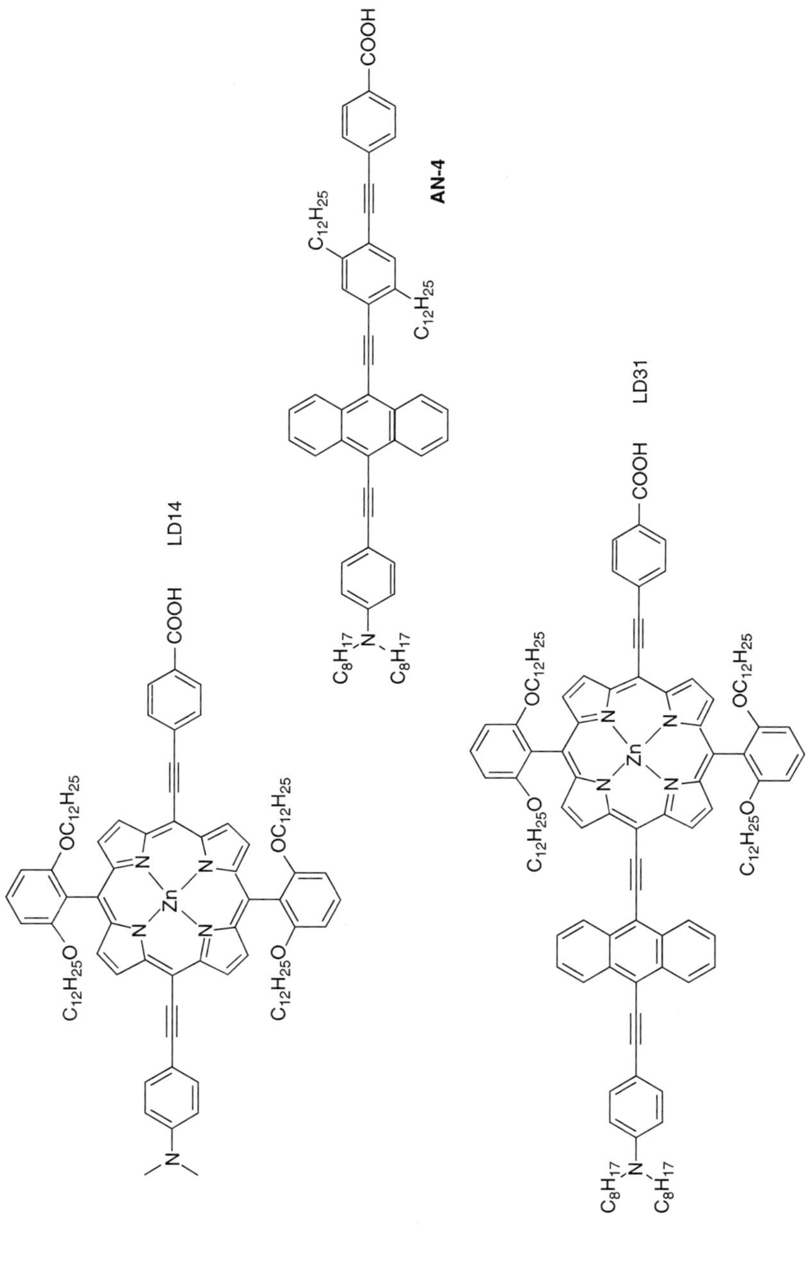

Figure 17. Molecular structures of **LD14**, **LD31** and **AN-4**.

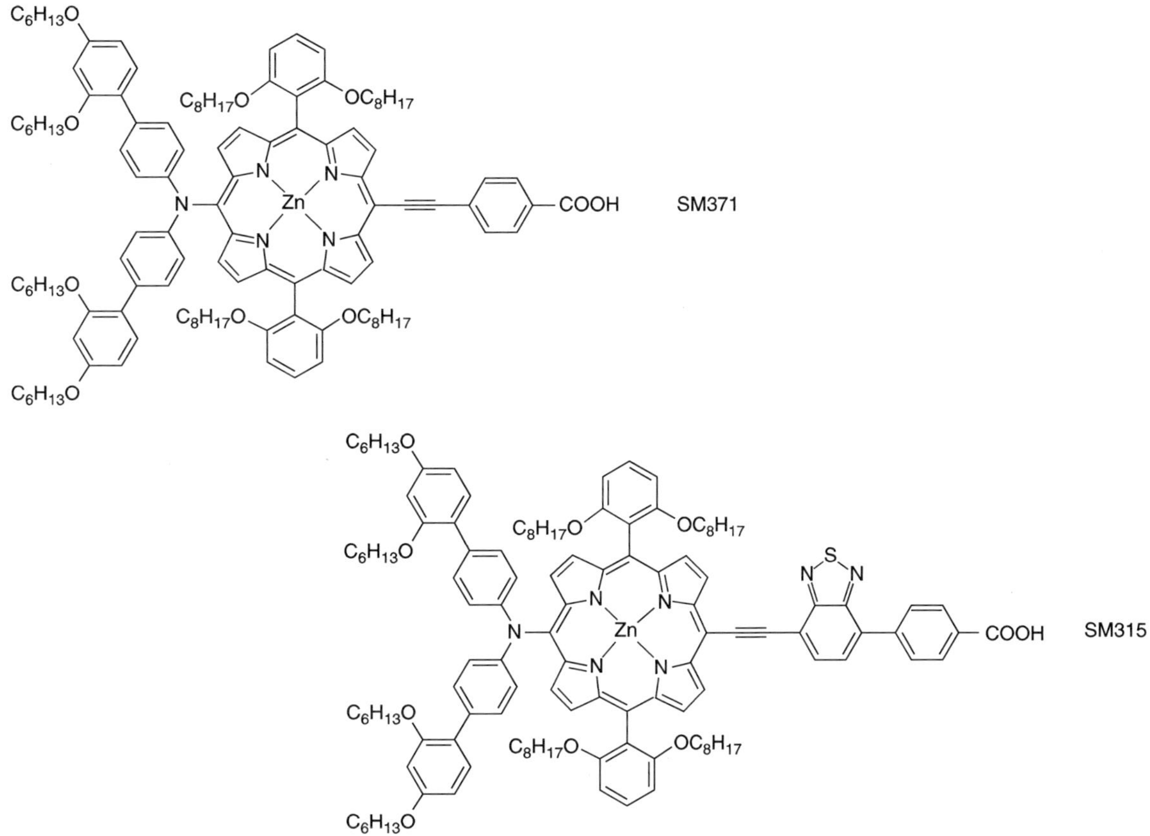

Figure 18. Molecular structures of **SM371** and **SM315**.

FF = 0.78) respectively using cobalt electrolyte. The higher Jsc obtained for **SM315** was due to the dramatically improved absorption properties that led to a high IPCE of 80% across the visible region (450 nm–750 nm). **SM371** has a 50 mV larger Voc under standard irradiation conditions compared to **SM315**. Moreover, it was shown that the electron lifetime is six times slower for the **SM371** dye. The slower recombination kinetics is likely to be due to the BDT unit which improves the excited state directionality and also prevents back electron transfer to the oxidized electrolyte.[52]

In Figure 19, we can observe the evolution of photovoltaic performances for DSSCs from 1991 to 2014.

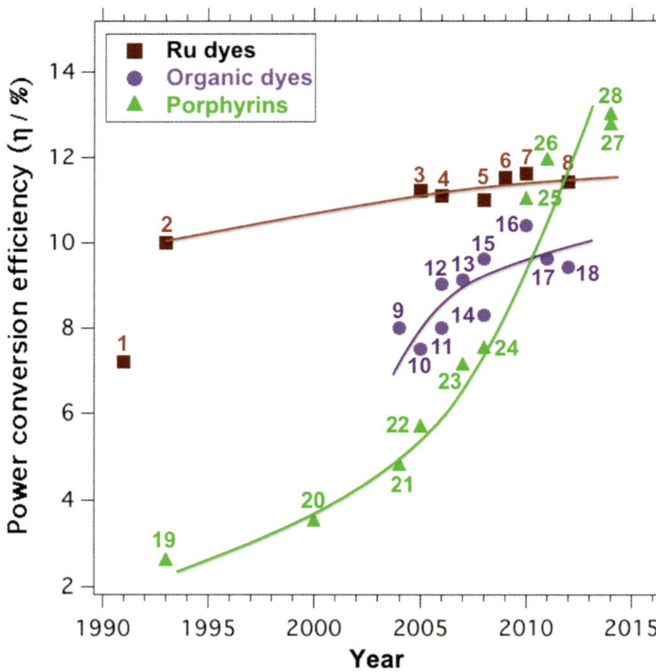

Figure 19. Progress on DSSC efficiency of the most relevant dyes including ruthenium complexes (**1–8**), organic dyes (**9–18**) and porphyrins (**19–28**). 1. Trinuclear trinuclear $RuL_2(\mu\text{-}(CN)ru(CN)L'_2)_2$[6]; **2**. N3[10]; **3**. N3[53]; **4**. N719[13]; **5**. C101[14]; **6**. CYC-B11[16]; **7**. C106[54]; **8**. Black dye[12]; **9**. Indoline dye[55]; **10**. NKX-2677[19]; **11**. JK2[56]; **12**. D149[9]; **13**. TA-St-CA[57]; **14**. MK-2[58]; **15**. D205[21]; **16**. C219[22]; **17**. Y123[26]; **18**. C218[12]; **19**. Cu-a-oxymesoisichlorin[39]; **20**. TCPP[40]; **21**. Zn-1a[59]; **22**. Zn-3[60]; **23**. GD2[20]; **24**. tda-2b-bd-Zn[61]; **25**. YD2[17]; **26**. YD2oC8[50]; **27**. GY50[52]; **28**. SM371.[52] (Reprinted from Dr. Lydia Cabau's Ph.D. thesis, 2014.)

II. Organic Solar Cells (OSC)

Another promising molecular solar cell that has be studied widely is organic solar cells (OSC). OSCs combine two organic materials — an electron donor or hole transport material (HTM), and an electron acceptor material or electron transport material (ETM) — which are "sandwiched" between two metal electrodes with different work functions (Figure 20). The photo-induced charges are separated at the interface between both types of organic materials and the free carriers are collected selectively at each electrode.

Electron-hole pairs (so called excitons) are generated upon irradiation of these devices. The lifetime of these excitons is short and they can only be transported a few nanometers (10–12 nm) depending on the nature of the organic material. When the exciton arrives at the interface of both organic materials, it separates into free carriers of different charge (so called polarons). These polarons must be transported before they can recombine to the contact electrodes (Figure 21).

Thus, the efficient formation of excitons as well as the optimization of exciton separation and polaron collection is key to achieving excellent device performance. A first approach is to select adequate donor and acceptor materials with sufficient energy onset to be able to separate efficiently charges at the interface. For this reason, the study of new materials and the morphology at the nanoscale have attracted much attention in recent years. The original OSC device was developed in the eighties by Tang (Figure 22) and yielded efficiencies as low as 1%[62] by using a bi-layer type device. In comparison, bulk-heterojunction solar cells, which involve the mixing of both types of organic materials, show efficiencies as high as 10% for single junction solar cells.

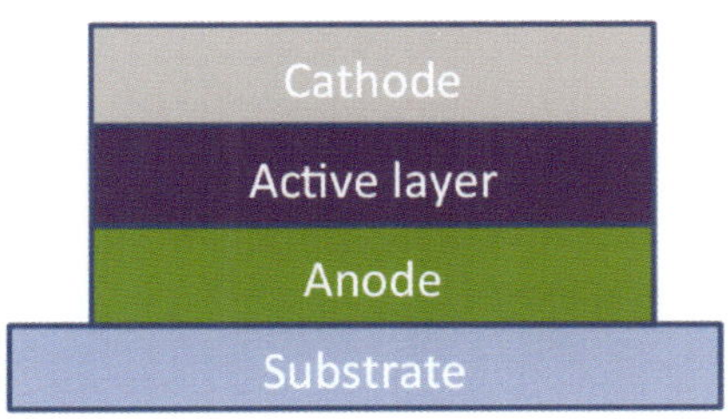

Figure 20. Schematic representation of a simple OSC device (reprinted from Dr. Lydia Cabau's Ph.D. thesis, 2014).

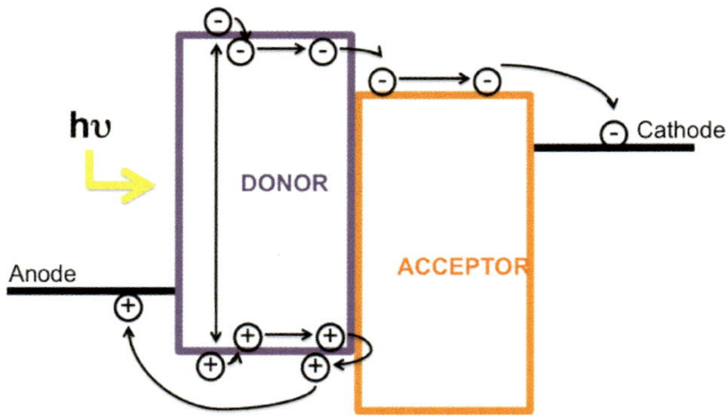

Figure 21. Schematic of the operating principle of an OSC (reprinted from Dr. Lydia Cabau's Ph.D. thesis, 2014).

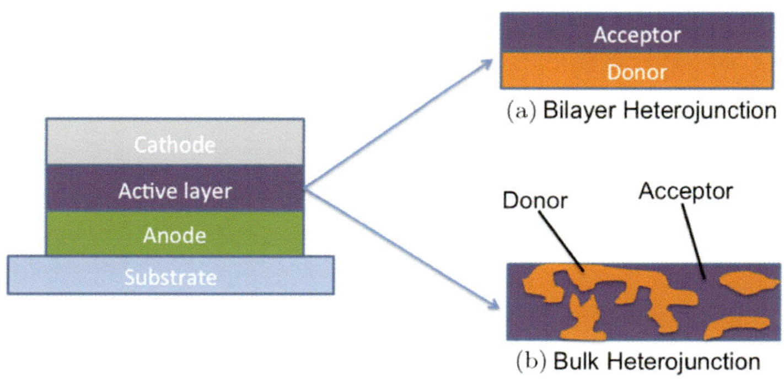

Figure 22. Architecture of a bilayer heterojunction (a) and a bulk heterojunction (b) (reprinted from Dr. Lydia Cabau's Ph.D. thesis, 2014).

A. Donor and Acceptor Materials

Here, in this section, we describe briefly the most relevant materials used in OSC.

1. *Electron Acceptor Materials*

Fullerenes and their derivatives are the dominant molecules used as electron acceptor materials in OSC. The use of this type of molecule is justified due to

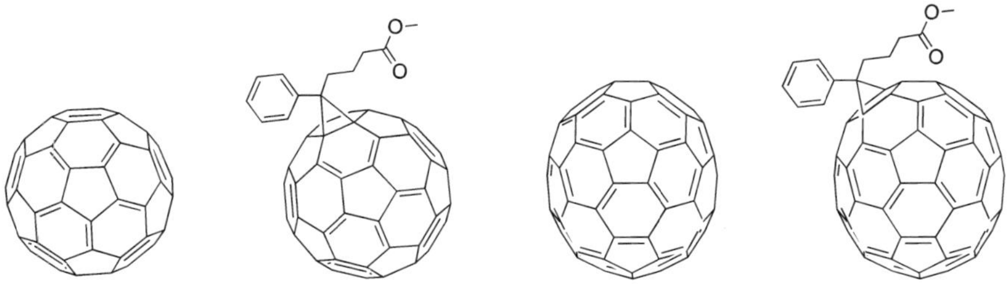

C_{60} PC$_{61}$BM C_{70} PC$_{71}$BM

Figure 23. Principal fullerenes used as acceptor moieties in OSC devices.

their strong ability to accept electrons from donor materials and also their electron mobility.[63] Derivatives of C_{60} and C_{70} such as PC$_{61}$BM and PC$_{71}$BM are the most commonly used in solution processed OPVs (Figure 23).

2. *Electron Donor Materials*

Electron donor materials have been readily explored as they are used as the main light-harvesting moiety in OSCs. Several types of molecules have been designed and synthesized for application in solution processed OSCs such as squaraines (SQ),[24,64–67] diketo-pyrrolopyrroles (DPP),[68–70] boron-dipyrromethene (BODIPY)[71] and also D-π-A dyes bearing triphenylamine units as secondary electron donors,[38,72,73] all with efficiencies ranging from 4% to 6% under standard illumination conditions of 1 sun (100 mW/cm^2 of sun simulated light 1.5 AM G spectrum).

Nowadays, the most efficient OSCs have been reported using small molecules in single junction devices. Bazan and Heeger have achieved an impressive 8.9% in this type of architecture with p-DTS(FBTTh$_2$)$_2$ (Figure 24).[74]

Alternatively, porphyrins (POR) have been also widely studied and used in many BHJ-OSC.[75–79] Peng and co-workers published in 2014 the highest efficiency solution-processed BHJ OSC based on a small molecule, in this case a porphyrin DPPEznP-O (Figure 25), with an efficiency of 7.23% (J_{sc} = 16.0 mA/cm^2, V_{oc} = 710 mV, FF = 0.63).[80]

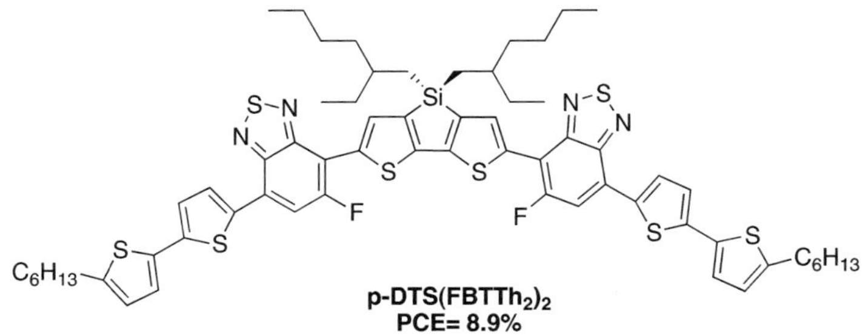

Figure 24. Molecular structure of p-DTS(FBTTh$_2$)$_2$.

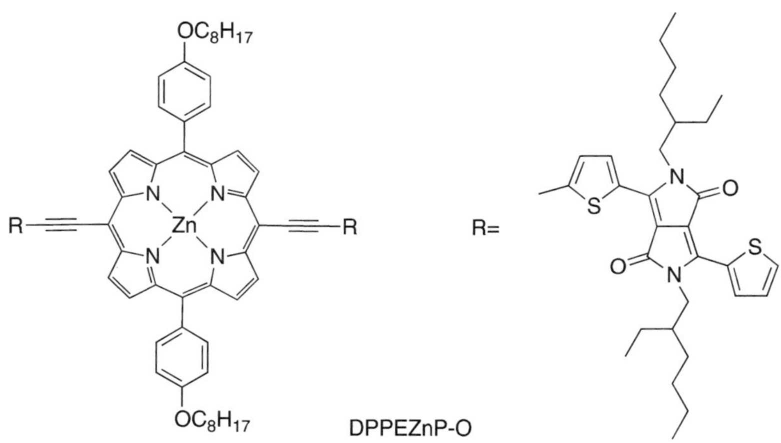

Figure 25. Molecular structure of DPPEznP-O.

III. References

1. Martinson, A. B. F.; Hamann, T. W.; Pellin, M. J.; Hupp, J. T. *Chem. Eur. J.* **2008**, *14*, 4458.
2. Fung, A. K. M.; Chiu, B. K. W.; Lam, M. H. W. *Water Res.* **2003**, *37*, 1939.
3. Zaban, A.; Ferrere, S.; Gregg, B. A. *J. Phys. Chem. B* **1998**, *102*, 452.
4. Hagberg, D. P.; Yum, J.-H.; Lee, H.; De Angelis, F.; Marinado, T.; Karlsson, K. M.; Humphry-Baker, R.; Sun, L.; Hagfeldt, A.; Grätzel, M.; Nazeeruddin, M. K. *J. Am. Chem. Soc.* **2008**, *130*, 6259.
5. Neale, N. R.; Kopidakis, N.; van de Lagemaat, J.; Grätzel, M.; Frank, A. J. *J. Phys. Chem. B* **2005**, *109*, 23183.
6. O'Regan, B.; Gratzel, M. *Nature* **1991**, *353*, 737.

7. Polson, M. I. J.; Medlycott, E. A.; Hanan, G. S.; Mikelsons, L.; Taylor, N. J.; Watanabe, M.; Tanaka, Y.; Loiseau, F.; Passalacqua, R.; Campagna, S. *Chem. Eur. J.* **2004**, *10*, 3640.
8. Hagfeldt, A.; Graetzel, M. *Chem. Rev.* **1995**, *95*, 49.
9. Ito, S.; Zakeeruddin, S. M.; Humphry-Baker, R.; Liska, P.; Charvet, R.; Comte, P.; Nazeeruddin, M. K.; Péchy, P.; Takata, M.; Miura, H.; Uchida, S.; Grätzel, M. *Adv. Mat.* **2006**, *18*, 1202.
10. Nazeeruddin, M. K.; Kay, A.; Rodicio, I.; Humphry-Baker, R.; Mueller, E.; Liska, P.; Vlachopoulos, N.; Graetzel, M. *J. Am. Chem. Soc.* **1993**, *115*, 6382.
11. Péchy, P.; Renouard, T.; Zakeeruddin, S. M.; Humphry-Baker, R.; Comte, P.; Liska, P.; Cevey, L.; Costa, E.; Shklover, V.; Spiccia, L.; Deacon, G. B.; Bignozzi, C. A.; Grätzel, M. *J. Am. Chem. Soc.* **2001**, *123*, 1613.
12. Han, L.; Islam, A.; Chen, H.; Malapaka, C.; Chiranjeevi, B.; Zhang, S.; Yang, X.; Yanagida, M. *Energy Environ. Sci.* **2012**, *5*, 6057.
13. Wang, Q.; Ito, S.; Grätzel, M.; Fabregat-Santiago, F.; Mora-Seró, I.; Bisquert, J.; Bessho, T.; Imai, H. *J. Phys. Chem. B* **2006**, *110*, 25210.
14. Gao, F.; Wang, Y.; Shi, D.; Zhang, J.; Wang, M.; Jing, X.; Humphry-Baker, R.; Wang, P.; Zakeeruddin, S. M.; Grätzel, M. *J. Am. Chem. Soc.* **2008**, *130*, 10720.
15. Chen, C.-Y.; Wu, S.-J.; Wu, C.-G.; Chen, J.-G.; Ho, K.-C. *Angew. Chem. Int. Ed.* **2006**, *45*, 5822.
16. Chen, C.-Y.; Wang, M.; Li, J.-Y.; Pootrakulchote, N.; Alibabaei, L.; Ngoc-le, C.-h.; Decoppet, J.-D.; Tsai, J.-H.; Grätzel, C.; Wu, C.-G.; Zakeeruddin, S. M.; Grätzel, M. *ACS Nano* **2009**, *3*, 3103.
17. Bessho, T.; Zakeeruddin, S. M.; Yeh, C.-Y.; Diau, E. W.-G.; Grätzel, M. *Angew. Chem. Int. Ed.* **2010**, *49*, 6646.
18. Clifford, J. N.; Yahioglu, G.; Milgrom, L. R.; Durrant, J. R. *Chem. Commun.* **2002**, 1260.
19. Hara, K.; Sato, T.; Katoh, R.; Furube, A.; Yoshihara, T.; Murai, M.; Kurashige, M.; Ito, S.; Shinpo, A.; Suga, S.; Arakawa, H. *Adv. Funct. Mat.* **2005**, *15*, 246.
20. Campbell, W. M.; Jolley, K. W.; Wagner, P.; Wagner, K.; Walsh, P. J.; Gordon, K. C.; Schmidt-Mende, L.; Nazeeruddin, M. K.; Wang, Q.; Grätzel, M.; Officer, D. L. *J. Phys. Chem. C* **2007**, *111*, 11760.
21. Ito, S.; Miura, H.; Uchida, S.; Takata, M.; Sumioka, K.; Liska, P.; Comte, P.; Péchy, P.; Grätzel, M. *Chem. Commun.* **2008**, 5194.
22. Zeng, W.; Cao, Y.; Bai, Y.; Wang, Y.; Shi, Y.; Zhang, M.; Wang, F.; Pan, C.; Wang, P. *Chem. Mater.* **2010**, *22*, 1915.
23. Feldt, S. M.; Gibson, E. A.; Gabrielsson, E.; Sun, L.; Boschloo, G.; Hagfeldt, A. *J. Am. Chem. Soc.* **2010**, *132*, 16714.
24. Bagnis, D.; Beverina, L.; Huang, H.; Silvestri, F.; Yao, Y.; Yan, H.; Pagani, G. A.; Marks, T. J.; Facchetti, A. *J. Am. Chem. Soc.* **2010**, *132*, 4074.
25. Bai, Y.; Zhang, J.; Zhou, D.; Wang, Y.; Zhang, M.; Wang, P. *J. Am. Chem. Soc.* **2011**, *133*, 11442.

26. Tsao, H. N.; Yi, C.; Moehl, T.; Yum, J.-H.; Zakeeruddin, S. M.; Nazeeruddin, M. K.; Grätzel, M. *ChemSusChem.* **2011**, *4*, 591.
27. Yum, J.-H.; Baranoff, E.; Kessler, F.; Moehl, T.; Ahmad, S.; Bessho, T.; Marchioro, A.; Ghadiri, E.; Moser, J.-E.; Yi, C.; Nazeeruddin, M. K.; Grätzel, M. *Nat. Commun.* **2012**, *3*, 631.
28. Harnessing Solar Energy in *Handbook of Porphyrin Science*; Kadish K. M., Smith K. M., Guilard R., Eds.; World Scientific Publishing 2014, Vol. 34.
29. Bonnett, R. *Chem. Soc. Rev.* **1995**, *24*, 19.
30. Hamblin, M. R.; Hasan, T. *Photochem. Photobiol. Sci.* **2004**, *3*, 436.
31. Henderson, B. W.; Dougherty, T. J. *Photochem. and Photobiol.* **1992**, *55*, 145.
32. Di Natale, C.; Paolesse, R.; Macagnano, A.; Mantini, A.; Goletti, C.; D'Amico, A. *Sens. Actuator B-Chem.* **1998**, *52*, 162.
33. Di Natale, C.; Salimbeni, D.; Paolesse, R.; Macagnano, A.; D'Amico, A. *Sens. Actuator B-Chem.* **2000**, *65*, 220.
34. Xu, Y.; Zhao, L.; Bai, H.; Hong, W.; Li, C.; Shi, G. *J. Am. Chem. Soc.* **2009**, *131*, 13490.
35. Choi, M.-S.; Yamazaki, T.; Yamazaki, I.; Aida, T. *Angew. Chem. Int. Ed.* **2004**, *43*, 150.
36. Gust, D.; Moore, T. A.; Moore, A. L. *Acc. Chem. Res.* **2009**, *42*, 1890.
37. Campbell, W. M.; Burrell, A. K.; Officer, D. L.; Jolley, K. W. *Coor. Chem. Rev.* **2004**, *248*, 1363.
38. Chiu, S.-W.; Lin, L.-Y.; Lin, H.-W.; Chen, Y.-H.; Huang, Z.-Y.; Lin, Y.-T.; Lin, F.; Liu, Y.-H.; Wong, K.-T. *Chem. Commun.* **2012**, *48*, 1857.
39. Kay, A.; Grätzel, M. *J. Phys. Chem.* **1993**, *97*, 6272.
40. Cherian, S.; Wamser, C. C. *J. Phys. Chem. B* **2000**, *104*, 3624.
41. Imahori, H.; Umeyama, T.; Ito, S. *Acc. Chem. Res.* **2009**, *42*, 1809.
42. Radivojevic, I.; Varotto, A.; Farley, C.; Drain, C. M. *Energy Environ. Sci.* **2010**, *3*, 1897.
43. Hsieh, C.-P.; Lu, H.-P.; Chiu, C.-L.; Lee, C.-W.; Chuang, S.-H.; Mai, C.-L.; Yen, W.-N.; Hsu, S.-J.; Diau, E. W.-G.; Yeh, C.-Y. *J. Mater. Chem.* **2010**, *20*, 1127.
44. Chang, Y.-C.; Wang, C.-L.; Pan, T.-Y.; Hong, S.-H.; Lan, C.-M.; Kuo, H.-H.; Lo, C.-F.; Hsu, H.-Y.; Lin, C.-Y.; Diau, E. W.-G. *Chem. Commun.* **2011**, *47*, 8910.
45. Lee, M. J.; Seo, K. D.; Song, H. M.; Kang, M. S.; Eom, Y. K.; Kang, H. S.; Kim, H. K. *Tetrahedron Lett.* **2011**, *52*, 3879.
46. Orbelli Biroli, A.; Tessore, F.; Pizzotti, M.; Biaggi, C.; Ugo, R.; Caramori, S.; Aliprandi, A.; Bignozzi, C. A.; De Angelis, F.; Giorgi, G.; Licandro, E.; Longhi, E. *J. Phys. Chem. C* **2011**, *115*, 23170.
47. Pasunooti, K. K.; Song, J.-L.; Chai, H.; Amaladass, P.; Deng, W.-Q.; Liu, X.-W. *J. Photochem. Photobiol. A Chem.* **2011**, *218*, 219.
48. Warnan, J.; Buchet, F.; Pellegrin, Y.; Blart, E.; Odobel, F. *Org. Lett.* **2011**, *13*, 3944.
49. Xiang, N.; Zhou, W.; Jiang, S.; Deng, L.; Liu, Y.; Tan, Z.; Zhao, B.; Shen, P.; Tan, S. *Sol. Energy Mater. Sol. Cells* **2011**, *95*, 1174.

50. Yella, A.; Lee, H.-W.; Tsao, H. N.; Yi, C.; Chandiran, A. K.; Nazeeruddin, M. K.; Diau, E. W.-G.; Yeh, C.-Y.; Zakeeruddin, S. M.; Grätzel, M. *Science* **2011**, *334*, 629.
51. Wang, C.-L.; Hu, J.-Y.; Wu, C.-H.; Kuo, H.-H.; Chang, Y.-C.; Lan, Z.-J.; Wu, H.-P.; Wei-Guang Diau, E.; Lin, C.-Y. *Energy Environ. Sci.* **2014**, *7*, 1392.
52. Mathew, S.; Yella, A.; Gao, P.; Humphry-Baker, R.; CurchodBasile, F. E.; Ashari-Astani, N.; Tavernelli, I.; Rothlisberger, U.; NazeeruddinMd, K.; Grätzel, M. *Nat. Chem.* **2014**, *6*, 242.
53. Nazeeruddin, M. K.; De Angelis, F.; Fantacci, S.; Selloni, A.; Viscardi, G.; Liska, P.; Ito, S.; Takeru, B.; Grätzel, M. *J. Am. Chem. Soc.* **2005**, *127*, 16835.
54. Yu, Q.; Wang, Y.; Yi, Z.; Zu, N.; Zhang, J.; Zhang, M.; Wang, P. *ACS Nano* **2010**, *4*, 6032.
55. Horiuchi, T.; Miura, H.; Sumioka, K.; Uchida, S. *J. Am. Chem. Soc.* **2004**, *126*, 12218.
56. Kim, S.; Lee, J. K.; Kang, S. O.; Ko, J.; Yum, J. H.; Fantacci, S.; De Angelis, F.; Di Censo, D.; Nazeeruddin, M. K.; Grätzel, M. *J. Am. Chem. Soc.* **2006**, *128*, 16701.
57. Hwang, S.; Lee, J. H.; Park, C.; Lee, H.; Kim, C.; Park, C.; Lee, M.-H.; Lee, W.; Park, J.; Kim, K.; Park, N.-G.; Kim, C. *Chem. Commun.* **2007**, 4887.
58. Wang, Z.-S.; Koumura, N.; Cui, Y.; Takahashi, M.; Sekiguchi, H.; Mori, A.; Kubo, T.; Furube, A.; Hara, K. *Chem. Mater.* **2008**, *20*, 3993.
59. Nazeeruddin, M. K.; Humphry-Baker, R.; Officer, D. L.; Campbell, W. M.; Burrell, A. K.; Grätzel, M. *Langmuir* **2004**, *20*, 6514.
60. Hara, K.; Wang, Z.-S.; Sato, T.; Furube, A.; Katoh, R.; Sugihara, H.; Dan-oh, Y.; Kasada, C.; Shinpo, A.; Suga, S. *J. Phys. Chem. B* **2005**, *109*, 15476.
61. Park, J. K.; Lee, H. R.; Chen, J.; Shinokubo, H.; Osuka, A.; Kim, D. *J. Phys. Chem. C* **2008**, *112*, 16691.
62. Tang, C. W. *Appl. Phys. Lett.* **1986**, *48*, 183.
63. He, Y.; Li, Y. *Phys. Chem. Chem. Phys.* **2011**, *13*, 1970.
64. Silvestri, F.; Irwin, M. D.; Beverina, L.; Facchetti, A.; Pagani, G. A.; Marks, T. J. *J. Am. Chem. Soc.* **2008**, *130*, 17640.
65. Viterisi, A.; Montcada, N. F.; Kumar, C. V.; Gispert-Guirado, F.; Martin, E.; Escudero, E.; Palomares, E. *J. Mater. Chem. A* **2014**, *2*, 3536.
66. Wei, G.; Wang, S.; Sun, K.; Thompson, M. E.; Forrest, S. R. *Adv. Energy Mater.* **2011**, *1*, 184.
67. Wei, G.; Xiao, X.; Wang, S.; Zimmerman, J. D.; Sun, K.; Diev, V. V.; Thompson, M. E.; Forrest, S. R. *Nano Lett.* **2011**, *11*, 4261.
68. Lee, O. P.; Yiu, A. T.; Beaujuge, P. M.; Woo, C. H.; Holcombe, T. W.; Millstone, J. E.; Douglas, J. D.; Chen, M. S.; Fréchet, J. M. J. *Adv. Mat.* **2011**, *23*, 5359.
69. Tamayo, A. B.; Dang, X.-D.; Walker, B.; Seo, J.; Kent, T.; Nguyen, T.-Q. *Appl. Phys. Lett.* **2009**, *94*.
70. Walker, B.; Tamayo, A. B.; Dang, X.-D.; Zalar, P.; Seo, J. H.; Garcia, A.; Tantiwiwat, M.; Nguyen, T.-Q. *Adv. Funct. Mat.* **2009**, *19*, 3063.
71. Rousseau, T.; Cravino, A.; Bura, T.; Ulrich, G.; Ziessel, R.; Roncali, J. *J. Mater. Chem.* **2009**, *19*, 2298.

72. Lin, H.-W.; Lin, L.-Y.; Chen, Y.-H.; Chen, C.-W.; Lin, Y.-T.; Chiu, S.-W.; Wong, K.-T. *Chem. Commun.* **2011**, *47*, 7872.
73. Lin, L.-Y.; Chen, Y.-H.; Huang, Z.-Y.; Lin, H.-W.; Chou, S.-H.; Lin, F.; Chen, C.-W.; Liu, Y.-H.; Wong, K.-T. *J. Am. Chem. Soc.* **2011**, *133*, 15822.
74. Kyaw, A. K. K.; Wang, D. H.; Wynands, D.; Zhang, J.; Nguyen, T.-Q.; Bazan, G. C.; Heeger, A. J. *Nano Lett.* **2013**, *13*, 3796.
75. Hatano, J.; Obata, N.; Yamaguchi, S.; Yasuda, T.; Matsuo, Y. *J. Mater. Chem.* **2012**, *22*, 19258.
76. Huang, Y.; Li, L.; Peng, X.; Peng, J.; Cao, Y. *J. Mater. Chem.* **2012**, *22*, 21841.
77. Li, L.; Huang, Y.; Peng, J.; Cao, Y.; Peng, X. *J. Mater. Chem. A* **2012**, *1*, 2144.
78. Sharma, G. D.; Daphnomili, D.; Biswas, S.; Coutsolelos, A. G. *Org. Electr.* **2013**, *14*, 1811.
79. Shi, S.; Wang, X.; Sun, Y.; Chen, S.; Li, X.; Li, Y.; Wang, H. *J. Mater. Chem.* **2012**, *22*, 11006.
80. Qin, H.; Li, L.; Guo, F.; Su, S.; Peng, J.; Cao, Y.; Peng, X. *Energy Environ. Sci.* **2014**, *7*, 1397.

3 Semi-Synthetic Responsive Matrices for Artificial Photosynthesis

R. Purchase,*,† R. Cogdell,‡ F. Breitling, V. Stadler,¶ N. van Hulst,|| G.-J. Kramer, A. Ramirez,** R. Zwijnenberg, †† A. Kallergi,†† J. B. de Baan,‡‡ I. Rudra§§ and H. J. M de Groot*,||||**

*Biophysical Organic Chemistry/Solid State NMR Leiden Institute of Chemistry, RA Leiden, The Netherlands
†Natural and Artifical Photosynthesis Research School of Chemistry, The Australian National University, Acton, ACT 2601, Australia
‡Institute of Molecular, Cell and Systems Biology, University of Glasgow, Glasgow G12 8TA, Scotland, UK
§Institute of Microstructure Technology, Karlsruhe Institute of Technology, Hermann-von-Helmholtz-Platz, 1 76344 Eggenstein-Leopoldshafen, Germany
¶PEPperPRINT GmbH Rischerstrasse, 12 69123 Heidelberg, Germany
||ICFO — the Institute of Photonic Sciences, The Barcelona Institute of Science & Technology, Av. Carl Friedrich Gauss, 3 08860 Castelldefels, Barcelona, Spain
**Copernicus Institute of Sustainable Development Heidelberglaan, 2 3584 CS Utrecht, The Netherlands
††Leiden University Centre for Arts in Society, 2300 RA Leiden, The Netherlands

‡‡Royal Academy of Art The Hague Prinsessegracht, 4 2514 AN The Hague The Netherlands
§§Computational Centre of Expertise, Shell India Market Pvt. Ltd., Plot No. 7, Bangalore Hardware Park, Devanahalli Industrial Park, Bande (Mahadeva) Kodigehalli, Bengaluru – 562149, India
||||groot_h@chem.leidenuniv.nl

Table of Contents

List of Abbreviations

Aa	amino acids
Bb	building blocks
cLIFT	laser induced forward transfer for combinatorial synthesis
DNA	desoxyribonucleic acid
EET	excitation energy transfer
FRET	Förster resonance energy transfer
LCA	life cycle assessment
LH2	light harvesting 2
MAS NMR	magic angle spinning NMR
NA	numerical aperture
NCAP	non-adiabatic conversion by adiabatic passage
NMR	nuclear magnetic resonance

POR	porphyrins
PS	polystyrene
PV	photovoltaic
PYR	styrylpyridinium salts
RC/LH1	reaction center/light harvesting 1 core complex
REDOR NMR	rotational-echo, double-resonance NMR
TRL	technology readiness level

Nature engineers responsive protein matrices to control pigments' properties for light energy conversion in photosynthesis, in contrast to the standard chemists' approach where the chemical properties of the dye molecules are modified and the matrix is ignored. Engineering the matrix offers very subtle ways to enhance the quantum efficiency of the breaking and making of chemical bonds in photochemistry and catalysis by non-adiabatic conversion of reactants into products. We present a roadmap for the development of "next-generation," high quantum efficiency artificial photosynthetic devices that convert reactants into products with long living twisting modes. These devices are based on learning from nature how photosynthetic proteins are activated for non-adiabatic conversion with high yield by structural control over collective modes in the thermal molecular dynamics. When driven far from equilibrium by optical excitation, energy transfer, or charge transfer, conformational twisting develops quantum states that are not eigenstates by the crossing of reactant and product energy levels. The off-diagonal elements in the Hamiltonian drive photochemical reactions in a lossless non-adiabatic conversion by adiabatic passage (NCAP) process where information from handedness in the structure is transiently converted into quantum information and back into structural information. Truncation of the Hamiltonian in the interaction frame of the nuclear motion associated with a long living vibration coupled to conformational twisting in a chiral responsive matrix enables coherent conversion.

The construction of proof-of-principle responsive matrix functional materials that implement this NCAP concept in biohybrid and bioinspired architectures for incorporation into artificial photosynthesis devices is currently emerging. We propose tailoring sturdy bacterial light-harvesting and reaction center complexes to enable the exploration of structure-dynamics relationships for the maximisation of functional quantum coherence induced by NCAP. With such a blueprint at hand, high-density protein-like arrays can be generated with high throughput peptide synthesis, assembled for bottom up construction of semisynthetic organic-inorganic hybrid architectures and screened for high yield, quantum-coherent enabled energy transfer and charge separation. To resolve the anisotropic motion from non-aligned enhanced twisting modes in the ground-state dynamic structure, and the spatial and temporal extent of coherences in the excited state, we advocate MAS NMR on isotope labelled preparations and ultra-fast optical spectroscopy. Input from experiments can guide the accurate multi-scale simulation of high yield light harvesting and charge separation to determine the optimal characteristics for complex bottom up construction of artificial photosynthesis device vehicles, such as size of the responsive matrix; the number, type and combination of chromophores; architecture; nature of the inter-chromophore electronic coupling; and means to promote robustness including withstanding oxidation.

The principle ethics of artificial photosynthesis is that it has to disappear into its practice of decentralized application. Environmental, humanities and design researchers can evaluate the environmental, energetic and economic consequences of artificial photosynthesis, envision the potential for systemic transformation in our relationship with energy and natural resources, and highlight the benefits and challenges we are likely to encounter on a path to large-scale deployment of responsive matrix technology in the densely populated environment.

I. Introduction

Around the world, large-scale deployment of wind and solar energy to generate renewable electricity is expected. However, to achieve a zero-carbon energy system, the world will need more than intermittent electricity. It will continue to rely massively on fuel for certain applications, such as heavy transport, aviation and iron ore reduction, and a long-term dependence on hydrocarbons is likely. Thus, the case for research and development into the field of artificial photosynthesis — systems to directly convert solar energy into fuel without oil or biomass intermediate — is strong. Furthermore, artificial photosynthesis, in contrast to other solar-to-fuel technologies, has the potential to be deployed in a small-scale, decentralized fashion. We have a vision of artificial photosynthesis becoming a common technology, and a significant contributor to the future energy resource mix. Although artificial photosynthesis is still a scientific challenge, all aspects of its deployment in urban environments, such as energy and resource/material availability, ecofriendliness and efficiency, and pathways from niche application to large-scale implementation and approval will need to be investigated. In this chapter, we present a roadmap for the development of "next-generation" artificial photosynthesis devices.

II. Nature's Paradigm

The early stages of photosynthesis can operate at high quantum efficiency across many different antenna and reaction center configurations. These natural systems are an inspiration for the next-generation of artificial photosynthesis devices and act as a blueprint for their construction. A key aspect of nature's approach is that biology engineers the matrix to control the properties of the pigments. This is in contrast to the standard chemists' approach of changing the properties of the pigment and ignoring the matrix. Engineering the dynamic structure of the matrix offers a whole new toolbox for subtle ways to enhance the outcome of both photochemistry and catalysis. This concept has become known as the responsive matrix.

A paradigm shift in artificial photosynthesis research can be achieved by engineering quantum coherent conversion at avoided crossings in chiral responsive matrices, similar to proteins found in nature. In photosynthetic complexes, the ubiquitous quantum instabilities that result from transient off-diagonal or non-adiabatic matrix elements in electron-nuclear Hamiltonians lead to resonant conversion of reactants into product states through vibronic coherence. Well out of equilibrium, chemical conversions can proceed in a deterministic fashion, transiently violating the limitations of the Carnot process. Quantum beats[1,2] with lifetimes of up to several ns have been observed in natural photosynthetic samples for both excitonic energy transfer[3] and charge separation[4,5] following optical excitation by means of 2D electronic spectroscopy. Here, we propose a much more general structure-function concept: non-adiabatic conversion by adiabatic passage (NCAP). Guided by our recent extensive analyses of semi-classical interconversion of electronic states with vibrational coherence in natural and artificial systems, we propose that photoexcitation of a responsive matrix leads to a conformational twisting response along a low frequency supramolecular mode with enhanced spectral density.[6,7] This leads to vibronic coupling of exciton levels and semi-classical coherent conversion of exciton states by resonance with a more rapid, second low frequency mode.[8] Although NCAP is a dissipative process out of equilibrium, it parallels the cross correlation detection of "NMR spin noise" without photoexcitation and "noisy light spectroscopy" with incoherent excitation that were recently taken to the level of 2D coherent spectroscopy.[9–12] During the dissipative adiabatic passage in a responsive matrix, near-unity resonant conversion from reactant |r> to product |p> is enabled by a transient semi-classical channel through the energy barrier (Figure 1). |r> and |p> are two pure states that describe, for example, antennae, charge separation or catalysts motifs during photoinduced excitation energy transfer, charge transfer or catalysis, respectively. Since resonators have a low impedance for the matching oscillation, the elec-

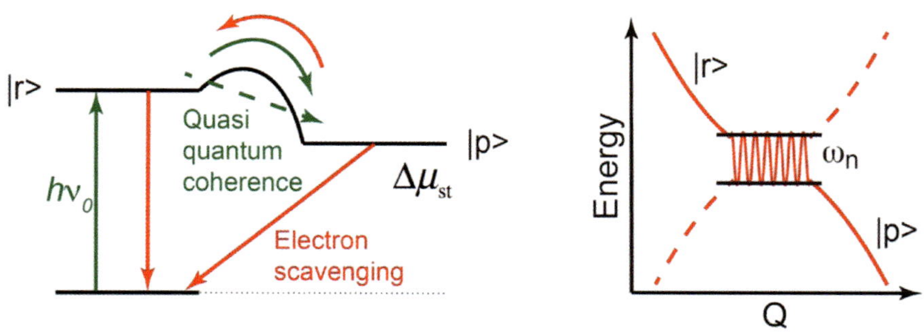

Figure 1. Conversion from a reactant state |r> to a product state |p>. Forward processes are shown in green and reverse processes in red. Semi-classical transfer from |r> to |p> proceeds with a channel through the barrier (green dashed arrow, left) by a conformational twisting response of the matrix for dynamic level crossing of |r> and |p>, and is driven by a ω_n twisting mode (right). This is in contrast to the Carnot process, which is shown as a hop over the barrier (green curved arrow, right). The Carnot process may be subject to back transfer (red curved arrow) and subsequent relaxation (red downwards arrow) or direct recombination as a result of electron scavenging on the nanoscale (red diagonal arrow).

tron-nuclear system can take up energy from the environment and give it back again, which leads to a robust mechanism for selection of a matching vibration.

Dynamic symmetry breaking is essential for such a vibrationally-coupled process, because without it, crossing of the electronic states is forbidden. We derived an analytic expression for the conversion by writing the equations of motion in the interaction frame of the nuclear circulation in terms of a Rabi oscillation with semi-classical angular momentum from molecular twisting in the chiral matrix, as opposed to spontaneous symmetry breaking, by selecting one of two counter rotating components representing a linear vibration.[8] Our perturbative approach to renormalize quantum and classical degrees of freedom along different directions in Hilbert space, follows a similar physical approach for describing the interaction of nuclear spins with macroscopic classical rotation.[13]

The physics of NCAP is fundamentally different from the current practice in quantum biology. It induces complete conversion by time reversal during adiabatic passage instead of stochastic decoherence. Adiabatic passage enables long-range transfer by recurrent deterministic coherence/decoherence and touches upon a fundamental property of quantum networks: systems in a pure state subject to non-adiabatic couplings self-select the fastest channel to the output. The dissipative character of NCAP provides a handle for kinetic control over the conversion process by trading time for energy efficiency. With minimal dissipation, slow and lossless conversion is realized, such as in rubisco;[14] while large dissipation leads to ultrafast conversion, as in the visual pigment rhodopsin.[15–18]

A principle challenge is thus to learn from nature and develop responsive matrices with quantum coherent kinetic control for high yield. We consider the absorption of ~90% of the incoming photons over as large as practicable part of the solar spectrum, an internal quantum efficiency of ~80%, and stability on the timescale of months a realistic target for practical implementation of photochemical conversion. The photon to product yield of the solar-to-fuel conversion technology will then be 90% × 80% = 72% of the thermodynamic limit. The higher the quantum efficiency, the less materials and surface area is needed. Current laboratory scale devices that produce hydrogen do not achieve anywhere near 72% external quantum efficiency and 80% photochemical yield.

Going from chemical yield to energy efficiency, the detailed balance energy conversion efficiency limit for near-quantitative conversion of photons to hydrogen is 42%, for a tandem absorber with cut off wavelengths of 700 and 1100 nm that drives water splitting and fuel production with two photons per electron transported into fuel. A realistic energy efficiency is thus 90% × 80% × 42% = 30%, which takes into account losses such as reflection at the surface, variation of the intensity over the optical spectrum during the day or between seasons, and a limited filling factor.[19] Most state of the art devices operate at around 5–10% solar-to-hydrogen efficiency.[20,21] There are a few very expensive devices that achieve around 15%.[22]

III. State of the Art

Current research strategies for converting sunlight to fuel with artificial photosynthesis are mostly heuristic and lead to similar scientific problems across a wide range of materials. Entropy and unavoidable irreversibility place a detailed balance thermodynamic limit on the efficiency of adiabatic photochemical energy conversion

$$\Delta\mu_{\mathrm{st}} = h\nu_0 - k_{\mathrm{B}}T\ln\frac{t}{\tau} \tag{1}$$

where $\Delta\mu_{st}$ is the chemical potential of energy storage and there is no $P\Delta V$ work involved from a volume change.[23] The first term in equation 1 is the energy difference for light absorption between the highest occupied molecular orbital and the lowest unoccupied molecular orbital. The second term is the mixing entropy for detailed balance between the concentrations of the photoproduct in the light and in the dark, where t is the lifetime of the storage reservoir, and τ is the excited state lifetime of the absorber. The longer the energy needs to be stored, the more mixing entropy needs to be generated in the chemical network to prevent light reactions running backwards and wasteful recombination. This is shown schematically in Figure 1.

When photochemical conversion proceeds near equilibrium (i.e. adiabatic) with a series of incoherent (i.e. Carnot type) conversion steps with intermediate dilution and concentration, as in PV coupled to electrolysis, the entropy and unavoidable irreversibility losses factorize rapidly. In contrast, the experimentalist designing an artificial photosynthetic system is faced with the choice of either ensuring that there is a big enough drop in energy built into the system to shift the photochemical equilibrium towards chemical storage and prevent light-driven reactions from running backwards during long storage times t leading to an energetically inefficient device; or, reducing the energy difference across the system, but then having a very poor quantum efficiency owing to recombination losses from higher probability back transfer followed by rapid fluorescence recombination with a rate $1/\tau$. In addition, for multi electron catalysis, charges need to be separated sufficiently on the nanoscale to prevent direct recombination by tunneling, but not too far apart to avoid unnecessary concentration losses. This requires matching of the tunneling distances with the storage times t for intermediate stages. Currently, there are no good solutions to overcome these two fundamental hurdles.

IV. Design Principles for Artificial Photosynthesis Based on Nature's Paradigm

The materials that we can use to mimic the responsive matrices that nature uses to enable high yield light reactions are limited to substances that can sustain correlated electrons and are capable of breaking and making chemical bonds with long-lived vibrational modes. Also, materials need to foster dynamic symmetry breaking in the NCAP mechanism. In natural photosynthesis, this is almost certainly provided by the intrinsic and induced chirality of the protein complexes and the embedded cofactors. Artificial materials that fulfil these criteria are, for instance, supramolecular systems, which, when biomimetically designed, are ready for quantum coherence. Paracrystalline or semicrystalline polymers with control over the dynamic structure,[24] too, may prove to be a fruitful avenue of investigation. There may be some parallels with the material requirements for quantum computing and spintronics, so some of the semiconducting materials used there may also be appropriate. While the use of conductors with free, uncorrelated electrons is problematic, materials with macroscopic quantum effects such as high-T_c superconductors may offer promises.

In nature, materials for light harvesting, charge separation, and catalysis are integrated in high-density arrays in the photosynthetic membrane. Such an arrangement enables high optical densities for efficient light absorption and efficient energy and electron transfer between neighboring components. Analogs of these high-density arrays have been successfully employed in artificial photosynthesis in the form of nanowire arrays for efficient light absorption.[25–27] Furthermore, high-density arrays are amenable to high-throughput manufacturing processes such as tiling and printing. High-density arrays are therefore another aspect of natural photosynthesis that should be mimicked in artificial systems.

V. Reverse Engineering Nature

Evolution has engineered the control of properties by photosynthetic matrices in biology by historical contingency over approximately three billion years. It is thus wise to use the biological matrix as a starting point and work from semi-synthetic materials, instead of *de novo* bottom up design of complex matrix systems with a controlled dynamic response for chemical conversion. With access to the universal principles that govern NCAP in natural systems, reverse engineering of natural photochemistry and catalysis in semi-synthetic materials using modified molecular building blocks and mimicking natural protein environments in supramolecular chemistry can rapidly become reality.

A variety of close-packed 2D arrays of photosynthetic complexes in native membranes can be used as a natural test bed for the generation, retention and quenching of coherences during intra- and inter-complex NCAP energy transfer and charge separation reactions. We will need natural photosynthetic samples that are robust, abundant, easily modified and in which quantum coherent behavior has been observed. The photosynthetic apparatus of photosynthetic purple bacteria is an example of such a system. A typical purple bacterial photosynthetic unit consists of reaction center/light harvesting 1 core complexes (RC/LH1) and light harvesting 2 complexes (LH2). The X-ray crystal structures of these complexes have been determined and are shown in Figure 2.[28,29–30] They are all constructed from the same basic design principle. They are circular oligomers of pairs of small, hydrophobic apoproteins. Each pair of apoproteins binds non-covalently a small

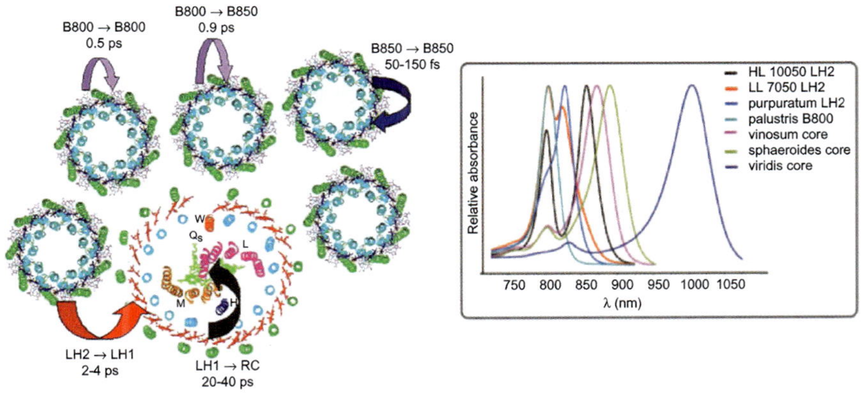

Figure 2. Membrane organization (left) and dispersion of the optical absorption for different LH species (right).

number of bacteriochlorophylls and carotenoids. The characteristic absorption spectra of these complexes are shown in Figure 2. The different pigment groups seen in the 3D structures in Figure 2 absorb light in separate regions of the spectrum. This makes functional studies straightforward in that each group can be specifically excited and their energy transfer and charge separation reactions clearly monitored. We know a great deal, therefore, about the basic structure and functional mechanisms of these purple bacterial complexes. The size and composition of the purple bacterial photosynthetic unit is controlled by the light intensity and wavelengths at which the bacterial cells are grown.[31,32] This gives exquisite control over which complexes are made and their relative arrangement in the native membranes.

With a view to artificially producing high-density arrays of photosynthetic units, and also with the aim of studying interactions between photosynthetic units, tiled arrays of purple bacterial photosynthetic units can be produced. Since purple photosynthetic bacteria are genetically amenable, mutant complexes and complexes tagged with DNA-binding zinc finger domains can readily be made so that they can be attached to DNA tiles in a sequence-specific manner.[33] This procedure is shown schematically in Figure 3. Photosynthetic complexes can be placed on these tiles with exquisite precision, enabling the investigation of the effects of distance (and hence coupling) between neighboring complexes upon both energy transfer efficiency and retention of coherences.

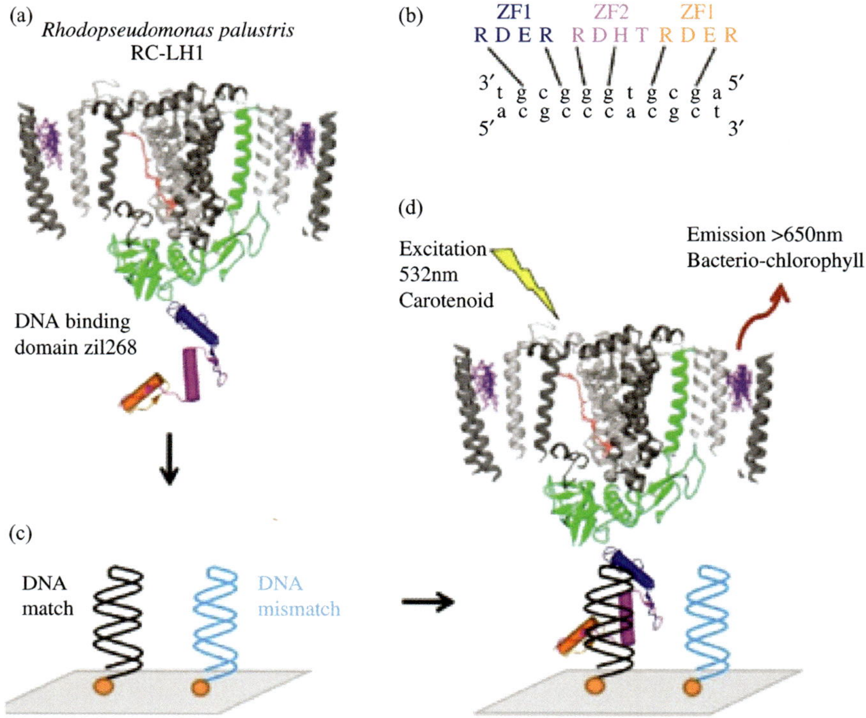

Figure 3. (a) Structure of the RC-LH1 from *Rhodopseudomonas palustris* (gray) with the H-subunit highlighted in green. (b) An 11 bp DNA zif268 recognition sequence. (c) Immobilization of cyclooctyne-tagged dsDNA to an azide-modified surface. (d) Excitation Energy Transfer (EET) was measured between the carotenoid pigment (in red) and bacteriochlorophyll (highlighted in purple) as a test for photosynthetic function. Reproduced from Henry *et al.*[33] with permission from The Royal Society of Chemistry.

Photosynthetic components are pre-programmed in the electronic ground state by their protein environment for optimal operation in the excited state. To guide the optimization of these processes in semi-synthetic systems, we will need to determine the global dynamics and local order parameters of pigments and amino acid side chains. MAS NMR is here the method of choice, since it enables sub-Å characterization of the dynamic structure.[34] In particular solid-state NMR relaxation and REDOR dipolar anisotropy measurements are excellent ways to obtain information about conformational twisting ground state dynamics. Purple bacteria can be grown on a variety of supplemented media to achieve isotope labeling to tailor samples for NMR analysis.

For the excited state, ultrafast optical spectroscopy is an excellent tool to elucidate the role of vibrations and the extent of coherences on the 10–100 nm level following coherent excitation or by noisy light spectroscopy.[35,36] Persistent coherences can also be addressed with single molecule detection in a confocal microscope with high NA (1.4-oil) excitation/detection and the sample in controlled atmospheric or liquid conditions. In this way, the fs-ps energy transfer path of the coherences and inter complex energy transfer can be traced. Localized coherent excitation enables the resolution of spatio-temporal energy transport on the quantum transfer network.[37,38] Such an experiment is shown conceptually in Figure 4.

To enhance our understanding of the structure-function relationships for NCAP, experiments should be complemented by theoretical methods. Guided by experimental data, modeling and simulations at various levels can be performed to inform the design process. Theoretical investigations should include quantum classical simulations of light harvesting and charge separation with NCAP.[6,7,39,40] Also, multi-scale modeling may yield insight into aspects of NCAP such as the size of the responsive

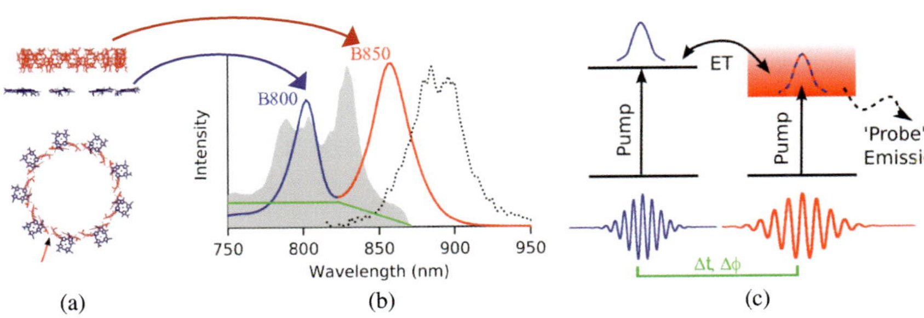

Figure 4. Concept of ultrafast phase-coherent excitation of individual light-harvesting complexes at room temperature. (a) Simplified structure of LH showing B800 (blue) and B850 (red) bands. (b) Absorption spectrum of B800 and B850, and the emission spectrum shown as dotted line. Broad band exciting laser spectrum (gray) and the applied spectral phase for pulse shaping (green line). (c) Concept of the experiment: interference of energy transfer (ET) and direct excitation is recorded in the fluorescent "probe" emission. Modified from Hildner *et al.*[37] with permission from AAAS.

matrix; the number, type and combination of chromophores; architecture (linear, cyclic, dendrimeric, lateral *vs.* layered); nature of the interchromophore electronic coupling (weak, strong); and means to promote robustness including withstanding oxidation.

Such an approach will enable the exploration of fundamental questions about energy transduction with NCAP and to use this improved understanding to boost quantum coherence with recurring adiabatic-passage events driven by interacting modes of molecular twisting. Questions such as the extent to which these coherences vary depending upon the size of the complexes and whether the coherences can be modulated by site-directed mutagenesis of apoproteins can be addressed. In this way, we can establish the physical principles that enable the maximisation of the functional role of NCAP in energy transduction in photosynthesis to efficiently concentrate energy for the long-term goal of fuel production. Study of the natural system will produce a blueprint for how to fabricate an artificial photosynthetic system with quantum coherences.

Once the "rules" by which NCAP operates in the natural system are established, this mechanism can be reverse-engineered in a more readily manufactured semi-synthetic material. High-throughput manufacturing techniques such as printing and tiling are good avenues to explore. In particular, there are two recently developed novel methods to synthesize and validate millions of different protein-like structures in unprecedented high-density array format. The one-cavity-one-peptide method enables the synthesis of millions of protein-like structures with built-in chromophores and electron relaying "stepping stones" at low cost.[41] A second technique uses Laser Induced Forward Transfer of materials for the combinatorial synthesis of high-density peptide arrays (cLIFT).[42] These techniques enable some 5000 to 10000 peptides per cm^2 in very high quality to be synthesized. Designer pigments synthesized from amino acid building blocks, such as porphyrin, styrylpyridinium, corrole, boron dipyrromethene or pyridinophane, can be introduced into the peptides arrays at different, well-controlled sites (i, k) within the growing peptide chain. Figure 5 shows the general formula of the desired constructs incorporating porphyrins (POR) or styrylpyridinium salts (PYR) prepared by the group of Balaban, who was a pioneer in this field. Different building blocks can be assembled for bottom up construction of semi-synthetic architectures.

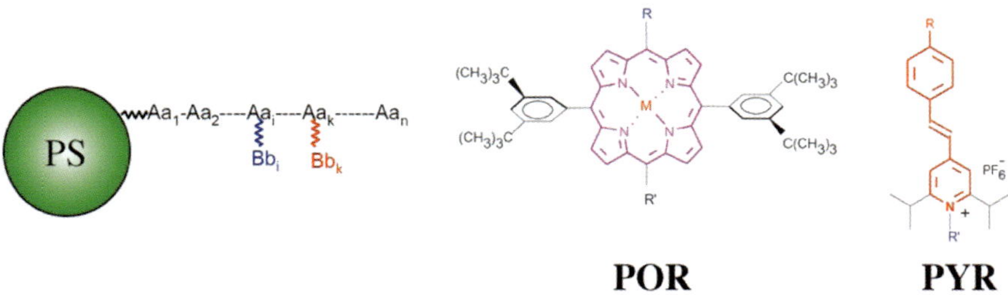

Figure 5. A peptide formed by *n* coupling steps of amino acids **Aa**. One or several building blocks **Bb** can be inserted at will, at any desired position within the peptide anchored on the solid support **PS**. The chromophoric system responsible for the photophysical properties is shown with coloured bonds while solubility inducing groups are black. M can be either 2H or various metals such as Zn, Cu, Ni, Mg, Cd, etc. The substituent R can be varied at will to both fine tune the absorption wavelength and redox potential while the blue substituents R′ serve as anchors for the automated solid phase synthesis of combinatorial peptides.

Furthermore, quite a few proteins and peptides are known to nucleate or bind to a diverse range of inorganic materials, e.g. ZnO. With these building blocks at hand, miniaturised self-assembled architectures can be constructed, e.g. by tiling different peptides that would nucleate different materials to DNA-based origami structures. Especially, two such building blocks fused together on a nanoscale distance should generate janus type particles made of two different semi-conductor materials that could serve as an artificial reaction center, which, again could be placed into close contact with semi-artificial or artificial LHC proteins (Figure 6).

As with the investigations of the natural system, quantum classical simulations, multi-scale modeling, and spectroscopic analyses with NMR and single molecule spectroscopy will guide the design of the system. Novel, semi-synthetic constructs will be screened to find molecule-surroundings that influence the spectroscopic behavior of built-in chromophores and, eventually, have semi-classical coherent conversion of reactants into products upon excitation with light, assisted by conformational twisting.

These techniques can be used to systematically scrutinize large numbers of peptides for quantum coherence and their capacity to nucleate and grow inorganic materials in array format (semiconductors, magnetite,

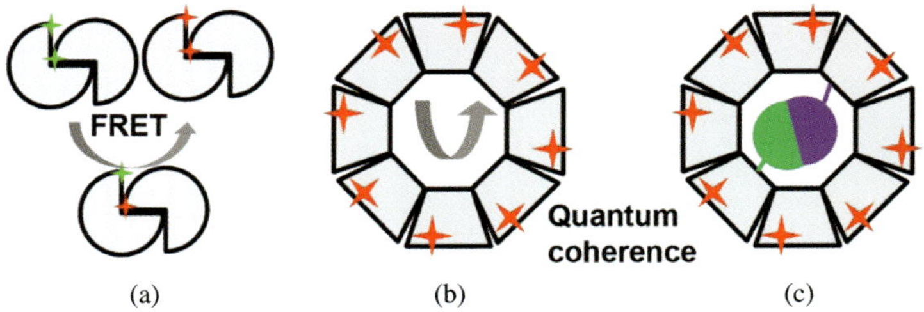

Figure 6. Screen for self-assembling protein-like structures. Peptides (> 40mers) are synthesized at high density in small cavities either with the one-cavity-one-peptide method (for initial screens), or with cLIFT (for systematic variations in the peptide's sequence), side chain protection groups are removed by harsh chemistry, the cavities are filled with a solvent (e.g. *n*-heptane), and sealed from each other, before the peptides are released by UV light into "their" cavity, where they self-assemble to dimers (a) or oligomers (b). Especially when "talking to each other" due to quantum coherence (one type of chromophore, see b), or by FRET (a mixture of two types of chromophores built into the peptides, see a), these oligomers drastically change their optical properties, which can be pre-screened with a high-resolution scanner equipped with three lasers and a large set of optical filters. Eventually, oligomerising peptides (doing quantum coherence) will be linked (by direct fusion, or by tiling them to DNA-origami structures) to additional peptides that nucleate inorganic material that could serve as an artificial reaction center (c).

insulators). With this step, proof of principle of the knowledge obtained from the natural system is provided. These protein-like materials explored with high-throughput peptide synthesis will be fundamentally different from the materials that are currently used in artificial photosynthesis research.

VI. The Societal Implications of Large-scale Decentralized Deployment of Artificial Photosynthesis

The development of responsive matrices for artificial photosynthesis aims to address a conundrum in solar energy supply. On the one hand there is biomass from natural photosynthesis, which provides 10% of the primary energy mix, is both affordable and sustainable, but is not scalable; it is by far not enough.[43,44] On the other hand, there is photovoltaics followed by electrolysis, which has been around for many decades, is in principle

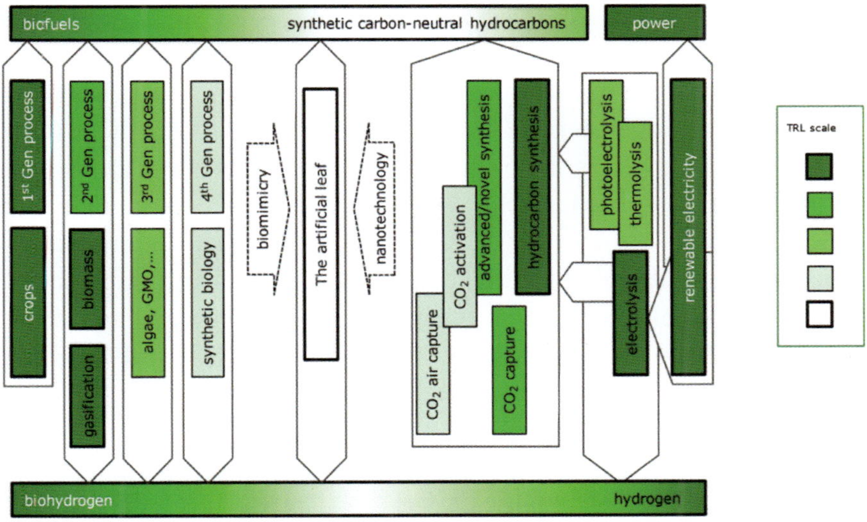

Figure 7. This schematic depicts the broad field of solar fuels, ranging from algal and other advanced biofuels, *via* synthetic biology to "the artificial leaf," onto 'conventional' conversion of renewable electricity, using electrolysis and synthesis with activated CO_2. This is a vibrant area of research. One can justifiably have high hopes for advances, because of the advances in life sciences (on the left) and nanotechnology (on the right). Artificial photosynthesis has the potential to combine the for all practical purposes "infinite" potential of solar energy, with the versatility of hydrocarbon fuels. At the highest level, the challenge is to find a technically novel optimum that combines the benefits of biofuels (i.e. a renewable hydrocarbon fuel) with the "unlimited" availability of renewable electricity. The colors are indicative of the so-called Technology Readiness Level (TRL) of the technology, running from dark green (mature technologies) to white, indicating technologies that are still conceptual or lab-scale.

scalable, but so far has not taken off for economic reasons. The challenge of converting hydrogen to a hydrocarbon fuel critically relies on the separate capture of CO_2 from air, so far a distant promise. This opens the opportunity space for more radically new technologies as proposed in this paper. The relationship between these advanced technologies and more conventional technologies is depicted in Figure 7. As is explained in the previous sections, the key lesson from nature is to avoid chains of Carnot cycles by integrating PV and electrochemistry, and use non-adiabatic (i.e. quasi quantum coherent) conversions that transiently violate detailed balance in semisynthetic NCAP responsive matrices. Artificial photosynthesis thus holds the potential to be both scalable and commercially viable as it takes the best aspects from biology and nanotechnology.

The Collingridge dilemma[45] states that in the early phases of development, a technology can still be changed, but the effects of the technology are hard to predict; when a technology is mature, its effects are clear, but it is difficult to change. By close collaboration with environmental, humanities and design researchers we hope that we can overcome the Collingridge dilemma by identifying, from the very beginning, semi-synthetic artificial photosynthesis technology development pathways that are not only technologically advanced, but also socially, environmentally and ethically compatible with large-scale urban deployment.

Life cycle assessment (LCA) is the prevailing quantitative framework for environmental assessment of energy and production systems. The combination of process-based life cycle inventories and input-output based models of the wider economy has been proposed to obtain both accurate and detailed descriptions of technologies while ensuring consistent and complete descriptions of the involved production network. This agnostic science approach ensures a basis for consistent environmental assessment across different technologies. For developing life cycle inventories of systems, several pieces of information are central: in particular, energy balances, and material and costs lists. Understanding the significance of different processes and emissions to different impact categories is at the heart of LCA. An understanding of how the impacts, and their uncertainties, accumulate throughout individual value chains, sub systems and network structures is best provided by application of methods such as structural path analysis. The lower the technological readiness of a technology, the greater is the uncertainty of its (future) performance and therefore of an LCA. Yet, the urgency of finding the most efficient solutions for high-yield, solar-energy conversion and storage in the context of the Paris agreement to establish a zero-emission energy system by 2045 and not later than 2065, with negative emissions thereafter, warrants that stakeholders are capable of deciding upon technology development paths with incomplete information. This is referred to as *ex-ante* technology life cycle assessment: the generation of a comprehensive assessment of opportunities and risks of technologies and their application, thereby providing knowledge suitable for actively shaping technological development. *Ex ante* LCA can be used to evaluate the environmental, energetic and economic consequences of artificial photosynthesis technology. It is crucial to the further development of this technology to establish the future success factors. When aspects of technological performance, economic viability and environmental impact are included in an integrated, early assessment of the technology, its future competitiveness against alternative low-carbon technologies can be evaluated.

Artificial photosynthesis is a technological response to our demand for renewable fuel. As such, it relates to our notions of nature, sustainability and the environment. Confronted with our environmental challenges, humanities research has embraced the subject of the environment, exposing the assumptions inscribed in our environmental thinking and challenging traditional notions of nature and the dichotomy of nature/technology. The now established field of environmental humanities, including practices such as eco-criticism and environmental ethics, is an essential resource for coping with our present environmental crisis, while transdisciplinary approaches are urged with regards to sustainable development and energy research.[46,47] Thanks to the humanities, we are getting more and more comfortable with the idea that humans are not fundamentally separate from nature, that nature is a socio-cultural construct, that the environment is in constant interaction with human technological activity and that even our cities and our urban environment are not necessarily in opposition to a "natural" ecosystem. At the same time, the framework of responsible innovation is gaining momentum in research and development.[48] It dictates continuous stakeholder involvement in responsive and reflective innovation.[49] Amongst various tools and methodologies involved, value sensitive design methodologies are attentive to the fact that our values are embodied in our technologies,[50] further justifying the significance of a philosophically informed conceptual investigation prior to the design process; yet, such methodologies may be inhibited by the dominant, and potentially outmoded, paradigms in our environmental thinking.

The ethics and aesthetics[51] of energy is that it has to disappear into its practice. Contrary to photovoltaics, artificial photosynthesis produces hydrogen or hydrocarbons in gaseous or liquid form that have to be stored locally or transported into a grid. Humanities researchers and industrial designers can work out a framework for understanding how the public may perceive (semi-synthetic) artificial photosynthesis, and interact with various stakeholders such as scientists, industry and the public to determine the design parameters required to make such a technology, on an unprecedented scale, desirable to the end user. We envision a deployment of artificial photosynthesis that is socially pioneering, socially responsible and one that fundamentally transforms our relationship with, and attitudes to, energy and natural resources. Such research may well include theoretical elaborations researched by means of design activities, such as envisioning and prototyping; and transform these concepts into tangible 3D artefacts, such as mock-ups. This approach safeguards that produced prototypes go beyond materialising current requirements, to become research artefacts that enable a critical reflection on the potential of artificial photosynthesis among all parties involved. Such research is relevant for developers of a technology such as artificial photosynthesis, as it promotes a consideration of technologies for renewable energy that is not agnostic to their broader societal implications but attentive to the fact that technologies are integrated into and interacting with socio-cultural systems.[52,53] For the discipline of the humanities, this research is relevant, as it advocates a dynamic role for the humanities as a force that can contribute to the shaping of a technology at an early stage. It seeks empowerment for the humanities to become a responsible shareholder in the societal debate about renewable energy resources and to critically support the implementation of technological innovation in a dynamic exchange between the humanities and science.

Together, these activities will enable us to envision the future large-scale implementation of artificial photosynthesis technologies; both its potential for systemic transformation as well as the challenges along the way. This will provide new perspectives on the societal impact of artificial photosynthesis, ease its introduction to the public and provide a means for disseminating information about future artificial photosynthesis technologies to the public and other stakeholders.

VII. Conclusions and Next Challenges

We offer a novel paradigm for artificial photosynthesis, based upon a new class of responsive matrix materials reverse engineered from nature, with many opportunities for scientific and applied research, and for innovation. When we have learned how to modulate the vibrational structure, established the symmetry breaking requirements for long-range quantum coherent energy transfer and applied it in a semi-synthetic material, this knowledge can be transferred and used non-heuristically to establish an entire new class of functional materials with superior properties at all scales, suitable for a broad range of energy conversion applications. This may prove to be an effective driver of all kind of catalytic conversion chemistries. Furthermore, the novel research methods advocated here, whereby science is combined with humanities and design, could well find broad application in the development of other future emerging technologies.

VIII. References

1. Chin, A. W.; Prior, J.; Rosenbach, R.; Caycedo-Soler, F.; Huelga, S. F.; Plenio, M. B. *Nat. Phys.* **2013**, *9*, 113.
2. Huelga, S. F.; Plenio, M. B. *Contemp. Phys.* **2013**, *54*, 181.
3. Chenu, A.; Scholes, G. D. *Annu. Rev. Phys. Chem.* **2015**, *66*, 69.
4. Novoderezhkin, V. I.; Romero, E.; van Grondelle, R. *Phys. Chem. Chem. Phys.* **2015**, *17*, 30828.
5. Romero, E.; Augulis R.; Novoderezhkin, V. I.; Ferretti, M.; Thieme, J.; Zigmantas, D.; van Grondelle, R. *Nat. Phys.* **2014**, *10*, 676.
6. Eisenmayer, T. J.; de Groot, H. J. M.; van de Wetering E.; Neugebauer, J.; Buda, F. *J. Phys. Chem. Lett.* **2012**, *3*, 694.
7. Monti, A.; Negre, C. F. A.; Batista, V. S.; Rego, L. G. C.; de Groot, H. J. M.; Buda, F. *J. Phys. Chem. Lett.* **2015**, *6*, 2393.
8. Purchase, R. L.; de Groot, H. J. M. *Interface Focus* **2015**, *5*, 20150014.
9. Bloch, F. *Phys. Rev.* **1946**, *70*, 460.
10. Chandra, K.; Schlagnitweit, J.; Wohlschlager, C.; Jerschow, A.; Müller, A. *J. Phys. Chem. Lett.* **2013**, *4*, 3853.
11. Coles, D. M.; Yang, Y.; Wang, Y.; Grant, R. T.; Robert A. Taylor, R. A.; Saikin, S. K.; Aspuru-Guzik, A.; David G. Lidzey, D. G.; Kuo-Hsiang Tang, J.; Smith, J. M. *Nat. Commun.* **2014**, *5*, 5561.
12. McCoy, M. A.; Ernst, R. R. *Chem. Phys. Lett.* **1989**, *159*, 587.
13. Boender, G. J.; Vega, S.; de Groot, H. J. M. *Mol. Phys.* **2011**, *95*, 921.
14. Yin, X.; Struik, P. C. *J. Exp. Bot.* **2015**, *66*, 6535.
15. Hahn, S.; Stock, G. *J. Phys. Chem. A*, **2001**, *105*, 2626.
16. Hoffman, D. P.; Mathies, R. A. *Acc. Chem. Res.* **2016**, *49*, 616.
17. Mathies, R. A. *Nat. Chem.* **2015**, *7*, 945.

18. Prokhorenko, V. I.; Nagy A. M.; Waschuk S. A.; Brown L. S.; Birge R. R.; Miller R. J. *Science* **2006**, *313*, 1257.
19. Hu, S.; Xiang, C.; Haussener, S.; Berger, A. D.; Lewis, N. S. *Energy Environ. Sci.* **2013**, *6*, 2984.
20. Abdi, F. F.; Han, L.; Smets, A. H. M.; Zeman, M.; Dam, B.; van de Krol, R. *Nat. Commun.* **2013**, *4*, 2195.
21. Verlage, E.; Hu, S.; Liu, R.; Jones, R. J. R.; Sun, K.; Xiang, C.; Lewis, N. S. Atwater, H. A. *Energy Environ. Sci.* **2015**, *8*, 3166.
22. Khaselev, O.; Turner, J. A. *Science* **1998**, *280*, 425.
23. de Groot, H. J. M. in "*Fundamentals of Materials for Energy and Environmental Sustainability*," Ginley, D. S; Cahen, D. eds. Cambridge University Press, USA, **2012**, pp. 365–378.
24. Collini, E.; Scholes, G. D. *Science* **2009**, *323*, 369.
25. Gao, L.; Cui, Y; Wang, J.; Cavalli, A.; Standing, A.; Vu, T. T. T.; Verheijen, M. A.; Haverkort, J. E. M.; Bakkers, E. P. A. M.; Notten P. H. L. *Nano Lett.* **2014**, *14*, 3715.
26. Hu, S., Chi, C.-Y.; Fountaine, K. T.; Yao, M.; Atwater, H. A.; Dapkus, P. D.; Lewis, N. S.; Zhou, C. *Energy Environ. Sci.* **2013**, *6*, 1879.
27. Yalamanchili, S.; Emmer, H. S.; Fountaine, K. T.; Chen, C. T.; Lewis, N. S.; Atwater, H. A. *ACS Photonics* **2016**, *3*, 1854.
28. McDermott, G.; Prince, S. M.; Freer, A. A.; Hawthornthwaite-Lawless, A. M.; Papiz, M. Z.; Gogdell, R. J. Isaacs, N. W. *Nature* **1995**, *374*, 517.
29. McLuskey, K.; Prince S. M.; Cogdell, R. J.; Isaacs, N. W. *Biochem.* **2001**, *40*, 8783.
30. Roszak, A. W.; Howard, T. D.; Southall, A. T.; Gardiner, A. T.; Law, C. J.; Isaacs, N. W.; Cogdell, R. J. *Science* **2003**, *302*, 1969.
31. Gardiner, A. T.; Cogdell, R. J.; Takaichi, S. *Photosynth. Res.* **1993**, *38*, 159.
32. Giraud, E.; Fardoux, J; Fourrier, N.; Hannibal, L.; Genty, B.; Bouyer, P.; Dreyfus, B.; Verméglio, A. *Nature* **2002**, *417*, 202.
33. Henry, S. L.; Withers, J. M.; Singh, I.; Cooper, J. M.; Clark, A. W.; Burley, G. A.; Cogdell, Richard J. *Org. Biomol. Chem.* **2016**, *14*, 1359.
34. Pandit, A.; Reus, M.; Morosinotto, T.; Bassi, R.; Holzwarth, A. R.; de Groot, H. J. *Biochim. Biophys. Acta* **2013**, *1827*, 738.
35. Brinks, D.; Hildner, R.; van Dijk E. M.; Stefani, F. D.; Nieder, J. B.; Hernando, J. van Hulst, N. F. *Chem. Soc. Rev.* **2014**, *43*, 2476.
36. Turner, D. B.; Arpin, P. C.; McClure, S. C.; Ulness, D. J.; Scholes, G. D. *Nat. Commun.* **2013**, *4*, 2298.
37. Hildner, R.; Brinks, D.; Nieder, J. B.; Cogdell, R. J.; van Hulst, N. F. *Science* **2013**, *340*, 1448.
38. Wientjes, E.; Renger, J.; Curto, A. G.; Cogdell, R. J.; van Hulst, N. F. *Nat. Commun.* **2014**, *5*, 4236.
39. Eisenmayer, T. J.; Buda, F. *Chemphyschem* **2014**, *15*, 3258.
40. Eisenmayer, T. J.; Lasave, J. A.; Monti, A.; de Groot, H. J. M.; Buda, F. *J. Phys. Chem. B* **2013**, *117*, 11162.

41. Nesterov-Müller, A.; Bykovskaya, V.; von Bojnicic-Kninski, C. M.; Löffler, F. F.; Popov, R.; Ridder, B.; Breitling, F.; Althuon, D. S. "*Ultra-hochdichte Oligomerarrays und Verfahren zu deren Herstellung*," Patent **2017**, DE102015117567A1, WO2017064165A1.
42. Loeffler, F. F.; Foertsch, T. C; Popov, R.; Mattes, D. S.; Schlageter, M.; Sedlmayr, M.; Ridder, B.; Dang, F.-X.; von Bojničić-Kninski, C.; Weber, L. K.; Fischer, A. Greifenstein. J.; Bykovskaya, V. Bischoff, F. R.; Meier, M. A. R.; Bräse, S.; Powell, A. K.; Balaban, T. S.; Breitling, F.; Nesterov-Mueller, A. *Nat. Commun.* **2016**, *7*, 11844.
43. Deng, Y. Y.; Koper, M.; Haigh, M.; Dornburg, V. *Biomass Bioenergy* **2015**, *74*, 253.
45. Buckley, J. A.; Thompson, P. B.; Whyte, K. P. *Technol. Soc.* **2017**, 48, 54.
44. Shell (2016) A better life with a healthy plant — pathways to net-zero emissions.
46. Palsson, G.; Szerszynski, B.; Sörlin, S.; Marks, J.; Avril, B.; Crumley, C.; Hackmann, H.; Holm, P.; Ingram, J.; Pardo B. M.; Weehuizen, R. *Environ. Sci. Policy* **2013**, *28*, 3.
47. Sorlin, S. *BioScience* **2012**, *62*, 788.
48. Von Schomberg, R. in: "*Responsible Innovation: Managing the Responsible Emergence of Science and Innovation in Society*", Owen, R.; Heintz. M.; Bessant, (eds.) John Wiley & Sons, Ltd, Chichester, UK, **2013**, pp. 51–71.
49. Stilgoe, J.; Owen, R.; Macnaghten, P. *Research Policy* **2013**, 42, 1568.
50. Friedman, B.; Kahn Jr., P. H.; Borning, A. in *Early Engagement and New Technologies: Opening up the Laboratory*, Doorn, N.; Schuurbiers, D.; van de Poel, I.; Gorman, M. E. (Eds.), Springer, **2013**, *16*, pp. 55–95.
51. Gaut, B. N. "*Art, Emotion and Ethics*", Oxford University Press, Oxford, **2007**.
52. Sovacool, B. K.; Gross, A. *Interface Focus* **2015**, *5*, 20140089.
53. Wüstenhagen, R.; Wolsink, M.; Bürer, M. J. *Energy Policy* **2007**, *35*, 2683.

4 Artificial Hemoprotein Assemblies in Development of Nanobiomaterials

Koji Oohora* and Takashi Hayashi**

Department of Applied Chemistry, Graduate School of Engineering, Osaka University, Suita, 565-0871, Japan
*oohora@chem.eng.osaka-u.ac.jp
**thayashi@chem.eng.osaka-u.ac.jp

Table of Contents

List of Abbreviations

RNA	ribonucleic acid
DNA	deoxyribonucleic acid
NMR	nuclear magnetic resonance
Cyt*b*	cytochrome b_{562}
Cyt*cb*	cytochrome cb_{562}
TEM	transmission electron microscopy
Cyt*c*	cytochrome *c*
SAXS	small angle X-ray scattering
Mb	myoglobin
ZnPP	zinc protoporphyrin IX
SEC	size exclusion chromatography
AFM	atomic force microscopy

I. Introduction

Biological systems utilize lipid membranes, RNA/DNA and protein as building blocks of self-assembling structures to achieve well-regulated functions, complex structures and dynamic molecular machines.[1–6] These self-assembling systems have inspired chemists to design various synthetic small molecules as building blocks for spontaneously assembled structures. Such "supramolecular polymers" demonstrate unique features generated by successive non-covalent interactions.[7–15] Furthermore, these systems have been applied to develop artificial supramolecular materials. Over the past two decades, artificial supramolecular assembling materials based on natural biomolecules have also been reported.[16] DNA has been a particularly well-studied building block used in development of supramolecular assemblies with the aim of constructing 2D and 3D structures and conjugates with other molecules.[17–19] This indicates that the relatively simple combination of the four nucleotide bases can provide a basis for rational design of well-defined structures. In contrast to simple DNA building blocks, protein building blocks have 20 amino acids with diverse chemical features. As a result, efforts to design protein assemblies are still developing. However, protein-inherent properties such as selective substrate recognition, storage, transport, and enzymatic reactivity will contribute to generation of unique protein-based functional biomaterials.[20,21]

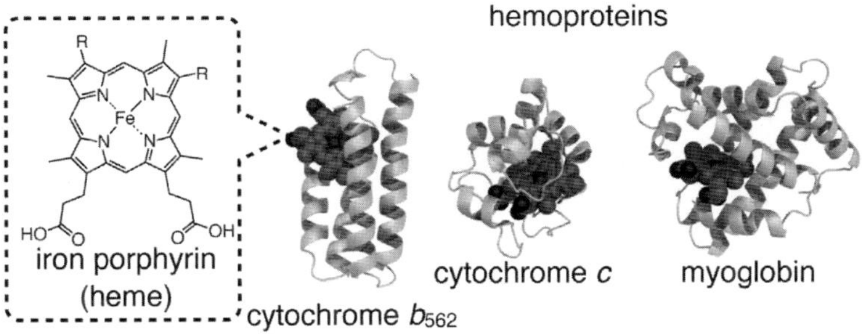

Figure 1. Molecular structure of heme and crystal structures of hemoproteins described in this chapter: cytochrome b_{562} (PDB ID: 256B), cytochrome *c* (PDB ID: 1HRC) and myoglobin (PDB ID: 2MBW).

In this context, advances in protein engineering have recently provided some examples of artificial protein assemblies.[21–28]

To generate an artificial supramolecular protein assembly, the utilization of inherent protein–protein interactions has been quite limited. This is caused by difficulties in fine-tuning of weak nonspecific protein–protein interactions such as electrostatic interactions, hydrogen bonding, and hydrophobic interactions. Undesirable interactions between protein surfaces can easily result in formation of undesired aggregates.[21] Although a few examples of development of artificial protein assemblies using these interactions have been demonstrated, known interfaces between protein units in crystals of native protein oligomers have been demonstrated as reliable locations for favorable protein-protein interactions for self-assembly.[29–31] In this context, a hemoprotein, which contains iron porphyrin (heme) as a cofactor, is an attractive protein building block due to its robustness and its unique interaction between the heme moiety and the protein matrix (Figure 1).[32,33] This chapter focuses on recent progress in development of artificial hemoprotein assembling systems.

II. Hemoproteins as Building Blocks for Supramolecular Assemblies

Hemoproteins provide important functions in biological systems including dioxygen transport/storage, electron transfer, gas sensing, and enzymatic oxidation/oxygenation.[34] Although, the same heme cofactor is employed by hemoproteins with different functions, the protein matrix of hemoproteins regulates the redox potential and reactivity of heme *via* coordination bonding, hydrogen bonding and electrostatic interactions. Over the past half-century, the crystal/NMR structures, functions, and physicochemical properties of many hemoproteins have been revealed, and these advances have encouraged us to modify the functions of hemoproteins by protein engineering in efforts to develop new biocatalysts and biomaterials.[35] Furthermore, several hemoproteins have been employed as building blocks of artificial protein assemblies. Due to their unique structures, functions and properties, hemoprotein assemblies are promising systems which are expected to provide the basis for new biomaterials with synergetic or cooperative effects derived from accumulated cofactor functions.[36,37]

III. Coordination Assembly

The coordination bonding interaction provides a useful driving force for oligomerization of proteins. Introduction of a metal ion-binding site onto protein surfaces has been investigated in efforts to achieve self-assembly.[38–44] Tezcan and coworkers have demonstrated a series of sophisticated hemoprotein assembling systems using the coordination bonding interaction (Figure 2a).[45–46] A variant of cytochrome b_{562} (Cyt*b*) known as cytochrome cb_{562} (Cyt*cb*), was employed in these investigations. Cyt*b* is one of the simplest known hemoproteins and has 106 amino acid residues which form a four-helix bundled structure with a removable heme *b* (iron protoporphyrin IX). Cyt*cb* is an engineered Cyt*b*, which has a covalently attached *c*-type heme in the intrinsic heme pocket linked to the protein *via* two thioether bonds with Cys residues.[47] The artificially engineered metal binding sites were created by site-directed mutagenesis at the protein surface and the addition of metal ions was found to spontaneously drive formation of the hemoprotein assembly *via* the coordination interaction. In the first report, two sets of two histidine residues located at the protein–protein interface in the crystal packing of Cyt*cb* were introduced into one of the four helices to provide metal binding sites. The oligomeric assembly in the solution state was characterized by analytical ultracentrifugation and the defined homotetrameric structure was revealed by X-ray crystallographic analysis. In the crystal structure, an unexpected coordination bonding interaction was found to include an aspartate residue instead of the introduced histidine

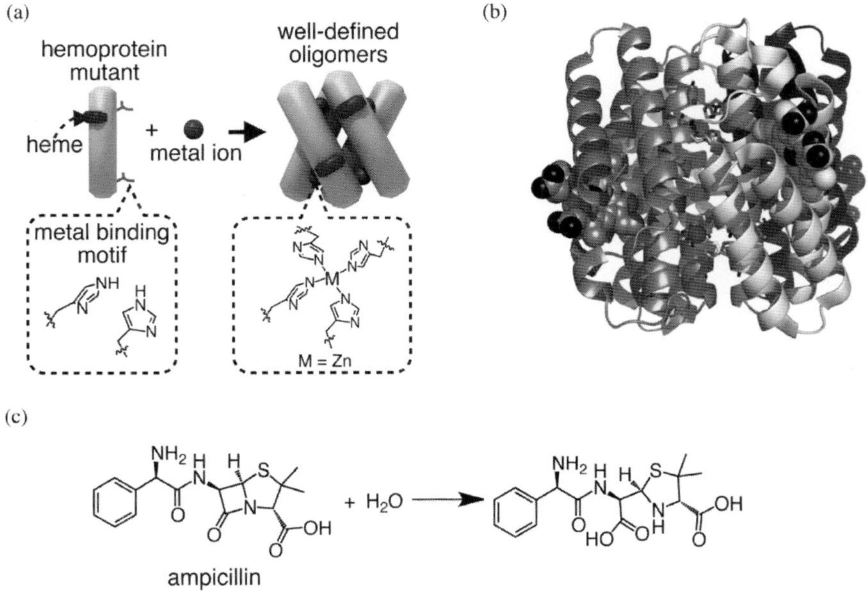

Figure 2. Coordination-driven hemoprotein assemblies. (a) Schematic representation of the assembly. (b) Crystal structure of engineered Cyt*cb* (PDB ID: 3HNJ). (c) β-lactamase catalyzed reaction.

residue. This structural information was utilized for further protein design to fine-tune the structure of the homotetrameric hemoprotein assembly (Figure 3b).[48] The first assembling system based on Cyt*cb via* a zinc ion-driven coordinating bonding interaction was a mixture of several semi-stable structures in solution state, whereas the defined homotetrameric structure in solution as well as in the crystals was obtained by fine-tuning the protein–protein interface. The zinc ion was repurposed to provide β-lactamase activity, which is unrelated to the native function of Cyt*cb*. β-lactamase activity against ampicillin was demonstrated to occur at the catalytic zinc sites (Figure 2c).[49] This artificial enzyme based on the tetramer of Cyt*cb* can be produced in the periplasm region of *Escherichia coli* and enables the bacteria to survive treatment with ampicillin. A mutant with higher catalytic activity was obtained by directed evolution *via* an *in vivo* selection strategy.[50]

A further investigation of the assembling interface resulted in micrometer-sized fibrous and sheet-like structures upon addition of metal

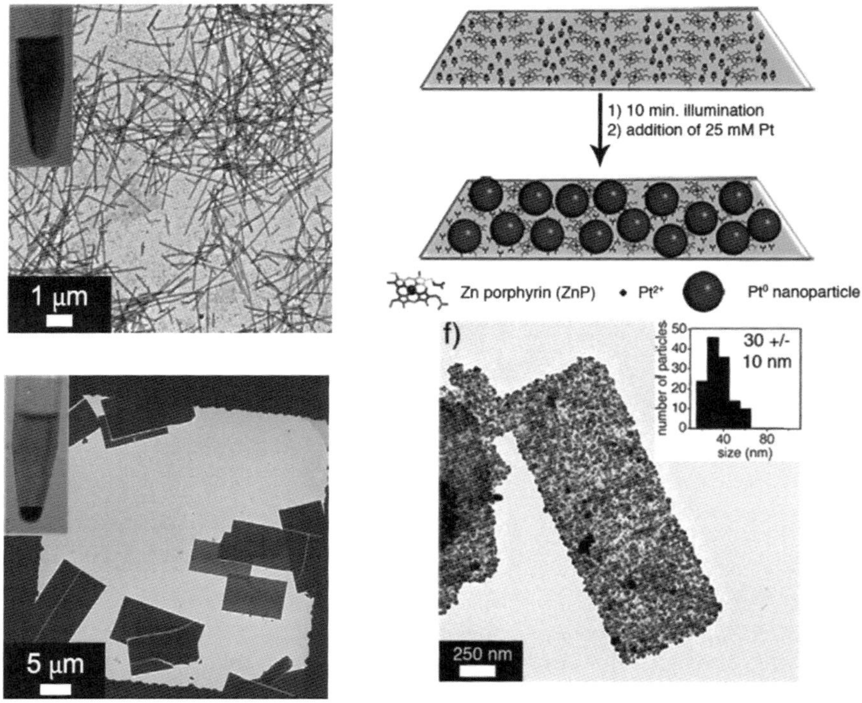

(a) (b)

Figure 3. (a) TEM-images of 1D- and 2D structures. Reproduced with permission from the Nature Publishing Group, Ref. [51] copyright 2012. (b) Photo-induced formation of Pt nanoparticles at Zn-substituted Cyt*cb* sheets. Reproduced with permission from National Academy of Sciences of the United States of America, Ref. [53] copyright 2014.

ions even under non-crystallizing conditions (Figure 3a).[51] The high-resolution TEM measurements show the periodic alignment of the building blocks in the fibers and sheets. The fibrous structure was developed into helical nanotubes with high flexibility, which is indicated by the Young's modulus estimated from the TEM images.[52] On the other hand, the sheet composed of zinc-substituted Cyt*cb* was found to be useful as the scaffold for photo-induced production of a Pt nanoparticle (Figure 3b).[55] Thus, the well-defined structure obtained through coordination was demonstrated to contribute to construction of artificial metalloenzymes and sub-micrometer-sized biomaterials.

IV. Domain Swapping

Domain-swapping of a hemoprotein (Figure 4a) has served as a unique strategy to construct an assembly. Hirota and coworkers have investigated a series of hemoprotein assemblies by domain swapping.[54] This group has focused on the polymerization mechanism of cytochrome *c* (Cyt*c*), a well-known electron transfer hemoprotein because it is involved in one of the most important events occurring during an early stage of apoptosis. The formation of kinetically stable oligomers of Cyt*c* at 4 °C was identified upon addition of ethanol to an aqueous solution of Cyt*c*. The isolated Cyt*c* dimer and trimer were characterized by X-ray crystallography and SAXS (small angle X-ray scattering) to reveal the assembly *via* swapping of the C-terminal helix (Figure 4b).

The polymers possessing higher molecular weights also have spectroscopic properties similar to those of the well-characterized dimer and trimer, suggesting that polymerization of Cyt*c* is caused by successive domain swapping. Data obtained from laser trapping experiments suggest that the domain-swapping event of Cyt*c* provides initial nucleation in formation of the amyloid fibril of Cyt*c*. Domain-swapped dimers were also found in assembly systems based on cytochrome c_{552},[55] a member of the Cyt*c* superfamily in the thermophilic organism *Hydrogenobacter thermophilus*. Interestingly, the overexpression system for production of cytochrome c_{552} in the periplasm region of *Escherichia coli* demonstrates

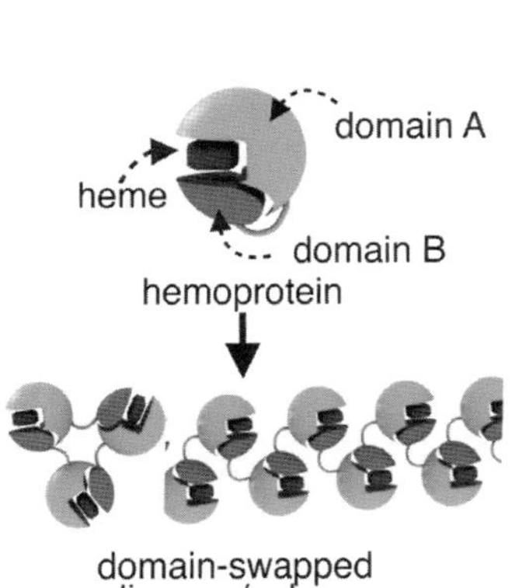

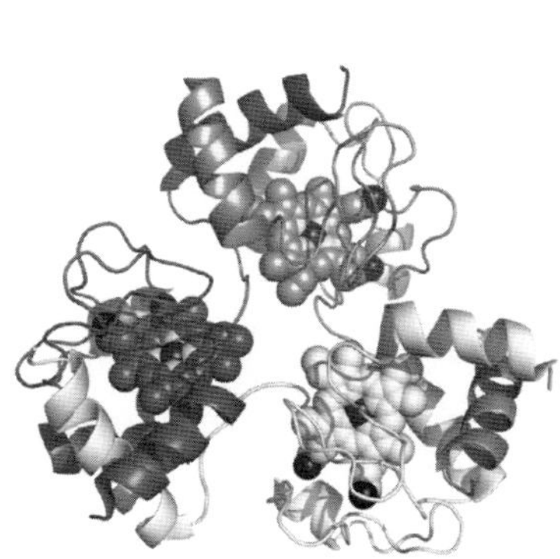

Figure 4. (a) Schematic representation of hemoprotein domain swapping. (b) Crystal structure of the domain-swapped trimer of Cyt*c* (PDB ID: 3NBT).

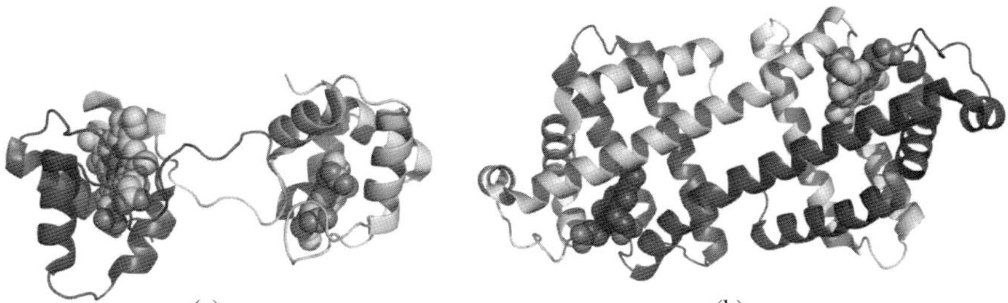

(a) (b)

Figure 5. (a) Crystal structure of the hetero dimer of chimeric cytochrome *c* proteins (PDB ID: 5XEC). (b) Crystal structure of the heterodimer of myoglobin mutants (PDB ID: 3WYO).

that polymerization of this protein occurs with domain swapping *in vivo*.[56] The amount of oligomer of cytochrome c_{552} increases as the concentration of the protein increases. A dimer formed from rational design of a chimeric protein of the Cyt*c* superfamily with domain swapping was demonstrated (Figure 5a).[57] In this system, cytochrome c_{551} from *Pseudomonas aeruginosa* was combined with cytochrome c_{552}. The two types of chimeric proteins of cytochrome c_{551} and cytochrome c_{552}, in which the N-terminal residue of one protein was conjugated with the C terminal residue of the other protein, were utilized to yield the heterodimer. Cyt*cb* also forms the domain-swapped dimer. In the crystal structure, a zinc ion cage was generated in the three domain-swapped dimers.[58] Myoglobin (Mb), an oxygen storage hemoprotein, from horse muscle was employed as the building block for a domain-swapped dimer.[59] The swapped Mb dimer was characterized by X-ray crystallography to reveal the formation of a new helix as a result of dimerization, whereas the affinity for dioxygen is similar to that of native Mb. The designed Mb mutants form the heterodimeric structure by domain-swapping.[60] The heterodimer contains the electrostatically complementary residues at the key helices during the process of domain swapping (Figure 5b). Further mutations introduced in the Mb heterodimer led to the creation of an engineered hemoprotein dimer with two different heme coordination environments, the native coordination structure and a bis-His ligated structure. Thus, precise crystal structures of domain-swapped hemoprotein oligomers will contribute not only to understanding diseases related to protein polymerization but also to the development of new strategies for design of artificial hemoprotein assemblies.

V. Amyloid-display

Amyloid fibrils are known as chemically and physically stable protein assemblies with β-sheet structures.[61,62] Barker and coworkers demonstrated an amyloid-display using $(SH3)_2$Cyt*b*, a recombinant fusion protein, in which Cyt*b* and an SH3 dimer domains are linked to construct a fibrous hemoprotein structure (Figure 6).[63] The SH3 dimer is a tandem repeat of 86 residues in the α-subunit of bovine phosphatidyl-3′-kinase which forms amyloid fibrils. The recombinant fusion protein assembly was observed by negatively stained TEM measurements, indicating the presence of a few fibrils with micrometer lengths and widths of approximately 8 nm. In addition, 200-nm helical pitches were also identified as characteristics of the SH3 dimer amyloid fibrils. The CD spectra of $(SH3)_2$Cyt*b* in the UV region indicate the presence of both α-helices and β-sheets in the folded cytochrome and amyloid moieties, respectively. A UV-vis spectrum of the fibrils indicates quantitative incorporation of heme into the apo form of the Cyt*b* moiety to form a six-coordinate low spin heme state. An NMR spectrum shows broad signals, suggesting that Cyt*b* has restricted motion on the amyloid fibrils. In another experiment, zinc protoporphyrin IX (ZnPP) was bound to the apo-form of the Cyt*b* moiety on the fibrils as a replacement for heme. Furthermore, the morphology of the amyloid fibrils was found to depend on the design of the monomer unit, and specifically the number of repeating moieties of the SH3 dimer and position of Cyt*b*.[64] In an effort to construct an electronic protein wire, the scanning tunneling

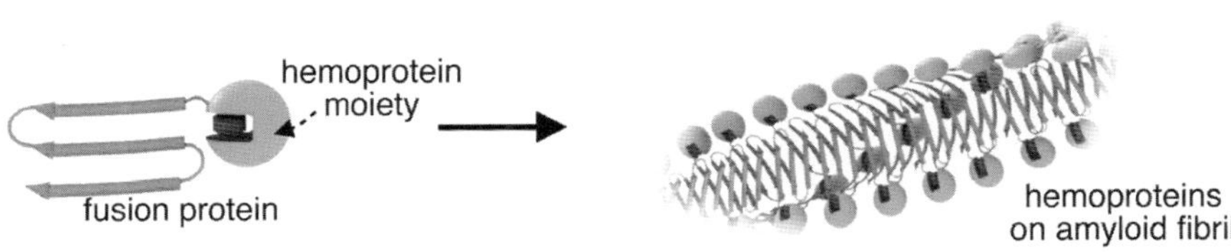

Figure 6. Schematic representation of amyloid-display using a fusion protein containing a hemoprotein moiety.

microscopic measurements for Cyt*b*-displayed amyloid fibrils demonstrated the presence of conductivity.[65]

VI. Heme–heme Pocket Interaction

Heme *b*, iron protoporphyrin IX, is easily removable from the heme pocket and can be re-inserted into the corresponding apoprotein to yield the reconstituted protein (Figure 7a).[35] Supramolecular hemoprotein assemblies using successive interprotein heme–heme pocket interactions have been demonstrated by our group (Figure 7b).[32,66,67] In the first example, the Cyt*b* mutant, Cytb^{H63C}, which has a cysteine residue on the protein surface, was designed and prepared. A thiol group of the cysteine residue on the protein surface was found to selectively react with iodoacetamide- or maleimide-tethered synthetic heme *via* a relatively hydrophilic flexible linker to produce a covalently heme-attached protein. The native heme cofactor was removed under acidic conditions and subsequent neutralization induced successive reconstitution of the modified hemoprotein to form the fibrous hemoprotein assemblies. Size exclusion chromatography (SEC) and atomic force microscopy (AFM) measurements were carried out on the assembling system to evaluate structures and thermodynamic behavior. The formation of the hemoprotein assembly was found to be thermodynamically reversible with the degree of polymerization depending on the monomer concentration, pH value and concentration of the additional heme or apoprotein species. UV-vis and CD spectra indicate the successful incorporation of heme into the apoprotein matrix, suggesting that each protein unit in the supramolecular assembly is correctly folded. Further addition of a synthetic heme triad into the hemoprotein linear assembly was found to lead to the formation of 2D-network structures on the graphite substrates, which were characterized by AFM.[67] Interesting assembling behavior with transitions between fiber- and micelle-like structures was confirmed when an azobenzene or stilbene moiety was employed as a linker between the protein and the heme moiety (Figure 7c).[68] Transient thermal stimuli provided to the fibrously assembled protein was found to induce formation of the spherical structure, which was maintained after cooling with slow conversion to the fibrous structure. The TEM image supports this anomalous transition. These findings indicate that the heme-mediated supramolecular assembling

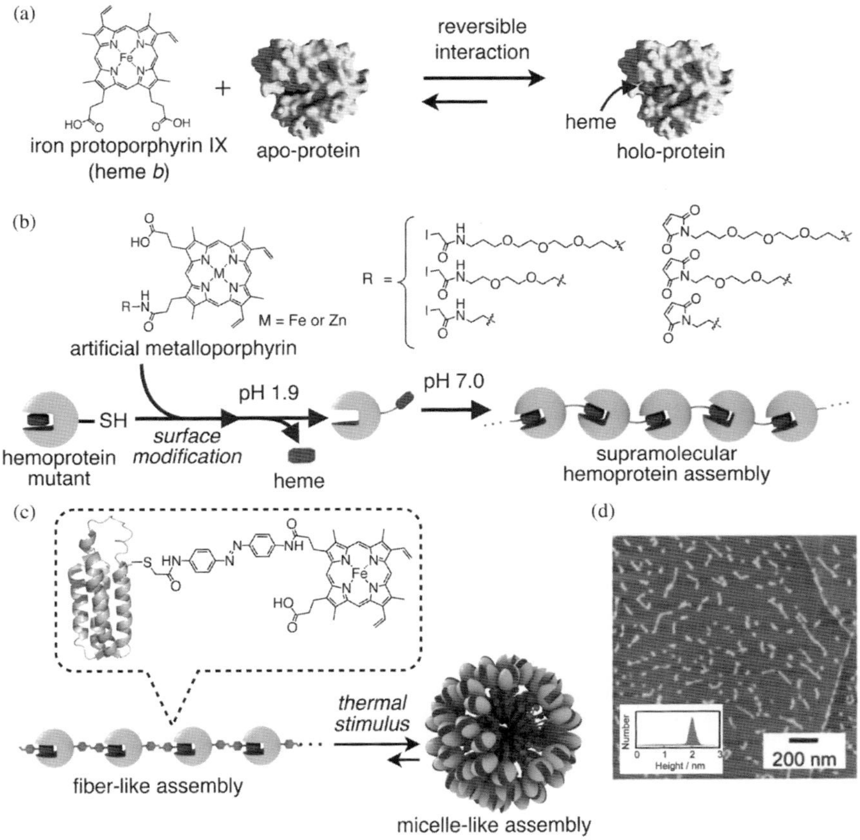

Figure 7. (a) Reversible heme binding in a hemoprotein. (b) Schematic representation of a supramolecular hemoprotein assembly formed *via* the heme–heme pocket interaction. (c) Unique dynamics of a supramolecular assembly of Cyt*b* formed *via* an azobenzene linker. Reproduced with permission from The Royal Society of Chemistry, Ref. [68] copyright 2017. (d) AFM image of a supramolecular Mb assembly. Reproduced with permission from The Royal Society of Chemistry, Ref. [69] copyright 2011.

systems based on hemoproteins are strongly modulated by the chemical structure of the linker moiety.

The strategy of using the interprotein heme–heme pocket interaction was applied to a Mb mutant, Mb^{A125C}, which has a cysteine residue at position 125.[69] A supramolecular Mb assembly was prepared using the same procedure used for Cyt*b*. AFM images of the Mb assembly are shown in Figure 7d. The system was found to form larger assemblies compared to that of Cyt*b* due to the higher affinity of heme for apoMb. This is consistent with the finding that the formation of the supramolecular assembly is thermodynamically controlled by the heme–heme pocket interaction. Addition of CO or CN^- as a sixth ligand stabilizes the Mb assembly against the unfavorable effect of dilution. It is known that native apoMb has greater affinity for heme in the presence of such a sixth ligand.[70] The intrinsic reactivity of Mb also causes a reaction of the linear Mb assembly with H_2O_2 to provide a three-dimensional cross-linked network due to coupling of free radicals generated at a tyrosine residue of surface of each protein unit. It was found that the dioxygen affinity is maintained after cross-linking, suggesting that the Mb assembly does not have a major influence on the native functions of the hemoprotein.

Another strategy for preparing a supramolecular hemoprotein assembly was demonstrated using two building blocks: a heme dimer and apoMb dimer.[71] This system also yields submicrometer-sized fibrous assemblies and the formation of the assemblies is strongly dependent on the design of the heme dimer and charges on the protein surface. Electrostatic repulsion between the building blocks causes small oligomers. Furthermore, the strategy using multiple building blocks was utilized to prepare an alternating aligned assembly of heterotropic proteins, where apoMb dimer, streptavidin and a biotin-linked heme derivative were employed as building blocks (Figure 8).[72] A co-assembly formed from the conjugate and two protein building blocks was characterized by AFM and SEC, confirming that the assembly has a sub-micrometer-sized fibrous structure. Alternating alignment of two different proteins was confirmed by further AFM measurements. These findings indicate that the specific interaction between the heme and the heme pocket is also useful for generating a heterotropic co-assembly.

A hemoprotein assembly was also immobilized on a gold electrode. To generate a photocurrent, a Zn-substituted Cyt*b*^{H63C} assembly was prepared and assembled with the gold electrode modified with a ZnPP derivative (Figure 9a).[73–75] AFM and quartz crystal microbalance measurements indicated that protein oligomers ranging in size from tetramers to heptamers were immobilized on the electrode. Interestingly, vertical fibrous assemblies on the electrode were observed in AFM

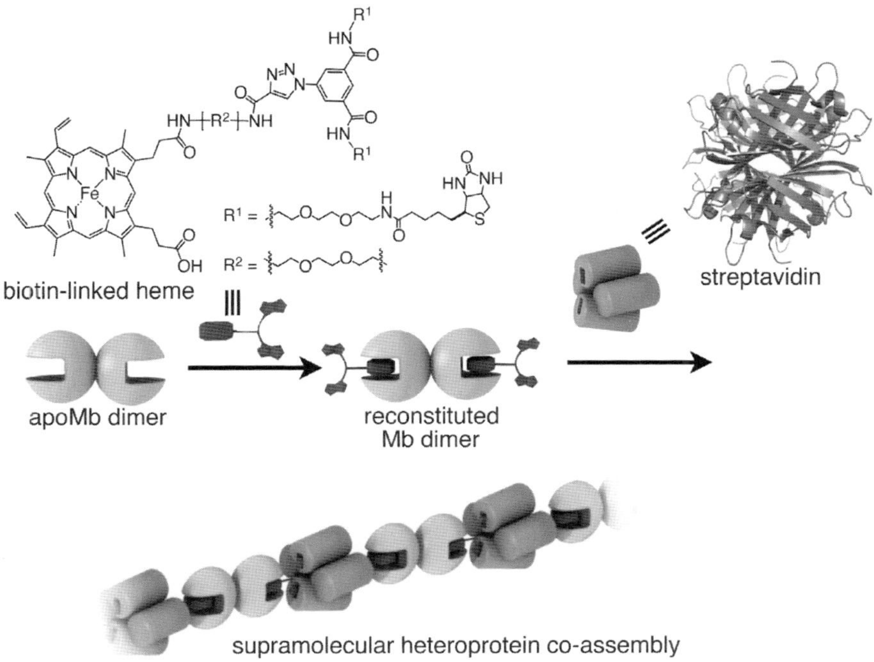

Figure 8. Supramolecular hetero-protein co-assembly of an apoMb dimer, streptavidin and a biotin-linked heme derivative. Reproduced with permission from Wiley-VCH, Ref. [72] copyright 2012.

images as shown in Figure 9b. Moreover, a cathodic current was observed upon irradiation with white light and the bias potential of –200 mV (*versus* Ag/AgCl) in the presence of methyl viologen as an electron acceptor. The efficiency was five times higher than that of the single layer of Cyt*b* with ZnPP on the modified electrode. The Zn-substituted hemoprotein assembly serves as an efficient photocurrent generation system.

VII. Conclusion and Outlook

The hemoprotein assembling systems described in this chapter are emergent topics for nanobiomaterials because several unique assembling strategies are available. The size-defined hemoprotein oligomers are

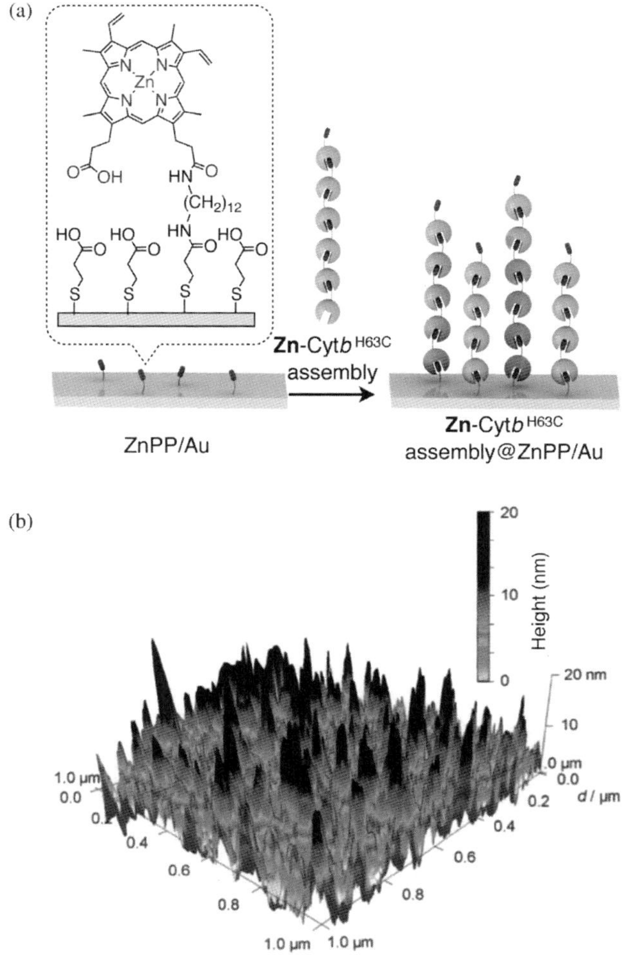

Figure 9. (a) Zn-substituted hemoprotein assembly on a ZnPP-modified gold electrode. (b) AFM image of the assembly on the electrode. Reproduced with permission from Wiley-VCH, Ref. [73] copyright 2012.

formed by metal-coordination between proteins or domain swapping. These structures are well characterized at atomic level resolution by X-ray crystallography. The defined structures are expanded to submicrometer-sizes, indicating that they can be used as useful scaffolds for biocatalysts and biomaterials. Supramolecular hemoprotein polymers such as fibrils or arrays, which are formed by amyloid-display or specific interactions between the heme and the heme pocket. These structures with submicrometer- or micrometer-sizes were clearly observed by AFM, TEM and/or SEM and applications as biowires for transfer of electrons and energy have been investigated. A variety of hemoprotein supramolecular assemblies with unique structures have been established. One of the next targets in this research field will be creation of hemoprotein assemblies with specific functions. Although several demonstrations of expected functions such as dioxygen binding, catalysis, and electron/energy transfer have been already reported, development of more innovative functions induced by synergistic effects of the assemblies has not yet been successful. Thus, further study of hemoprotein assembling systems is required to prepare assemblies which confer new chemical properties with the objective of constructing new biocatalysts and biomaterials. Since metalloporphyrins are known to possess unique physicochemical properties and reactivities, the creation of sophisticated assemblies based on hemoproteins is expected to lead to the development of smart biocatalysts, biomaterials and biodevices involving transport or storage of gas molecules/ions with the cooperative binding behaviors, stimuli-responsive materials responding to specific chemicals, enzyme-like catalysts with oligomerization-regulated activity, and self-assembled nano-wires providing efficient multiple electron transfer.

VIII. References

1. Whitesides, G. M.; Mathias J. P.; Seto, C. T. *Science* **1991**, *254*, 1312.
2. Kinbara, K.; Aida, T. *Chem. Rev.* **2005**, *105*, 1377.
3. Erickson, H. P.; Anderson, D. E.; Osawa M. *Microbiol. Mol. Biol. Rev.* **2010**, *74*, 504.
4. Sidhu, S. S. *Biomol. Eng.* **2001**, *18*, 57.
5. Sachse, C.; Chen, J. Z.; Coureux, P. D.; Stroupe, M. E.; Fandrich, M.; Grigorieff, N. *J. Mol. Biol.* **2007**, *371*, 812.
6. Harrison P. M.; Arosio, P. *Biochim. Biophys. Acta* **1996**, *1275*, 161.
7. Fouquey, C.; Lehn, J.-M.; Levelut, A.-M. *Adv. Mater.* **1990**, *2*, 254.
8. Lehn, J.-M. *Supramolecular Chemistry* 1995, VCH, Weinheim.
9. Greef, T. D. A.; Smulders, M. J.; Wolffs, M.; Schenning, A. P. H. J.; Sijbesma, R. P.: Meijer, E. W. *Chem. Rev.* **2009**, *109*, 5687–5754.
10. Dankers P. Y. W.; Meijer, E. W. *Bull. Chem. Soc. Jpn.* **2007**, *80*, 2047.
11. Aida, T.; Meijer E. W.; Stupp, S. I. *Science* **2012**, *335*, 813.

12. Castellano, R. K.; Rudkevich D. M.; Rebek, Jr., J. *Proc. Natl. Acad. Sci. USA* **1997**, *94*, 7132.
13. Hajjaj, F.; Yoon, Z. S.; Yoon, M. C.; Park, J.; Satake, A.; Kim D.; Kobuke, Y. *J. Am. Chem. Soc.* **2006**, *128*, 4612.
14. Michelsen U.; Hunter, C. A. *Angew. Chem., Int. Ed.* **2000**, *39*, 764.
15. Miyauchi, M.; Takashima, Y.; Yamaguchi, H.; Harada, A. *J. Am. Chem. Soc.* **2005**, *127*, 2984.
16. Mirkin C. A.; Niemeyer, C. M. *Nanobiotechnology II*, 2007, Wiley- VCH, Weiheim.
17. Seeman, N. C. *Nature* **2003**, *421*, 427.
18. Wilner, O. I.; Weizmann, Y.; Gill, R.; Lioubashevski, O.; Freeman R.; Willner, I. *Nat. Nanotechnol.* **2009**, *4*, 249.
19. Lo, P. K.; Karam, P.; Aldaye, F. A.; McLaughlin, C. K.; Hamblin, G. D.; Cosa G.; Sleiman, H. F. *Nat. Chem.* **2010**, *2*, 319.
20. Lutolf, M. P.; Hubbell, J. A. *Nat. Biotechnol.* **2005**, *23*, 47.
21. Yeates, T. O.; Padilla, J. E. *Curr. Opin. Struct. Biol.* **2002**, *12*, 464.
22. Grueninger, D.; Treiber, N.; Ziegler, M. O.; Koetter, J. W.; Schulze, M. S.; Schulz, G. E. *Science* **2008**, *319*, 206.
23. Bromley, E. H.; Channon, K.; Moutevelis, E.; Woolfson, D. N. *ACS Chem. Biol.* **2008**, *3*, 38.
24. Matsuura, K.; Watanabe, K.; Matsuzaki, T.; Sakurai, K.; Kimizuka, N. *Angew. Chem., Int. Ed.* **2010**, *49*, 9662.
25. Sicard, D.; Cecioni, S.; Iazykov, M.; Chevolot, Y.; Matthews, S.; Praly, J.-P.; Souteyrand, E.; Imberty, A.; Vidal, S.; Phaner-Goutorbe, M. *Chem. Commun.* **2011**, *47*, 9483.
26. Reynhout, I. C.; Cornelissen, J. J. L. M.; Nolte, R. J. M. *J. Am. Chem. Soc.* **2007**, 129, 2327.
27. Kulkarni, S.; Schilli, C.; Grin, B.; Müller, A. H. E.; Hoffman, A. S.; Stayton, P. S. *Biomacromolecules* 2006, 7, 2736.
28. Müller, M.; Petkau, K.; Brunsveld, L. *Chem. Commun.* **2011**, *47*, 310.
29. King, N. P.; Sheffler, W.; Sawaya, M. R.; Vollmar, B. S.; Sumida, J. P.; Andrė, I., Gonen, T.; Yeates, T. O.; Baker, D. *Science* **2012**, *336*, 1171.
30. King, N. P.; Bale, J. P.; Sheffler, W.; McNamara, D. E.; Gonen, S.; Gonen, T.; Yeates, T. O.; Baker, D. *Nature* **2014**, *510*, 103.
31. Gonen, S.; DiMaio, F.; Gonen, T.; Baker, D. *Science* **2015**, *348*, 1365.
32. Oohora, K.; Onoda, A.; Hayashi, T. *Chem. Commun.* **2012**, *48*, 11714.
33. Oohora, K.; Hayashi, T. *Curr. Opin. Chem. Biol.* **2014**, *19*, 154.
34. Messerschmidt, A.; Huber, R.; Poulos, T.; Wieghardt, K. *Handbook of Metalloproteins Vol. 1.,* **2001**, Wiley, New York.
35. Hayashi, T. Generation of functionalized biomolecules using hemoprotein matrices with small protein cavities for incorporation of cofactors *in Coordination Chemistry in Protein Cages: Principles, Design, and Applications*, eds. Ueno, T.; Watanabe, Y. **2014**, Wiley, Hoboken, pp. 87–110.

36. Lai, Y.-T.; Cascio, D.; Yeates, T.O. *Science* **2012**, *336*, 1129.
37. Sinclair, J. C.; Davies, K. M.; Vénien-Bryan, C.; Noble, M. E. M. *Nat. Nanotech.* **2011**, *6*, 558.
38. Burazerovic, S.; Gradinaru, J.; Pierron, J.; Ward, T. R. *Angew. Chem. Int. Ed.* **2007**, *46*, 5510.
39. Munch, H. K.; Heide, S. T.; Christensen, N. J.; Hoeg-Jensen, T.; Thulstrup, P. W.; Jensen, K. J. *Chem. Eur. J.* **2011**, *17*, 7198.
40. Biswas, S.; Kinbara, K.; Oya, N.; Ishii, N.; Taguchi, H.; Aida, T. *J. Am. Chem. Soc.* **2009**, *131*, 7556.
41. Pires, M. M.; Chmielewski, J. *J. Am. Chem. Soc.* **2009**, *131*, 2706.
42. Laganowsky, A.; Zhao, M.; Soriaga, A. B.; Sawaya, M. R.; Cascio, D.; Yeates, T. O. *Protein Sci.* **2011**, *20*, 1876.
43. Bogdan, N. D.; Matache, M.; Roiban, G.-D.; Dobrotă, C.; Meier, V. M.; Funeriu, D. P. *Biomacromolecules* **2011**, *12*, 3400.
44. Huard, D. J. E.; Kane, K. M.; Tezcan, F. A. *Nat. Chem. Biol.* **2013**, *9*, 169.
45. Salgado, E. N.; Radford, R. J.; Tezcan, F. A. *Acc. Chem. Res.* **2010**, *43*, 661.
46. Salgado, E. N.; Faraone-Mennella, J.; Tezcan, F. A. *J. Am. Chem. Soc.* **2007**, *129*, 13374.
47. Faraone-Mennella, J.; Tezcan, F. A.; Gray, H. B.; Winkler, J. R. *Biochemistry* **2006**, *45*, 10504.
48. Salgado, E. N.; Ambroggio, X. I.; Brodin, J. D.; Lewis, R. A.; Kuhlman, B.; Tezcan, F. A. *Proc. Natl. Acad. Sci. USA* **2010**, *107*, 1827.
49. Song W. J.; Tezcan F. A. *Science* **2016**, *346*, 1525.
50. Song, W. J.; Yu, J.; Tezcan, F. A. *J. Am. Chem. Soc.* **2017**, *39*, 16772.
51. Brodin, J. D.; Ambroggio, X. I.; Tang, C.; Parent, K.; Baker, T. S.; Tezcan, F. A. *Nat. Chem.*, **2012**, *4*, 375.
52. Brodin, J. D.; Smith, S. J.; Carr, J. R.; Tezcan F. A. *J. Am. Chem. Soc.* **2015**, *137*, 10468.
53. Brodin, J. D.; Carr, J. R.; Sontz, P. A.; Tezcan, F. A. *Proc. Natl. Acad. Sci. USA* **2014**, *111*, 2897.
54. Hirota, S.; Hattori, Y.; Nagao, S.; Taketa, M.; Komori, H.; Kamikubo, H.; Wang, Z.; Takahashi, I.; Negi, S.; Sugiura, Y.; Kataoka, M.; Higuchi, Y. *Proc. Natl. Acad. Sci. USA* **2010**, *107*, 12854.
55. Hayashi, Y.; Nagao, S.; Osuka, H.; Komori, H.; Higuchi, Y.; Hirota, S. *Biochemistry* **2012**, *51*, 8608.
56. Hayashi, Y.; Yamanaka, M.; Nagao, S.; Komori, H.; Higuchi, Y.; Hirota, S. *Sci. Rep.* **2015**, *6*, 19334.
57. Zhang, M.; Nakanishi, T.; Yamanaka, M.; Nagao, S.; Yanagisawa, S.; Shomura, Y.; Shibata, N.; Ogura, T.; Higuchi, Y.; Hirota, S. *Chem. Bio. Chem.* **2017**, *18*, 1712.
58. Miyamoto, T.; Kuribayashi, M.; Nagao, S.; Shomura, Y.; Higuchi, Y.; Hirota, S. *Chem. Sci.* **2015**, *6*, 7336.

59. Nagao, S.; Osuka, H.; Yamada, T.; Uni, T.; Shomura, Y.; Imai, K.; Higuchi, Y.; Hirota, S. *Dalton. Trans.* **2012**, *41*, 11378.
60. Lin, Y.-W.; Nagao, S.; Zhang, M.; Shomura, Y.; Higuchi, Y.; Hirota, S. *Angew. Chem. Int. Ed.* **2015**, *54*, 511.
61. Baxa, U.; Speransky, V.; Steven, A. C.; Wickner, R. B. *Proc. Natl. Acad. Sci. USA* **2002**, *99*, 5253.
62. Sondheimer, N.; Lindquist, S. *Mol. Cell* **2000**, *14*, 163.
63. Baldwin, A. J.; Bader, R.; Christodoulou, J.; MacPhee, C. E.; Dobson, C. M.; Barker, P. D. *J. Am. Chem. Soc.* **2006**, *128*, 2162.
64. Forman, C. J.; Nickson, A. A.; Anthony-Cahill, S. J.; Baldwin, A. J.; Kaggwa, G.; Feber, U.; Sheikh, K.; Jarvis, S. P.; Barker, P. D. *ACS Nano* **2012**, *6*, 1332.
65. Forman, C. J.; Wang, N.: Yang, Z. Y.; Mowat, C. G.; Jarvis, S.; Durkan, C.; Barker, P. D. *Nanotehcnology* **2013**, *24*, 175102.
66. Kitagishi, H.; Oohora, K.; Yamaguchi, H.; Sato, H.; Matsuo, T.; Harada, A.; Hayashi, T. *J. Am. Chem. Soc.* **2007**, *129*, 10326.
67. Kitagishi, H.; Kakikura, Y.; Yamaguchi, H.; Oohora, K.; Harada, A.; Hayashi, T. *Angew. Chem. Int. Ed.* **2009**, *48*, 1271.
68. Oohora, K.; Onuma, Y.; Tanaka, Y.; Onoda, A.; Hayashi, T. *Chem. Commun.* **2017**, *53*, 6879.
69. Oohora, K.; Onoda, A.; Kitagishi, H.; Yamaguchi, H.; Harada, A.; Hayashi, T. *Chem. Sci.* **2011**, *2*, 1033.
70. Hargrove, M. S.; Wilkinson, A. J.; Olson, J. S. *Biochemistry* **1996**, *35*, 11300.
71. Onoda, A.; Takahashi, A.; Oohora, K.; Onuma, Y.; Hayashi, T. *Chem. Biodivers.* **2012**, *9*, 1684.
72. Oohora, K.; Burazerovic, S.; Onoda, A.; Wilson, Y. M.; Ward, T. R.; Hayashi, T. *Angew. Chem. Int. Ed.* **2012**, *51*, 3818.
73. Onoda, A.; Kakikura, Y.; Uematsu, T.; Kuwabata, S.; Hayashi, T. *Angew. Chem. Int. Ed.* **2012**, *51*, 2628.
74. Kakikura, Y.; Onoda, A.; Kubo, E.; Kitagishi, H.; Uematsu, T.; Kuwabata, S.; Hayashi, T. *J. Inorg. Organomet. Polym.* **2013**, *23*, 172.
75. Onoda, A.; Kakikura, Y.; Hayashi, T. *Dalton Trans.* **2013**, *42*, 16102.

5 Hydrogenases: From Biomimetic to Bioinspired Models

Dumitru Sirbu,*[,†,‡] Tatiana Straistari[†] and Andrew C. Benniston*

*Molecular Photonics Laboratory, School of Chemistry, Newcastle University, Newcastle upon Tyne, NE1 7RU, UK
[†]Institute of Chemistry, Academy of Sciences of Moldova, Academiei str., Chisinau, MD-2028, Republic of Moldova
[‡]dumitru.sirbu@ncl.ac.uk

Table of Contents

List of Abbreviations

[Fe]-H_2ase	iron only containing hydrogenase
[FeFe]-H_2ase	iron-iron containing hydrogenase
[NiFe]-H_2ase	nickel-iron containing hydrogenase
CEC	chemical reaction-electron transfer-chemical reaction
CECE	chemical reaction-electron transfer-chemical reaction-electron transfer
DFT	density functional theory
DMF	dimethylformamide
ECCE	electron transfer-chemical reaction-chemical reaction-electron transfer
ECEC	electron transfer-chemical reaction-electron transfer-chemical reaction
EECC	electron transfer-electron transfer-chemical reaction-chemical reaction
EECEC	electron transfer-electron transfer-chemical reaction-electron transfer-chemical reaction
EPR	electron paramagnetic resonance
H_2ase	hydrogenase
Methenyl-H_4MPT^+	methenyl-tetrahydromethanopterin cation
Methylene-H_4MPT^+	methylene-tetrahydromethanopterin cation

TFA	trifluoroacetic acid
TOF	turnover frequency
TON	turnover number

I. Introduction

An energy efficient and robust method for the production of dihydrogen by the reduction of water offers a way to generate a highly combustible fuel or feedstock for chemical reactions.[1,2] Even though the electrolysis of water for dihydrogen production is a well-established technology it has yet to be adopted on a sizeable scale.[3] Fossil fuels combustion to generate the electricity required to produce H_2 is counterproductive, yet electrolysis offers a way of storing the energy from renewable sources to overcome problems with intermittent supply of e.g., wind and solar power.[4] However, there is a significant cost barrier to the widespread use of electrolysis to generate hydrogen and much of this cost comes from the price of the electrode materials, which are usually based on precious metals.[5] Even though proton reduction at platinum is facile the scarcity of the metal has fuelled research efforts to develop cheaper and more abundant materials. The two main problems to overcome is the large overpotential that hampers the use of, for example, glassy carbon electrodes and the catalyst stability during prolonged reaction time.[6]

II. Enzymatic Hydrogen Production

The first report of the enzymatic activity for hydrogen production was published in 1931, and subsequently used for reducing a variety of substrates.[7] The enzymes, hydrogenases (H_2ases), are found in archaea and bacteria where they play the crucial role in microbial energy metabolism thanks to their capacity to catalyze the reversible oxidation of molecular hydrogen.

$$H_2 \rightleftarrows 2\,H^+ + 2\,\bar{e}$$

Proton reduction (H_2 evolution) is the last step of pyruvate fermentation and H_2 oxidation is coupled to reduction of dioxygen, nitrate, sulphate, fumarate and carbon dioxide. While the function of these enzymes has been known for over a century, elucidation of the structure of their active sites as well as the explanation of the mechanisms by which they operate have required much more time. It was only in 1956, that the presence of iron was first found in H_2ases.[8] During the 1970s, using electron paramagnetic resonance (EPR) spectroscopy, it was shown that these enzymes contain ferredoxin similarly to iron-sulfur clusters.[9] It took until 1980, when Lancaster[10] and Thauer[11] teams detected the presence of the nickel in certain H_2ases. Nowadays, depending on the structure of the active sites, H_2ases are commonly grouped into three main classes: [NiFe]-H_2ases,[12] [FeFe]-H_2ases[13] and [Fe]-H_2ases.[14] Most of the reported H_2ases belong to the first two classes.

A. Nickel-Iron Containing Hydrogenases, [NiFe]-H_2ases

The first crystal structure of [NiFe]-H_2ases was obtained from *Desulfovibrio gigas* by Fontecilla-Camps and co-workers.[12] It is described as a heterodimeric periplasmatic protein consisting of two subunits of 28 kDa and 60 kDa, while the active site contains one Ni-atom and one Fe-atom. Two forms of the active sites are known. In the oxidized form the Fe-atom and the Ni-atom are 2.9 Å from each other and bridged by two S–Cys and one O-atom from a hydroxo or oxo ligand (X in Figure 1a). In the reduced form the Fe-atom and the Ni-atom are bridged only by two S–Cys ligands with a relatively short Ni–Fe distance of 2.5–2.6 Å (Figure 1b).

[NiFeSe]-H_2ases are a special subgroup of the [NiFe]-H_2ases in which one of the terminal cysteine ligands bound to Ni-atom is replaced by a selenocysteine (SeCys). The active site of the [NiFeSe]-H_2ases is

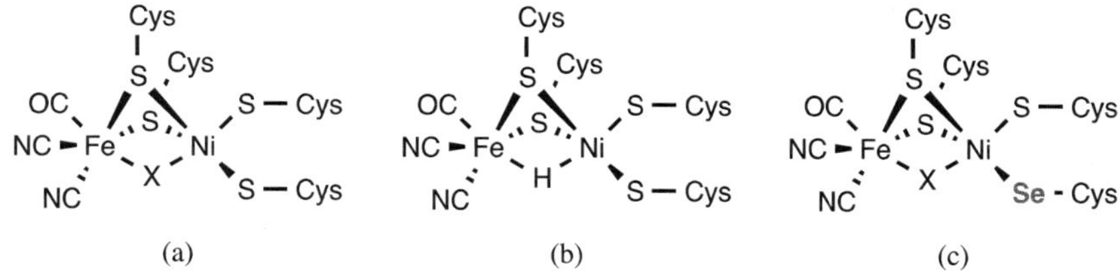

(a) (b) (c)

Figure 1. The active site structure of the oxidized (a) and reduced (b) forms of [NiFe]-H_2ases and of the [NiFeSe]-H_2ases in the oxidized form (c).

basically the same as in the [NiFe]-H_2ases. The [NiFeSe]-H_2ase structure was determined in enzymes isolated from *Desulfovibrio baculatum*[15,16] and *Desulfovibrio vulgaris H.*[17] The Ni-atom is coordinated by three sulfurs from cysteine and one selenium from selenocysteine, two of the cysteine ligands are linked to the Fe center. The coordination sphere of the Fe-atom is completed with three diatomic ligands (cyanide and carbon monoxide) and the sixth coordination position is occupied by an hydroxo or oxo bridge in the oxidized form, therefore the Fe(II) atom is in octahedral coordination geometry (Figure 1c).

B. Iron-Iron Containing Hydrogenases, [FeFe]-H_2ases

One of the first structures of the [FeFe]-H_2ases was obtained from studies of enzymes isolated from *Clostridium pasteurianum*[18] and *Desulfovibrio desulfuricans.*[13] These hydrogenases are predominantly involved in hydrogen production. The [FeFe] cluster consists of two octahedral coordinated Fe-atoms containing five non-protein carbonyl (CO) and cyano (CN) ligands stabilizing their low-spin and oxidation state.[19] The coordination sphere is completed by the bridging dithiolate cofactor contributing as an effective proton relay (Figure 2).[20]

C. Iron Only Containing Hydrogenases, [Fe]-H_2ases

The [Fe]-H_2ase named also H_2-forming methyltetrahydromethanopterin dehydrogenase, which was discovered in 1990, has been labeled as an "iron-sulfur cluster-free hydrogenase" because it does not contain iron-sulfur clusters.[21] Their biological function is to catalyze the reduction of the methenyl-tetrahydromethanopterin cation (methenyl-H_4MPT^+) into

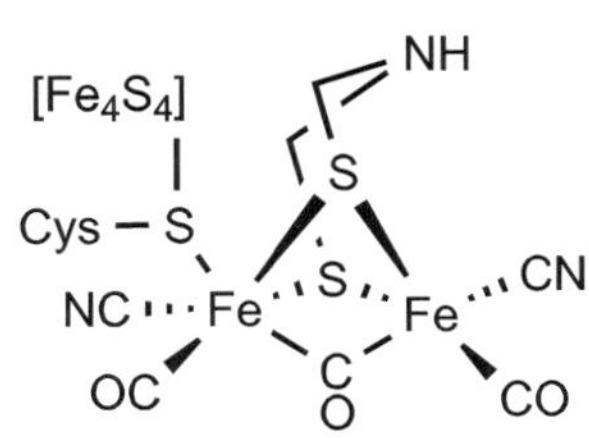

Figure 2. The active site structure of the [FeFe]-H_2ases.

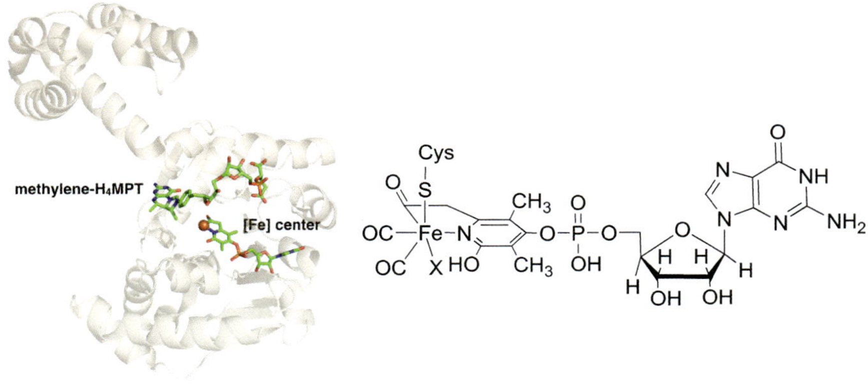

(a) (b)

Figure 3. (a) X-ray crystallographic structure of the monomeric *Methanocaldococcus jannaschii* [Fe]-hydrogenase with methylene-H_4MPT (PDB 3H65);[23] (b) The molecular structure of the Fe-GP cofactor.

methylene-tetrahydromethanopterin (methylene-H_4MPT^+) with H_2.[22] [Fe]-H_2ases contain two moles of iron per mole of homodimer of 75 kDa. The low spin Fe-atom is not redox active as the catalysis mechanism does not involve iron centered redox steps. Thus, it is also called "metal-free hydrogenase." The iron center contains one S–Cys ligand, one acyl substituted guanylylpyridinol-based chelating ligand (GP) and two CO ligands in a *cis* arrangement (Figure 3). The acyl unit was previously thought to be a carboxylate moiety. The pyridinol hydroxyl group is proposed to act as a pendant base.

D. Common Features and Comparison of the [NiFe]- and [FeFe]-H_2ases

Although structure and composition of hydrogenases are significantly different, the [NiFe]-H_2ase enzyme still strongly resembles the [FeFe] analog in various aspects. Both enzymes possess proton and dihydrogen channels, which deliver protons or dihydrogen to the active site and allow the subsequent release of dihydrogen or protons. Another common feature is the presence of three successive iron-sulfur clusters which allow the transfer of electrons (Figure 4), as opposed to [Fe]-H_2ase in which electrons are directly delivered to the active site.

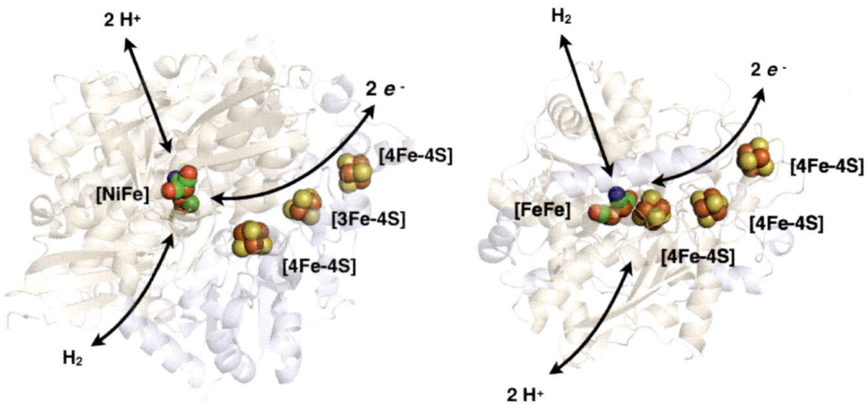

Figure 4. X-ray crystallographic structures of the [NiFe]-hydrogenase from *Dv*MF (left) (PDB 1WUH)[24] and of the [FeFe]-hydrogenase from *Dd* (right) (PDB 1HFE)[13] showing the proton and dihydrogen channels and the iron-sulfur clusters involved in the electron transfer.

Regarding the metallic centers, the common feature of these enzymes is: (1) the presence of several thiolate ligands from cysteine or dithiolate cofactor and (2) the organometallic nature of their active sites. Indeed, the presence of CN^- and CO ligands in the catalytic active site of an enzyme is without precedent in biology. In both cases the metals are redox active ($Ni^{II/III}$ and $Fe^{I/II}$) within the catalytic cycle.

E. Redox Active Species and Mechanism

In biological conditions (P = 0.1 bar of H_2, 30°C, pH 7), the [NiFe]-H_2ases catalyze very efficiently the $2H^+/H_2$ interconversion. Numerous works summarized the large amount of spectroscopic data, including infrared, crystallography and paramagnetic parameters used to identify and classify the catalytic intermediates of [NiFe]-H_2ases.[25–28] Kinetic and spectroscopic studies of the catalytic activity of [NiFe]-H_2ases allowed the characterization of different intermediate states of the enzyme.[29] EPR studies have revealed the existence of (1) three paramagnetic Ni-states (Ni-A, Ni-B and Ni-C) and (2) two silent Ni-states: Ni-SI_a (SI for silent and "a" for active) and the reduced state Ni-R (R for reduced). The presence of CN^- and CO ligands on the Fe-atom have allowed for a better characterization by following the vibrational bands of the diatomic ligands in IR spectra. These different states have been classified into three groups: (1) inactive states (the most oxidized) subdivided in unready states (Ni-A, Ni-SU) and (2) ready states (Ni-B, Ni-SI) according to their slow or fast kinetics in the reductive activation of hydrogen, and (3) the active states (Ni-SI_a, Ni-C and Ni-R) directly involved in the catalytic cycle of the enzyme. Despite the enormous amount of data regarding the [NiFe]-H_2ases catalysis mechanism, obtained experimentally and through theoretical studies, we still lack conclusive experimental evidence for the site of dihydrogen binding, although most models indicate binding at the Ni center.[30,31]

Binding and heterolytic dihydrogen cleavage at the Ni ion was proposed by several groups to occur through a μ-hydride intermediate after a proton transfer event.[32–34] At the same time a competing theory proposes dihydrogen binding and activation at the Fe center, where the Ni is modelled as high spin.[35] This is proposed to be followed by a bridging Ni–H–Fe hydride intermediate, as suggested by spectroscopic and molecular modelling data.[36,37] On the basis of spectroscopic studies a mechanism was later proposed in which a Cys bound to Ni-center plays the role of a general base in H_2 activation by [NiFe]-hydrogenases (Figure 5).[38]

Armstrong's group reported that the role of general base in H_2 activation could be played by the suspended guanidine N-atom from arginine R509 located at 4.5 Å from the metal centers (Figure 6).[39] Most evidences on catalysis intermediates relies on studies of [NiFe(Se)]-H_2ases subgroup, thereby more insights might emerge from ongoing research on parent [NiFe]-H_2ases enzymes.

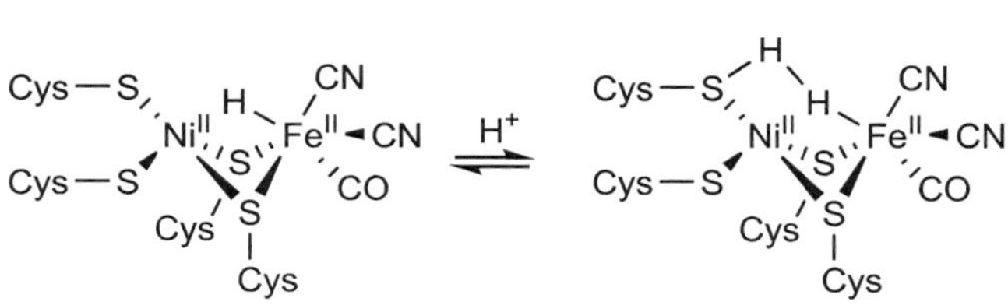

Figure 5. Proposed H_2 activation step in [NiFe]-H_2ases employing cysteine as a general base.

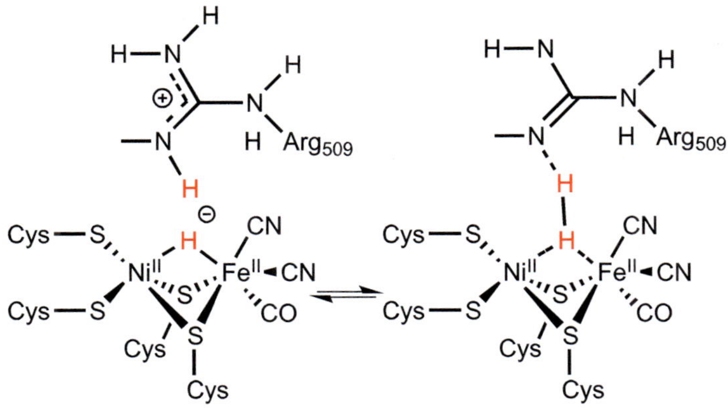

Figure 6. Proposed H_2 activation step in [NiFe]-H_2ases based on suspended guanidine as a general base.

III. Biomimetic Models of Hydrogenases

The studies on the structure and the operation mechanism of H_2ases led to an intense research of [NiFe]-H_2ases synthetic models. Synthesis of biomimetic models of [NiFe]-H_2ases active sites is important from two perspectives: (1) from a structural point of view, because their structures were unlike any previously known compound, and more importantly since they use abundant metals (i.e. nickel and iron) and (2) due to their high effectiveness and selectivity for the production or the oxidation of hydrogen. These catalysts could play a leading role in hydrogen based technologies, including fuel cells and photo-, electro-chemical production of hydrogen.

A. Biomimetic *versus* Bioinspired Models

According to Vullev[40] "biomimesis and bioinspiration" carry different denotations and connotations, and as such, their implications for science and engineering also differ. The term biomimetic (or bio-imitating), introduced in the 1970s, is defined as imitations of biological systems to create new synthetic compounds.[41] It may be a compound that mimics a biological material in its structure or function (e.g., molecular recognition hosts, artificial enzymes, medicinal chemistry targets, light antennae) but also a laboratory activity designed to imitate the natural mechanism (e.g., synthesis of natural product, chiral catalysis).

Bioinspired systems do not have evident resemblance to the parent biological analogs but use similar ideas and principles. The benefit of not being restricted to the natural design is the possibility to add structural elements and thus properties that are not accessible for natural models. As result the bioinspired approach has the potential to outperform the efficiency of biomimetic or biological systems.

B. Biomimetic Models of [NiFe]-H_2ases

Rather than completing a story or "raising more questions than it answers" the Volbeda *et al.* crystal structure of [NiFe]-H_2ase provides a focus and motivation for synthesizing interesting heterometallic complexes as structural and catalytic models.[12] The intricate chemistry required to correctly arrange coordination metal centers in structural and functional models of [NiFe]-H_2ase for H_2 oxidation or evolution is a special topic of study. In the reproduction of active sites of hydrogenase several aspects are taken into account: the metal centers, coordination environment of the metal and the electrochemistry of the ligand.

A year after the structure determination of the [NiFe]-H_2ases, the first crystal of a thiolate bridged [NiFe] complex was isolated.[42] The structure of the dimeric NiFe complex (Figure 7) shows a resemblance to the [NiFe]-H_2ase active site, although the Ni–Fe distance of 3.1 Å is longer than that in the [NiFe]-H_2ases structure. Shortly another binuclear thiolate bridged Ni–Fe–nitrosyl complex was characterized[43] as shown in Figure 7. In this complex, the distance between Ni–Fe centers is shorter (2.8 Å) than in active center of [NiFe]-H_2ase. But this compound was unstable in the electrochemical study. The first structure that closely reproduces the

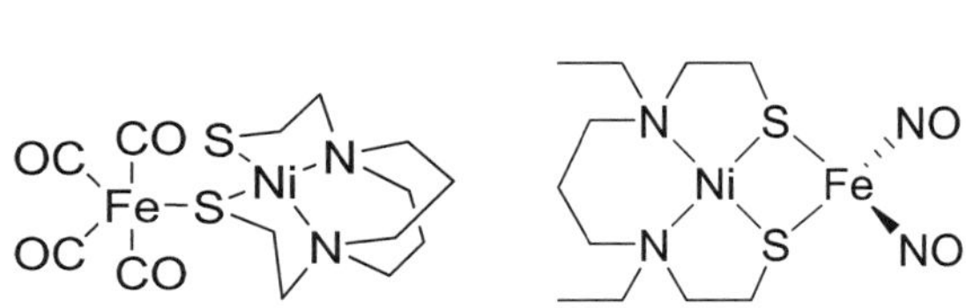

Figure 7. Structure of dimeric [NiFe] complexes containing (a) monothiolate and (b) dithiolate bridged nickel and iron atoms.

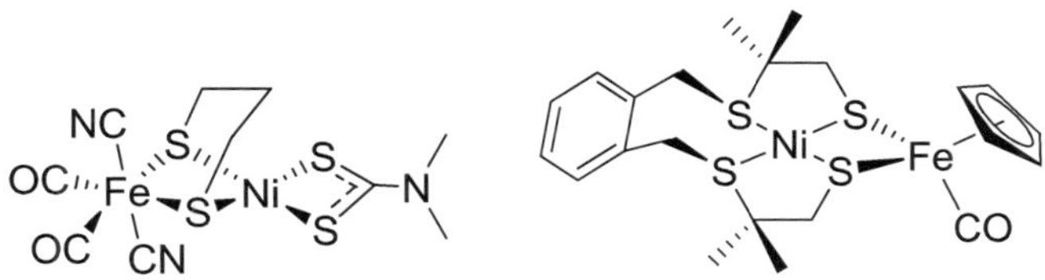

Figure 8. Molecular structure of thiolate-based [NiFe]-organometallic complexes.

coordination environment of the active site of [NiFe]-H_2ase, in terms of cyanide and carbonyl units on the Fe atom but also tetrathio-coordinated environment of Ni (Figure 8), was synthesized by Tatsumi's group.[44] In the reported complex the two thiolate S atoms link Ni and Fe, the Ni atom is tetrahedrally coordinated with two additional S atoms by chelating to a dithiocarbamate and the Fe atom is in an octahedral geometry by coordinating to two cyanide and two carbon monoxide ligands. Unfortunately, these models showed no catalytic activity. Artero's group reported a [NiFe]-H_2ase model organometallic complex (Figure 8), with a tetradentate ligand providing S_4-coordination at the Ni ion with two Ni–Fe bridging sulphur atoms and a carbonyl along with cyclopentadienyl ligand on the iron center.[45] The distance between the Ni–Fe atoms is 2.96 Å. The complex partially mimics the structure of the [NiFe] catalytic site but most importantly reproduces the enzyme's activity.

The first dinuclear [NiFe]-model complex (Figure 9b) mimicking the catalysis intermediates of [NiFe]-H_2ases (Figure 9a), through protonation of the Cys–SH group was reported in 2012.[46] The neutral compound is reversibly protonated by using a strong acid. The protonation takes place at the terminal thiolate unit coordinated to nickel, but also has a significant impact on the electronic structure of the iron center. Both the neutral and protonated forms reveal a short Ni–Fe distance of 2.4 Å and a bonding interaction was suggested between the metal ions.

Another approach to the synthesis of H_2ase mimicking complexes consists of using the abiological ligands, e.g., phosphine coordinating atoms. Molecular systems with S-bridged Ni–Fe that contains phosphine based ligands (Figure 10) have been developed by Evans's group.[47,48] Only partial similarity with the natural enzyme can be noticed. The Ni ion is P_2S_2Cl five-coordinated in a square pyramidal geometry with the

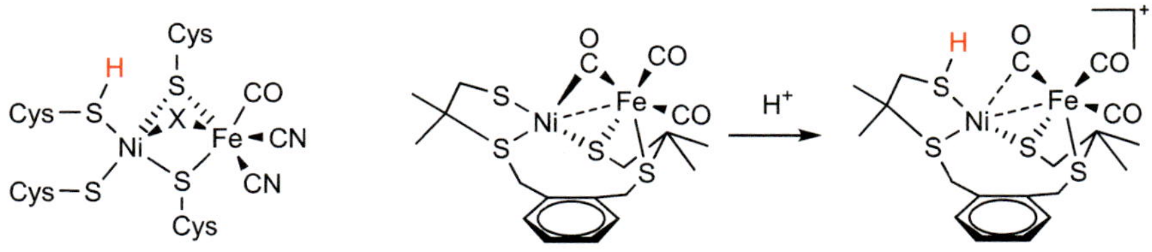

Figure 9. (a) Molecular structure of the active site of the [NiFe]-H_2ase in the protonated oxidized inactive ready state (Ni–B, X = OH^-) and reduced active state (Ni–C, X = H^-); (b) Protonation of the synthetic [NiFe] model.

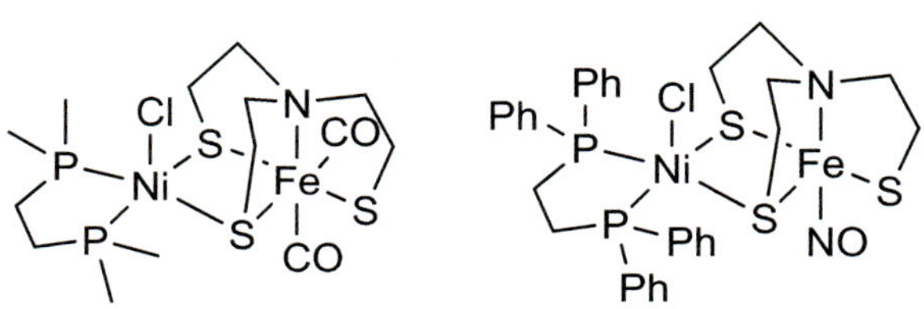

Figure 10. Molecular structure of diphosphine-based [NiFe]-H_2ases.

chloride ligand in the apical site, and P_2S_2 atoms in the basal plane. The iron atom is five or six coordinated through the S, N and C atoms. No electrochemical description is available for the reported structures.

Several works by Rauchfuss *et al.* present NiFe complexes with P_2S_2 four-coordinate nickel ion.[49–51] Even though the compound contains abiological phosphine-based ligands, a close mimic of the [NiFe]-H_2ase was reported, and most importantly catalysis showed similar bidirectional properties of hydrogen evolution/oxidation. Electrochemical H_2 oxidation to protons occurs in the presence of a base, while strong acids induce H_2 release. Coordination of borane units to the Fe–CN centers was used to both suppress side-reactions at the Fe–CN centers and promote decarbonylation that leads to reaction with H_2 to form a hydride species (Figure 11a), similarly to the reaction of the native hydrogenase. The Fe–Ni distance of 2.55 Å reported for the Ni-R state is very close to the values of 2.59 and 2.57 Å found for the reported hydride states. However, the downside of the phosphine ligands is that they cannot act as intramolecular bases as compared to the sulphur-based ligands in the native enzyme. It should also be mentioned that the redox reaction is iron centered, as opposed to the nickel-based process in the natural enzyme.

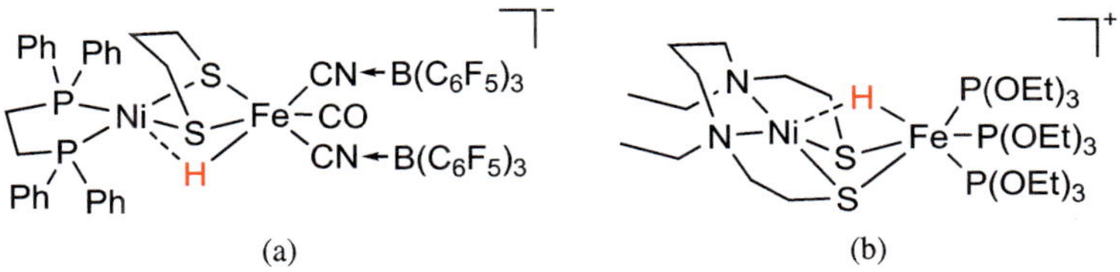

Figure 11. Molecular structure of phosphine containing [NiFe]-H_2ase models showing the hydride intermediate species.

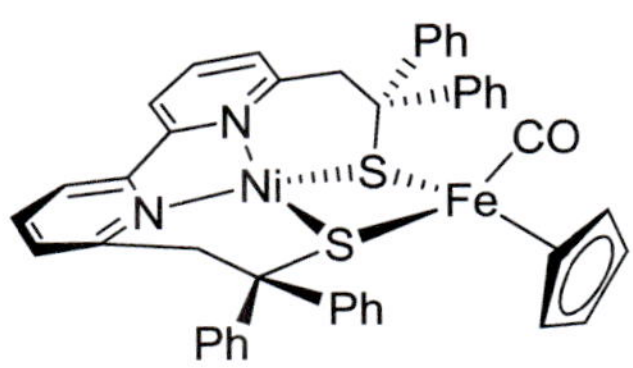

Figure 12. The [NiFe]-H_2ase model reported by Duboc and Artero.[53]

Another [NiFe]-based example of bidirectional catalytic behavior was reported by using a triphosphine-Fe based complex,[52] and showed both H_2 evolution and oxidation mimicking the parent enzyme. Two nitrogen atoms are complemented by two sulfur Ni–Fe bridging atoms for a tetradentate N_2S_2 coordination of Ni ion (Figure 11b). A somewhat longer Ni–Fe distance of 2.8 Å was found for this H_2ase model. Heterolytic H_2 cleavage requires a strong base (methanolate). Upon removal of a coordinated solvent the complex reacts with molecular hydrogen to form a Ni–H–Fe bridging hydride intermediate that releases a proton when oxidized with methyl viologen, and thus closes the catalytic cycle. On the other hand, release of H_2 requires a strong acid.

The teams of Duboc and Artero have reported a very interesting H_2ase structural and functional biomimetic complex (Figure 12), that for the first time for [NiFe] based models display reactivity at the Ni site, thus mimicking the natural enzyme mechanism.[53] Furthermore, the Fe modulates the electronic properties of the N_2S_2 coordinated Ni ion thus enhancing its reactivity. Two important catalytic intermediates were detected and characterized for this complex that replicate structural and electronic features of the Ni-L and Ni-R states of the natural enzyme. The [FeNi] model displays high rates of electrocatalytic hydrogen evolution turnover frequency (TOF) of 250 s^{-1} at 10 mM concentration in mildly acidic acetonitrile solutions.

C. [NiRu]-Based Biomimetic Models of [NiFe]-H_2ases

In the context to develop new compounds as efficient and/or stable catalysts for hydrogen conversion, a vision in this direction is the substitution of iron with another electron-rich metal center. Not surprisingly researchers attempted to use Ru that was known to act as catalyst for many other chemical reactions. Thus, several groups reported on the synthesis and catalytic properties of [NiRu]-models for [NiFe]-H_2ases.

One of the first functional bioinspired Ru-based model of [NiFe]-H_2ases was reported in 2006.[54] The nickel ion is coordinated by four sulfur atoms forming a square-planar environment (Figure 13). The ruthenium ion is connected to nickel through two bridging sulphur atoms. The distance between the two metal atoms is 3.2 Å, longer than Ni–Fe distances observed in enzymes. The same group later published other Ni–Ru complexes by changing the ligands, but keeping the same sulphur-based bridging mode.[55–58]

Catalytic properties were investigated in DMF solutions with triethylammonium as the proton source. Electrochemical studies assisted by theoretical calculations established the mechanism consisting of the initial formation of a metal-hydride which further reacts bimolecularly to undergo reductive elimination of hydrogen. The heterolytic pathway is also taken into consideration, consisting of protonation of the intermediate hydride to evolve hydrogen in a monomolecular reaction, followed by a one-electron reduction event. Hence H_2 evolution either follows a CEC or a CECE mechanism (E is an electron transfer step and C is a chemical step).[57,59] The introduction of a cyclopentadienyl ligand to the ruthenium ion helps decrease the electrochemical overpotential and the turnover rates.

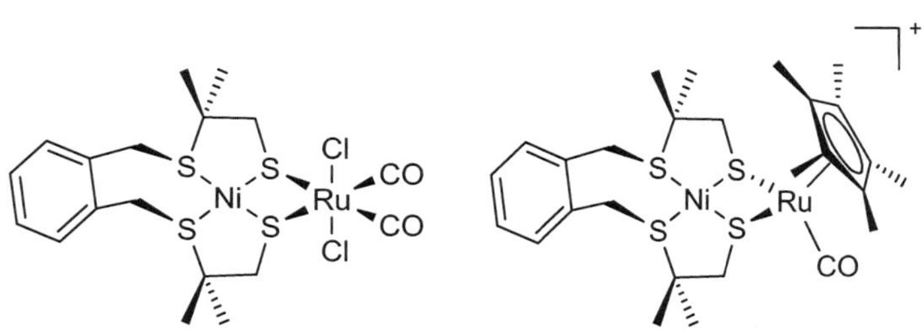

Figure 13. Molecular structures of some [NiRu]-based complexes mimicking the [NiFe]-H_2ases.

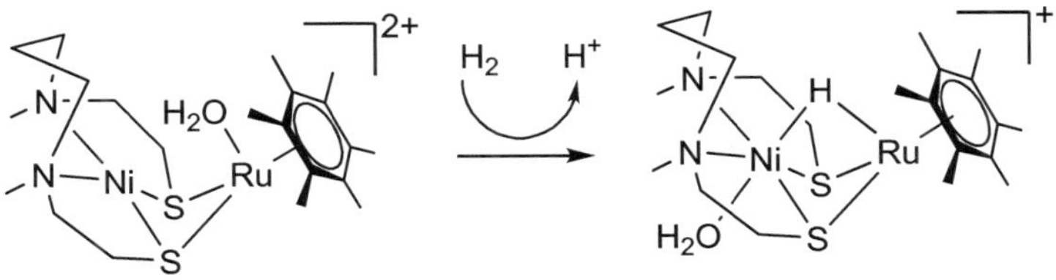

Figure 14. Formation of [Ni-H-Ru] hydride complex by reaction with dihydrogen.

Another group has reported a [NiRu] with a reasonably similar structure, but using a N_2S_2 tetradentate ligand to coordinate the Ni ion in a square-planar geometry (Figure 14). The metal hydride [Ni–H–Ru] species was isolated in the aqueous media under a dihydrogen atmosphere (0.1 MPa) and the structure determined by neutron diffraction analysis.[60] Formation of the hydride is accompanied by Ni(II, S = 0) coordinating a water molecule and the bridging hydride, thus transforming to Ni(II, S = 1) in a pseudo-octahedral coordination geometry. This research supports the formation of a [Ni–H–Ru] intermediate as part of the above-mentioned mechanism proposed by Fontecave and Artero.[57,59]

D. Biomimetic Models of [FeFe]-H_2ases

The first generation of binuclear Fe–Fe complexes (Figure 15) modeling the [FeFe]–H_2ase tried to copy the ligand environment of the natural analogs by including carbon monoxide/cyanide ligands and a dithiolate bridging ligand.[61–63] The intermetallic distance of 2.6 Å in these systems is close to that determined in [FeFe]-H_2ase.

A large number of similar compounds have quickly emerged after these first examples, with huge impact on understanding the structure of the natural enzymes. Yet, limited catalytic performance was found mostly due to the use of the Fe(I)–Fe(I) and even Fe(I)–Fe(0) redox levels causing large overpotentials, while the natural enzyme is known to work at very low potentials through Fe(II) –Fe(I) active species.

Substitution of one carbon on the dithiolate bridge by a heterogeneous atom was realised by several groups and showed enhancement of catalytic properties. The oxadithiolate model complexes of the active site of [FeFe]-hydrogenases (Figure 16) have been reported by the Hu research

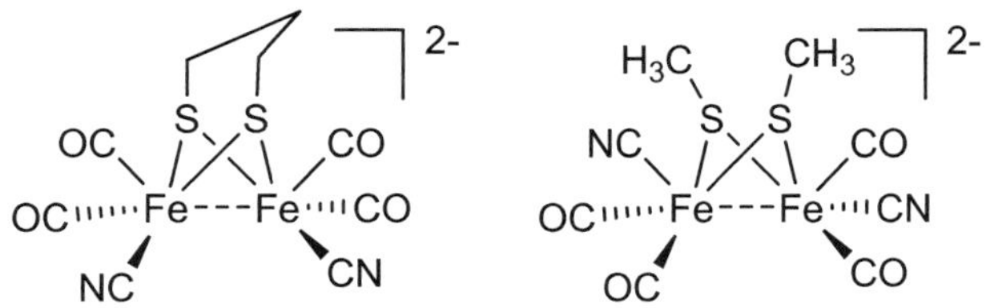

Figure 15. Molecular structures of some of the first biomimetic models of [FeFe]-H_2ases.

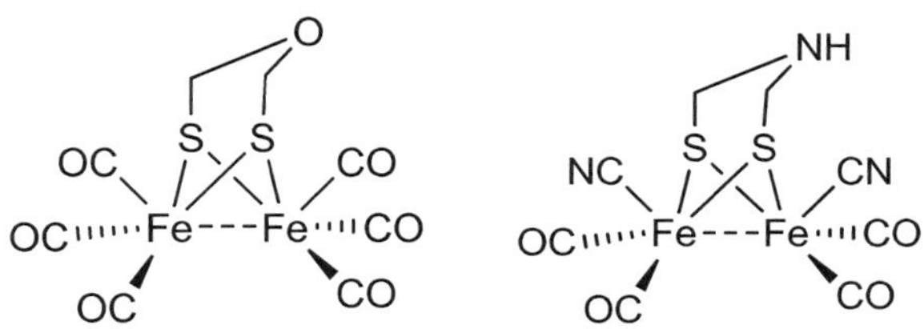

Figure 16. Molecular structures of [FeFe]-H_2ases with a modified dithiolate group.

group.[64,65] For this series of compounds the electrochemical investigations showed high catalytic activity in proton reduction in the acetic acid media and the EECC mechanism was proposed. Azadithiolato diiron complexes represent another approach in the synthesis of biomimetic models of [FeFe]-H_2ases.[66] The presence of N–H bond in the axial position leads to increased electron-density at the iron. Electrochemical studies showed that the protonation of the amine group can assist the formation of the H–H bond to produce dihydrogen during the catalysis at low potentials.[67]

The first [2Fe3S] clusters containing carbon monoxide (Figure 17) were described in 2001, with the Fe–Fe distance of 2.5 Å in the metal bonding range.[68] In the presence of two equivalents of cyanide ligand a metastable CO bridged di-cyanide species is obtained. In the presence of isocyanide ligands, the more stable diferrous dithiolate complexes are formed, containing μ-CO ligands with the Fe ion in the octahedral geometry thus structurally resembling the oxidized form of the [FeFe]-H_2ase.[69]

Not many examples are known for the [FeFe]-H_2ase with redox active units attached for mimicking the electron transfer relay. The first report described a [4Fe4S] cluster attached to the $[Fe_2(\mu\text{-}SR)_2(CO)_6]$ catalysis

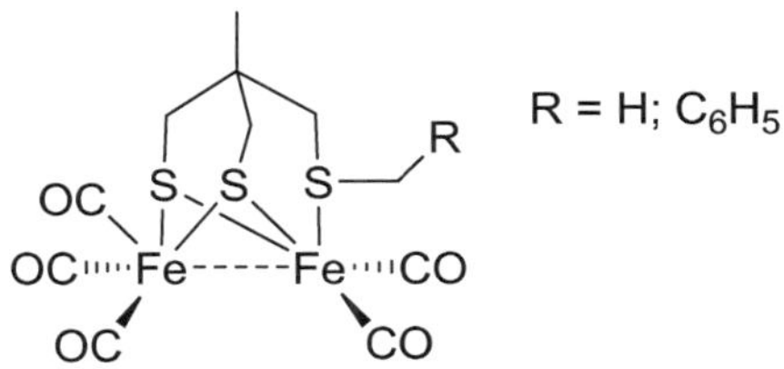

Figure 17. The molecular structure of the [2Fe3S] based biomimetic model of [FeFe]-H_2ase.

subsite.[70] Although the system works as a catalyst, large overpotentials are required thus limiting its use. Several other examples were published but very few showed functional importance of the appended unit.[71,72] The best example consists of a ferrocene moiety connected to a [FeFe] subsite. It was shown that the ferrocene unit can provide the second positive charge required for H_2 oxidation from the ferricinium cation obtained at mild over potentials.

IV. Bioinspired Models of Hydrogenases

Research on the biomimicry of the natural H_2ase enzymes played a crucial role in understanding and learning from Nature; ways were developed to sustain an efficient catalysis of hydrogen evolution from water by keeping low overpotentials, high frequency rates, robustness (though under specific conditions) and crucially by using largely available non-noble metals. The next emerging step is to apply all this knowledge to produce new synthetic catalysts with no direct structural resemblance to the natural enzymes, but with catalytic activity surpassing that of hydrogenases.

This section aims to give a review of literature on the metal complexes as bioinspired models of hydrogenases for the generation of dihydrogen from protons and electrons.

A. Bioinspired Nickel Complexes

One of the first works on nickel complexes shown to catalyze hydrogen evolution from water was based on cyclam derived compounds studied in the laboratory of Sauvage (Figure 18).[73] Turnover numbers (TON)

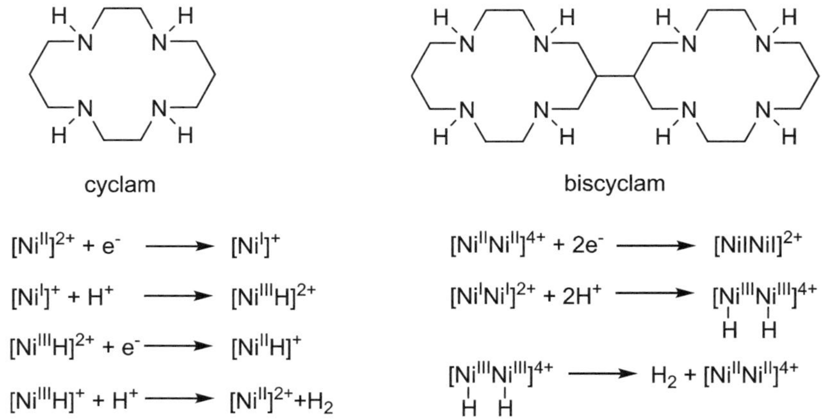

Figure 18. Structure of cyclam ligands and the catalysis mechanism for both nickel complexes.

up to 100 were reached for nickel complexes by measuring the amount of hydrogen produced in the course of electrolysis in water at pH 7. It is intriguing that the efficiency of the dinuclear compound as an electrocatalyst for water reduction to hydrogen is significantly larger than that of the mononuclear compound, for an identical analytical concentration of nickel. The heterolytic catalysis mechanism was proposed for the monomer consisting of an ECEC sequence, namely each electron transfer step is followed by a proton transfer, while an initial double reduction followed by double protonation (EECC) was suggested for the dimer (Figure 18). Formation of dihydride is proposed to be responsible for the better performance of the dimer. Interestingly under a CO_2 atmosphere formation of carbon monoxide is much more favored compared to dihydrogen formation.

The NiP_4 molecular system as functional model of H_2ase has been extensively studied by DuBois and co-workers. One of the first studied models $[M(diphosphine)_2]^{2+}$, where M = Ni, Pt, activates hydrogen to form the hydrides in the presence of bases.[74] Later the $[Ni(diphosphine)_2]^{2+}$ complexes containing ligand-appended nitrogen bases ($[Ni(PNP)_2]^{2+}$) were reported.[75] These systems were found to catalyze the electrochemical oxidation of hydrogen. The electrochemical studies confirm the heterolytic activation of hydrogen through the intermediary species $[HNi(PNP)(PNHP)]^{2+}$. The two hydrogens undergo rapid intramolecular

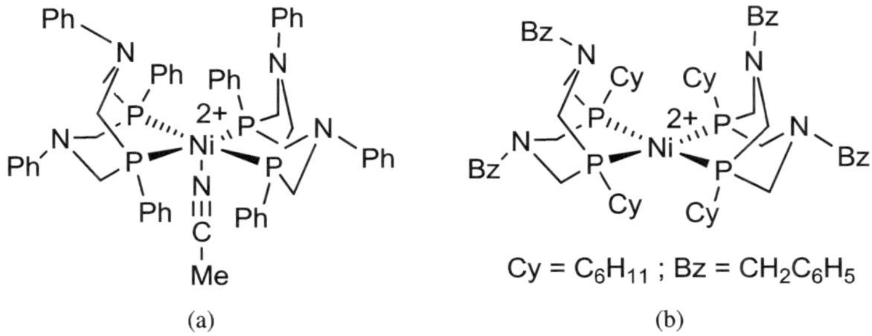

Figure 19. Structure of (a) the $[Ni(P^{Ph}{}_2N^{Ph}{}_2)_2(CH_3CN)]^{2+}$ cation, and (b) the $[Ni(P^{Cy}{}_2N^{Bz}{}_2)_2]^{2+}$ cation.

and intermolecular exchange with each other and protons in solution, similarly to the natural [FeFe]-H_2ase.

Later the same group reported a series of Ni(II) complexes based on cyclic diphosphine ligands (Figure 19) which catalyze both electrochemical hydrogen evolution and oxidation.[76] Pendant bases are incorporated in the ligand backbone and play an important role in activation and cleavage of the H–H bond.

Electrochemical studies showed similar properties for both compounds. However, electrocatalysis studies revealed that the fully phenyl substituted complex (Figure 19a) is an electrocatalyst for proton reduction in acidic solutions while the cyclohexyl/benzyl analog (Figure 19b) catalyzes electrochemical hydrogen oxidation in basic solution.

The TOF decreases drastically from 350 s^{-1} (300 mV overpotential in solution of protonated DMF) to 5 s^{-1} when the amine substituent is changed from phenyl to benzyl, showing how more basic amines disfavor the catalytic proton reduction. By introducing onto the phenyl ring various substituents with different electron-withdrawing or donating capability (OMe, Me, $CH_2P(O)(OEt)_2$, H, Br, CF_3) a pronounced effect on the redox potentials of the $Ni^{II/I}$ and $Ni^{I/0}$ was noticed, and respectively catalytic activity for proton reduction. Thus it was shown that pendant amines in the second coordination sphere can facilitate heterolytic dihydrogen cleavage or evolution, intra- and inter-molecular proton transfers and their coupling with the electron-transfer process.[77] Catalysts containing pendant amines usually show much lower overpotentials and much higher catalytic rates as compared to analogous complexes lacking the pendant amines.

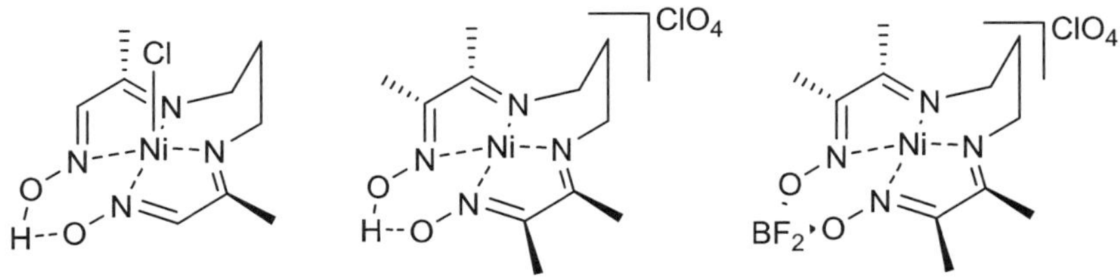

Figure 20. Structure on Ni complex with glyoxime ligands.

Examples of catalysts for electrochemical proton reduction by using organometallic complexes of Ni are also known based on tridentate P_2O pincer ligands.[78] Moderate overpotentials of 0.35 V were reported along with a TOF of 209 s^{-1}.

Electrocatalysts based on the nickel complexes with diimine-dioxime ligands (Figure 20) were reported to be active for protons reduction.[79] Bulk electrolysis experiments of H_2 evolution at a graphite rod electrode from various acids in acetonitrile catalyzed by nickel complexes showed production of hydrogen with a turnover number (TON) of up to 20 (per 3 h) and a Faradaic yield in the range of 30–70%. The presence of a proton site is believed to help reducing the overpotential. A bimetallic homolytic pathway is proposed based on spectroscopic observations.

An interesting approach was reported of using chelating ether groups as the second coordination site that can bind alkali metal cations to form Lewis acids in close vicinity to the redox-active metal center (Figure 21) and thereby promote proton reduction efficiency.[80] The main catalytic site is based on a salicylaldimine Ni complex while the introduction of sulfonate units allows for electrocatalysis of H_2 evolution in neutral and acidic aqueous solutions.

A number of papers were reported on H^+/H_2 interconverting Ni complexes based on thio-ligands with a N_xS_y coordination mode.[81–83] Although electrocatalysis experiments were not performed complexes were shown to activate the dihydrogen molecule and catalyze H/D exchange in the same way as hydrogenases do. Phenol units seem to be critical for this process so that even removal of one of the two phenols (Figure 22) stops the H/D exchange, thus a cooperative hydrogen bonding mechanism of dihydrogen activation is likely.

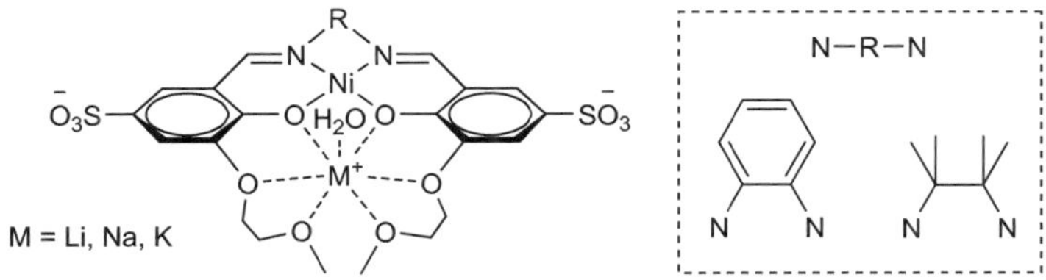

Figure 21. Molecular structure of salicylaldimine based Ni complexes with pendant ether groups.

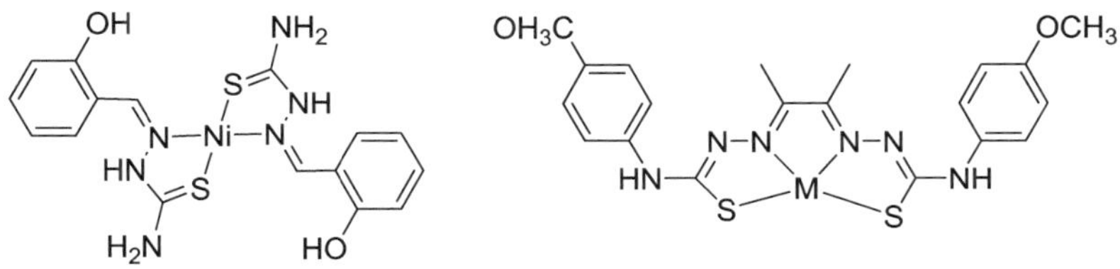

Figure 22. Ni complexes based on S/N containing ligands.

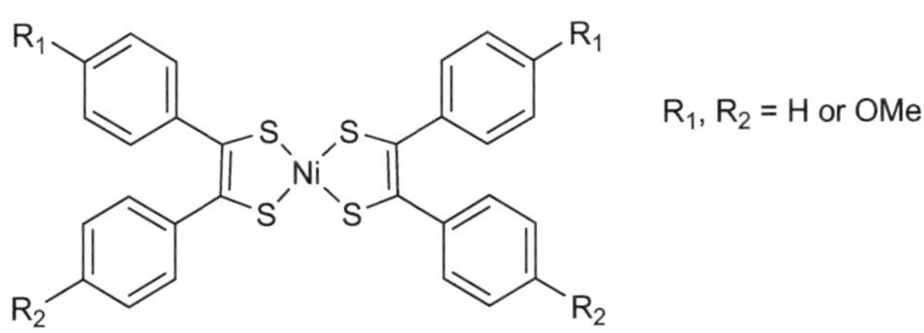

Figure 23. Molecular structure of dithiolene based Ni complexes.

It was shown that Ni complexes with N_2S_2 bis-thiosemicarbazone based ligand (Figure 22) can reduce protons at very low overpotentials (350 mV) and good turnover efficiencies. Molecular modeling suggests a ligand-metal synergistic catalysis mechanism that involves first ligand-based reduction and protonation steps followed by metal-centered processes.[84]

In contrast to the traditional metal-hydride mediated catalysis several examples show that catalysts with ligand-centered proton-transfer and electron-transfer events are a viable alternative. Neutral and monoanionic nickel complexes with 1,2-dithiolene ligands (Figure 23) were used as proton reducing catalysts in DMF with trifluoroacetic acid (TFA) as the proton source.[85] Computational modeling suggests ligand-based EECC or EECEC mechanisms depending on the acid concentration and the substituents on the phenyl rings.

B. Bioinspired Cobalt Complexes

The ability of the CoN_4 complexes to electrocatalyze the reduction of protons was discovered by chance. In the 1980s, Fisher and Eisenberg were investigating the electrocatalytic properties of a cobalt complex with a macrocyclic diene-N_4 ligand (Figure 24) for the reduction of CO_2 in water/acetonitrile solutions.[86] The products of the electrocatalytic reaction were determined to be carbon monoxide and dihydrogen in equimolar ratio with a TON of 7.8 and 9, respectively, per hour.

Much later several research groups focused their attention on bisglyoximato cobalt complexes (cobaloximes). In 2005 and 2007, Artero and Fontecave investigated the cobaloxime complex (Figure 25a) and its BF_2

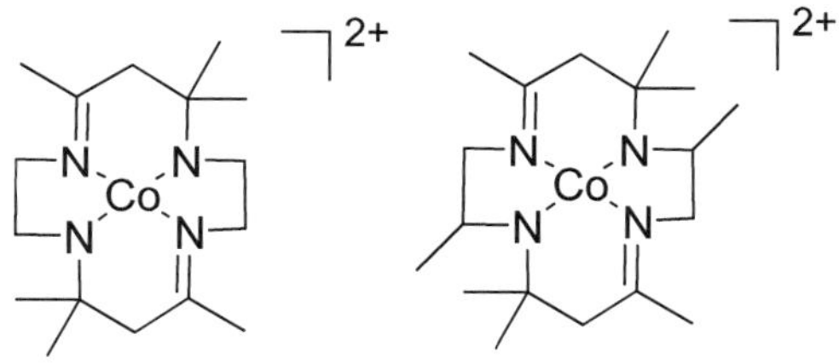

Figure 24. Molecular structures of some of the first cobalt complexes used for the reduction of protons.

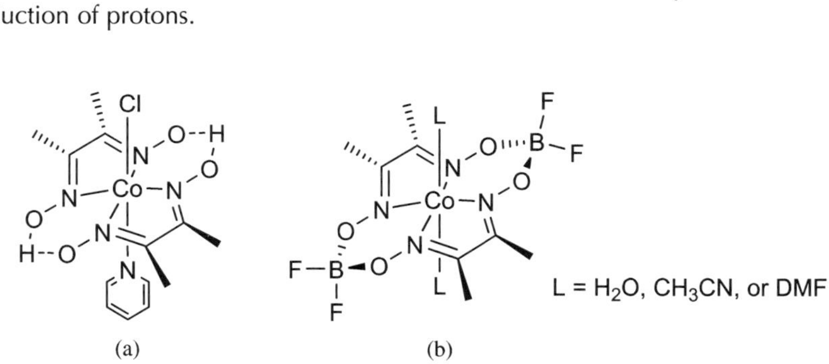

Figure 25. Molecular structures of cobaloxime complexes.

coordinated derivative (Figure 25b) as functional models of hydrogenases.[87,88]

Several cobaloximes have been examined as electrocatalysts for hydrogen evolution from DMF solution of triethylammonium chloride. During the controlled-potential electrolysis TONs reaching 100 were achieved and cyclic voltammogram simulations were consistent with a heterolytic catalysis mechanism. Both the equatorial glyoxime ligand and the axial pyridine ligand have a significant influence on the catalytic activity of the complexes. Introduction of electron donating or withdrawing substituents in the axial pyridine ligand significantly increased (respectively decreased) the catalytic rate of the electrocatalysts. The complexes containing difluoroboryl-bridges were investigated for protons electroreduction in both DMF and acetonitrile using different acids as proton source. Better performance was obtained as compared to parent cobaloximes. Depending on the strength of the acid used and the potential applied, three different pathways for hydrogen production are discussed. Namely EECC, ECEC and ECCE sequences, with heterolytic (most likely in strong acids) or homolytic hydrogen evolution, are possible in weak, intermediate and strong acids respectively. The best turnover frequency of 92 h^{-1} after 30 min at –0.5 V *versus* Ag/AgCl was obtained when p-cyanoanilinium tetrafluoroborate was used as the acid, making this class of compounds one of the most promising for proton reduction applications.

Several mechanisms for proton reduction promoted by cobaloxime complexes have been proposed in the literature, however some details remain a topic of intense debate, largely owing to the difficulty of detection of some key reactive intermediates such as the protonated species.[89]

Substitution of phenyl to methyl on the glyoxime ligand (Figure 26) resulted in smaller overpotential but lower electrocatalytic activity was also noticed.[90,91] Interestingly indication of hydrogen oxidation catalysis by cobaloximes was also shown.

Although encouraging studies show the great potential cobaloximes, as most other molecular catalysts they suffer from degradation in prolonged experiments, most likely as result of acid catalyzed hydrolysis of the ligand, thus reducing their application potential. A way to solve this problem consists of depositing them on surfaces. Several attempts were made, including immobilization through axial coordination of cobalt onto

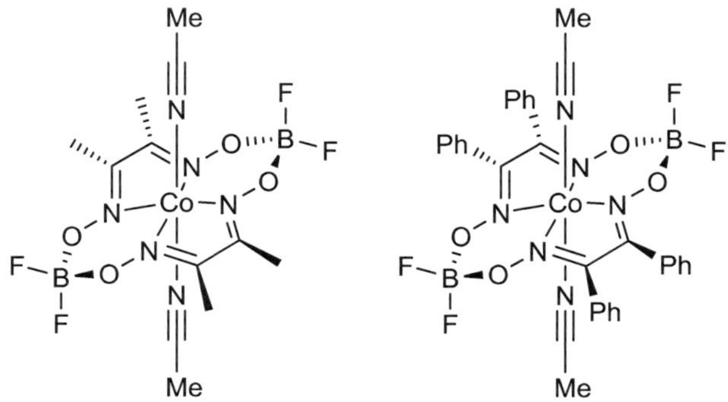

Figure 26. Other cobalt tetraimino complexes.

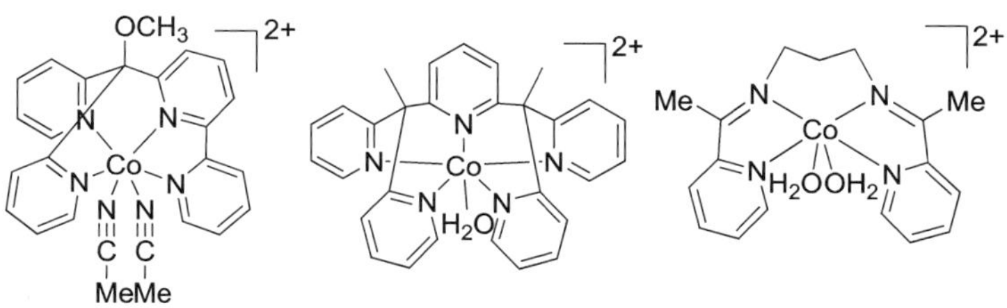

Figure 27. Molecular structures of cobalt complexes containing polypyridyl ligands.

polymer-modified nanostructured indium tin oxide electrode[92] and pyridine-coated carbon nanotubes,[93] or through diimine–dioxime ligand covalent link to multiwalled carbon nanotubes,[94] giving although somehow contradictory still very promising results (e.g., 55,000 TON was reached).

Cobalt polypyridine complexes (Figure 27) showed promising results for hydrogen evolution from neutral water with mostly quantitative Faradaic yields.[95,96] They were found to efficiently reduce protons for 60 h without major signs of degradation, reaching a TON of 55,000. Yet a different approach for the TON calculation was used preventing from accurate comparison to other electrocatalysts. A higher overpotential and a 87% Faradaic yield was found for the cobalt complex with the iminopyridine-based ligand (Figure 27).[97]

Cobalt corroles are another promising class of compounds for catalytic protons reduction (Figure 28).[98,99] The use of pendant acid/base units

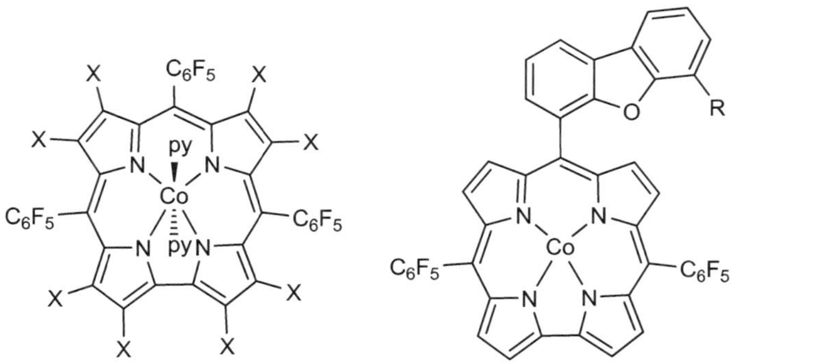

X = H, F, Cl, Br R = Br, COOH, $PO(OH)_2$, $CH_2PO(OH)_2$

Figure 28. Molecular structure of cobalt corrole catalysts.

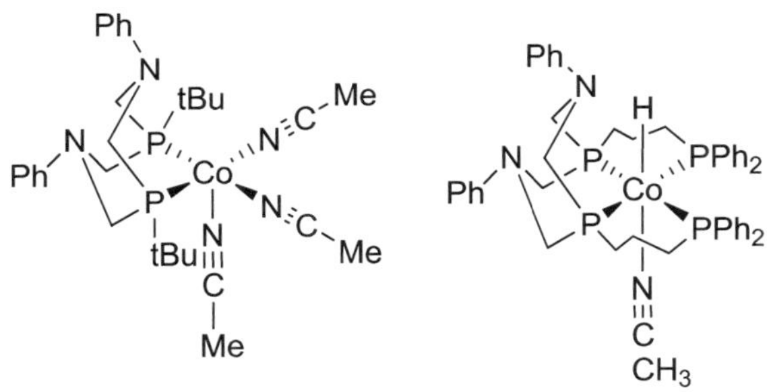

Figure 29. The diphosphine and tetraphosphine cobalt complexes.

allowed overpotential reduction to 450 mV in neutral aqueous solutions; in case of methylenephosphonic acid the group can enhance the hydrogen evolution reaction by increasing the effective local concentration of protons.

A number of cobalt complexes with ligands containing phosphine coordination atoms are known (Figure 29).[100–103] The P_4-tetradentate ligand (Figure 29) showed electrocatalysis for hydrogen production in DMF solutions with a TOF of 980 s^{-1} but at a high overpotential of 0.930 V. Interestingly addition of water increases the TOF to 18,000 s^{-1}. Mechanistic studies suggest that dihydrogen evolution can occur by protonation of the HCo^{II} hydride, while the pendant amines facilitate proton delivery to the reaction center.

C. Bioinspired Complexes with Other Metals

In order to diversify the classes of compounds that are capable of catalyzing protons reduction, complexes of various other metals (e.g., Cu, Pd, Mn) with an immense variety of ligands were probed as hydrogen evolution candidates.

Porphyrin complexes were known for a long time to catalyze proton reduction reaction, after the first studies in Savéant's group.[104,105] The Fe^{III}-tetraphenylporphyrin (Figure 30, left) was found to electrochemically produce dihydrogen in quantitative Faradaic yields from DMF solutions in the presence of weak acids, yet with large overpotentials. Two mechanisms were taken into consideration after formation of the dianion $Fe^{0}TPP^{2-}$; protonation of the porphyrin ring to give the phlorin reactive intermediate or formation of a Fe^{II} hydride intermediate that upon reaction with a proton would produce dihydrogen. As no catalysis was observed for CuTPP the former pathway was ruled out.

We have shown that Cu and Pd porphyrin complexes (Figure 30, right) can be activated for electrocatalysis of protons reduction in DMF with weak acids by introducing ferrocene units.[6] Unlike the other example the metal ion is not redox active in the catalytic cycle. DFT calculations show that sterically bulky ferrocenes distort the porphyrin at the *meso*-position thus facilitating formation of the intermediate phlorin anion determined spectroscopically. Another benefit is a decrease in the overpotential. Later the phlorin intermediate was proposed to form by proton-coupled electron

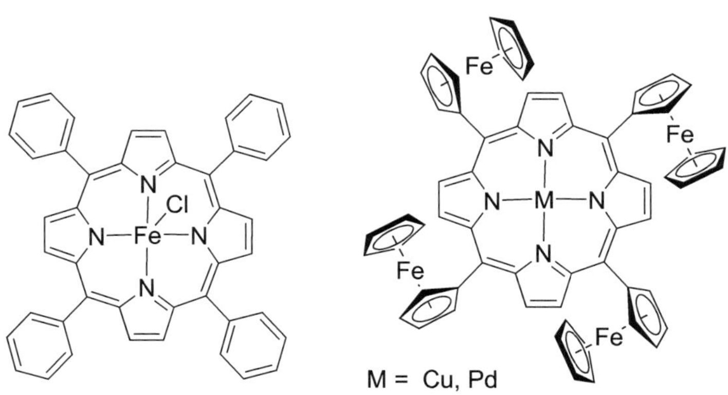

Figure 30. Molecular structure of the porphyrin catalysts.

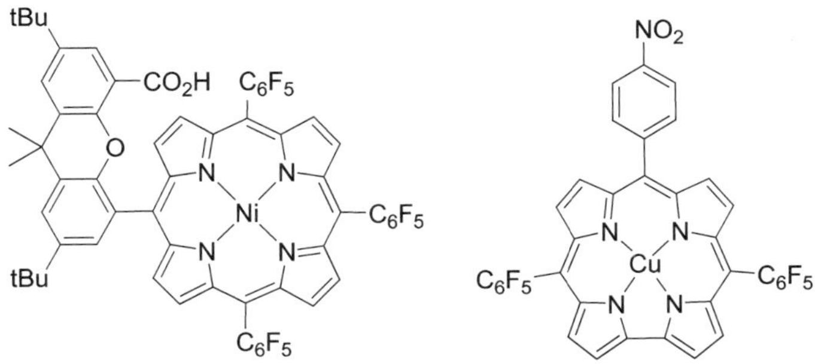

Figure 31. Molecular structure of the porphyrin and corrole-based catalysts.

transfer during electrocatalytic hydrogen evolution by using a Ni complex of a porphyrin with a pendant carboxylic acid (Figure 31).[106] The copper complex of the structurally related 5,15-pentafluorophenyl-10-(4-nitrophenyl)corrole (Figure 31) also showed electrocatalytic properties towards hydrogen evolution at relatively mild overpotentials.[107] The theoretical and experimental studies indicate an EECC mechanism of hydrogen evolution based on the usual metal hydride intermediate pathway.

A dinuclear Ni–Mn complex (Figure 32) was proposed as bioinspired model of [NiFe]-H_2ases by catalyzing hydrogen evolution from TFA solutions, yet with relatively large overpotential of 0.86 V.[108] The catalysis cycle intermediate, based on a bridging hydride ligand between Ni and Mn, has a structural resemblance to the Ni–C active state of NiFe hydrogenases.

The thiolato dimanganese hexacarbonyl anion (Figure 32) with some structural resemblance to the active site of [FeFe]-hydrogenase, showed electrocatalytic activity towards proton reduction from TFA solutions in acetonitrile.[109] High turnover frequency values were reported in controlled conditions at a moderate overpotential of 0.61 V.

The copper complex of a 2-methylpyridinyl substituted ethylenediamine (Figure 33) was reported in 2014 as a protons reduction catalyst at moderate overpotentials (420 mV), and almost quantitative Faradaic yields in acidic aqueous solutions, with a catalysis rate of over 10,000 s^{-1}.[110] The initially proposed mechanism for H_2 evolution involved two proton-coupled reduction steps, but later DFT calculations were used to develop an alternative mechanism based on coupling of a Cu(II)-hydride and a pendant pyridinium unit.[111]

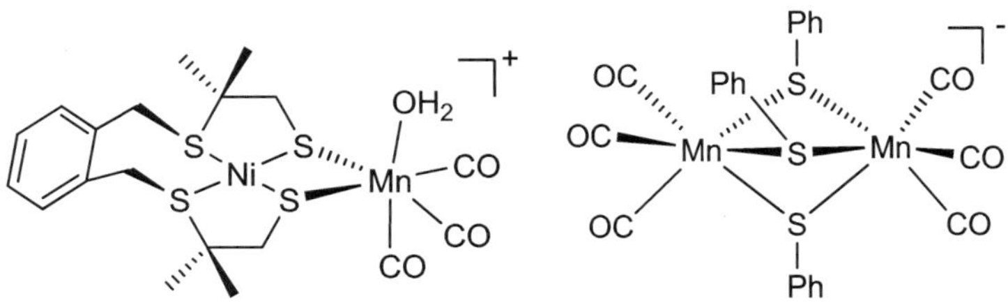

Figure 32. Molecular structure of manganese containing complexes.

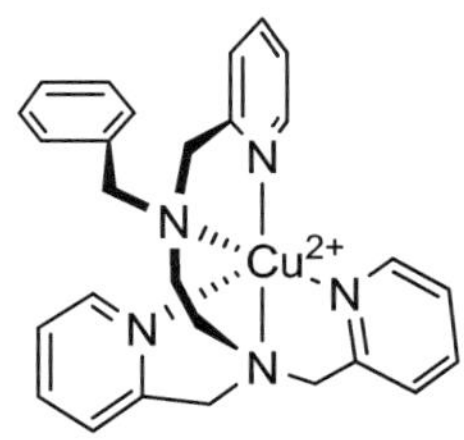

Figure 33. Molecular structure of copper-based proton reduction catalysts.

The reaction starts with a proton-coupled electron transfer to generate a Cu^{I}–pyridinium species, in which the proton is in a fast equilibrium between two pyridine groups. The following proton-coupled electron transfer leads to the formation of a Cu^{II}-hydride intermediate followed by dihydrogen formation by the coupling of the Cu^{II}-hydride and the proton from the pyridinium unit. Thus, the pendant pyridine helps lower the overpotential for H_2 formation.

Knowledge obtained from these studies provides guidelines for the future design of new catalysts of hydrogen evolution reaction based on non-biological elements and structures unusual for nature.

V. Conclusions

During the past decades, a vast progress has been made into the understanding of the structure and function of hydrogenases, and development of synthetic models reaching or to a certain extent overpassing the natural systems. Determination of active intermediates and the tuning of thermodynamic energies in the catalytic cycle, use of pendant acid and bases to control the proton movement and stabilizing of the high energy intermediates, as well as understanding the steric factors helps building increasingly more efficient platforms for the generation of molecular hydrogen.

High structural versatility and the "mastery" to finely control the local catalytic environment, along with the ability of advanced spectroscopic and computational study of the catalysis mechanism make them worthy of being investigated further, and yet, molecular catalysts have several major problems. The main issue is their poor robustness as compared to materials. With the above mentioned we nevertheless believe that combination of highly active and selective molecular catalysts and functional surfaces, thus increasing their robustness, might be the way forward to overcome the challenging task and make molecular catalysts truly useful for real applications.

VI. References and Citations

1. Bard, A. J.; Fox, M. A. *Acc. Chem. Res.* **1995**, *28*, 141.
2. Lewis, N. S.; Nocera, D. G. *Proc. Natl. Acad. Sci.* **2006**, *103*, 15729.
3. Holladay, J. D.; Hu, D. L.; Wang, Y. K. *Catal. Today* **2009**, *139*, 244.
4. Koroneos, C.; Dompros, A.; Roumbas, G.; Moussiopoulos, N. *Int. J. Hydrogen Energ.* **2004**, *29*, 1443.
5. Carmo, M.; Fritz, D. L.; Mergel, J.; Stolten, D. A. *Int. J. Hydrogen Energ.* **2013**, *38*, 4901.
6. Sirbu, D.; Turta, C.; Gibson, E. A.; Benniston, A. C. *Dalton Trans.* **2015**, *44*, 14646.
7. Stephenson, M.; Stiklans, L. H. *Biochem. J.* **1931**, *25*, 205.
8. Peck, H. D.; San Pietro, A.; Gest, H. *Proc. Natl. Acad. Sci. USA.* **1956**, *42*, 13.
9. Adams, M. W.; Mortenson, L. E.; Chen, J. S. *Biochim. Biophys. Acta.* **1980**, *594*, 105.
10. Lancaster, J. R. *FEBS Lett.* **1980**, *115*, 285.
11. Albracht, S. P.; Graf, E. G.; Thauer, R. K. *FEBS Lett.* **1980**, *140*, 311.
12. Volbeda, A.; Charon, M. H.; Piras, C.; Hatchikian, E. C.; Frey, M.; Fontecilla-Camps, J. C. *Nature*, **1995**, *373*, 580.
13. Nicolet, Y.; Piras, C.; Legrand, P.; Hatchikian, E. C.; Fontecilla-Camps, J. C. *Structure* **1999**, *7*, 13.
14. Peters, J. W.; Lanzilotta, W. N.; Lemon, B. J.; Seefeldt, L. C. *Science* **1998**, *282*, 1853.
15. Garcin, E.; Vernède, X.; Hatchikian, E. C.; Volbeda, A.; Frey, M.; Fontecilla-Camps, J. C. *Structure* **1999**, *7*, 557.
16. Volbeda, A.; Amara, P.; Iannello, M.; de Lacey, A. L.; Cavazza, C.; Fontecilla-Camps, J. C. *Chem. Commun.* **2013**, *49*, 7061.

17. Marques, M. C.; Coelho, R.; Pereira, I. A. C.; Matias, P. M. *Int. J. Hydrogen Energ.* **2013**, *38*, 8664.
18. Lemon, B. J.; Peters, J. W. *Biochemistry* **1999**, *38*, 12969.
19. Silakov, A.; Reijerse, E. J.; Albracht, S. P. J.; Hatchikian, E. C.; Lubitz, W. *J. Am. Chem. Soc.* **2007**, *129*, 11447.
20. Berggren, G.; Adamska, A.; Lambertz, C.; Simmons, T. R.; Esselborn, J.; Atta, M.; Gambarelli, S.; Mouesca, J. M.; Reijerse, E. J.; Lubitz, W.; Artero, V.; Fontecave, M. *Nature* **2013**, *498*, 66.
21. Zirngibl, C.; Hedderich, R.; Thauer, R. K. *FEBS Lett.* **1990**, *261*, 112.
22. Afting, C.; Hochheimer, A.; Thauer, R. K. *Arch. Microbiol.* **1998**, *169*, 206.
23. Hiromoto, T.; Warkentin, E.; Moll, J.; Ermler, U.; Shima, S. *Angew. Chem., Int. Ed.* **2009**, *48*, 6457.
24. Ogata, H.; Hirota, S.; Nakahara, A.; Komori, H.; Shibata, N.; Kato, T.; Kano, K.; Higuchi, Y. *Structure* **2005**, *13*, 1635.
25. Lubitz, W.; Ogata, H.; Rüdiger, O.; Reijerse, E. *Chem. Rev.* **2014**, *114*, 4081.
26. Vincent, K. A.; Parkin, A.; Armstrong, F. A. *Chem. Rev.* **2007**, *107*, 4366.
27. Siegbahn, P. E. M.; Tye, J. W.; Hall, M. B. *Chem. Rev.* **2007**, *107*, 4414.
28. Shafaat, H. S.; Rüdiger, O.; Ogata, H.; Lubitz, W. *Biochim. Biophys. Acta Bioenerg.* **2013**, *1827*, 986.
29. Lubitz, W.; Reijerse, E.; van Gastel, M. *Chem. Rev.* **2007**, *107*, 4331.
30. Thauer, R. K.; Kaster, A. K.; Goenrich, M.; Schick M.; Hiromoto, T.; Shima, S. *Annu. Rev. Biochem.* **2010**, *79*, 507.
31. Cramm, R. J. *Mol. Microbiol. Biotechnol.* **2009**, *16*, 38.
32. Fan, C.; Teixeira, M.; Moura, J.; Moura, I.; Huynh, Boi H.; Le Gall, J.; Peck, H. D.; Hoffman, B. M. *J. Am. Chem. Soc.* **1991**, *113*, 20.
33. Bruschi, M.; Tiberti, M.; Guerra, A.; de Gioia, L. *J. Am. Chem. Soc.* **2017**, *136*, 1803.
34. Wu, H.; Hall, M. B. *C. R. Chim.* **2008**, *11*, 790.
35. Keith, J. M.; Hall, M. B. *Inorg. Chem.* **2010**, *49*, 6378.
36. Foerster, S.; Gastel, M.; Brecht, M.; Lubitz, W. *J. Biol. Inorg. Chem.* **2005**, *10*, 51.
37. Kampa, M.; Lubitz, W.; van Gastel, M.; Neese, F. *J. Biol. Inorg. Chem.* **2012**, *17*, 1269.
38. Peters, J. W.; Schut, G. J.; Boyd, E. S.; Mulder, D. W.; Shepard, E. M.; Broderick, J. B.; King, P. W.; Adams, M. W. W. *BBA — Molecular Cell Research* **2015**, *1853*, 1350.
39. Evans, R. M.; Brooke, E. J.; Wehlin, S. A. M.; Nomerotskaia, E.; Sargent, F.; Carr, S. B.; Phillips, S. E. V.; Armstrong, F. A. *Nat. Chem. Biol.* **2015**, *12*, 46.
40. Vullev, V. I. *J. Phys. Chem. Lett.* **2011**, *2*, 503.
41. Kupchan, S. M.; Schubert, R. M. *Science* **1974**, *185*, 791.
42. Lai C. H.; Reibenspies J. H.; Darensbourg M. Y. *Angew. Chem. Int. Ed.* **1996**, *35*, 2390.
43. Osterloh, F.; Saak, W.; Haase, D.; Pohl. S. *Chem. Commun.* **1997**, *10*, 979.

44. Li, Z. L.; Ohki, Y.; Tatsumi, K. *J. Am. Chem. Soc.* **2005**, *127*, 8950.
45. Canaguier, S.; Field, M.; Oudart, Y.; Pecaut, J.; Fontecave, M.; Artero, V. *Chem. Commun.* **2010**, *46*, 5876.
46. Weber, K.; Kramer, T.; Shafaat, H. S.; Weyhermuller, T.; Bill, E.; van Gastel, M.; Neese F.; Lubitz, W. *J. Am. Chem. Soc.* **2012**, *134*, 20745.
47. Davies, S. C.; Evans, D. J.; Hughes, D. L.; Longhurst, S.; Sanders, J. R. *Chem. Commun.* **1999**, *19*, 1935.
48. Smith, M. C.; Barclay, J. E.; Cramer, S. P.; Davies, S. C.; Gu, W.-W.; Hughes, D. L.; Longhurst, S.; Evans, D. J. *Dalton Trans.* **2002**, *13*, 2641.
49. Barton, B. E.; Whaley, C. M.; Rauchfuss, T. B.; Gray, D. L. *J. Am. Chem. Soc.* **2009**, *131*, 6942.
50. Barton, B. E.; Rauchfuss, T. B. *J. Am. Chem. Soc.* **2010**, *132*, 14877.
51. Manor, B. C.; Rauchfuss, T. B. *J. Am. Chem. Soc.* **2013**, *135*, 11895.
52. Ogo, S.; Ichikawa, K.; Kishima, T.; Matsumoto, T.; Nakai, H.; Kusaka, K.; Ohhara, T. *Science* **2013**, *339*, 682.
53. Brazzolotto, D.; Gennari, M.; Queyriaux, N.; Simmons, T. R.; Pecaut, J.; Demeshko, S.; Meyer, F.; Orio, M.; Artero, V.; Duboc, C. *Nat. Chem.* **2016**, *8*, 1054.
54. Oudart, Y.; Artero, V.; Pecaut, J.; Fontecave, M. *Inorg. Chem.* **2006**, *45*, 4334.
55. Oudart, Y.; Artero, V.; Pécaut, J.; Lebrun, C.; Fontecave, M. *Eur. J. Inorg. Chem.* **2007**, *18*, 2613.
56. Oudart, Y.; Artero, V.; Norel, L.; Train, C.; Pécaut, J.; Fontecave, M. *J. Organomet. Chem.* **2009**, *17*, 2866.
57. Canaguier, S.; Vaccaro, L.; Artero, V.; Ostermann, R.; Pecaut, J.; Field, M. J.; Fontecave, M. *Chem. Eur. J.* **2009**, *15*, 9350.
58. Canaguier, S.; Fontecave, M.; Artero, V. *Eur. J. Inorg. Chem.* **2011**, *7*, 1094.
59. Canaguier, S.; Fourmond, V.; Perotto, C. U.; Fize, J.; Pecaut, J.; Fontecave, M.; Fielde, M. J.; Artero, V. *Chem. Commun.* **2013**, *49*, 5004.
60. Ogo, S.; Kabe, R.; Uehara, K.; Kure, B.; Nishimura, T.; Menon, S. C.; Harada, R.; Fukuzumi, S.; Higuchi, Y.; Ohhara, T.; Tamada, T.; Kuroki, R. *Science* **2007**, *316*, 585.
61. Schmidt M.; Contakes, S. M.; Rauchfuss, T. B. *J. Am. Chem. Soc.* **1999**, *121*, 9736.
62. Le Cloirec, A.; Best, S. P.; Borg, S.; Davies, S. C.; Evans, D. J.; Hughesa, D. L.; Pickett, C. J. *Chem. Commun.* **1999**, *22*, 2285.
63. Lyon, E. J.; Georgakaki, I. P.; Reibenspies, J. H.; Darensbourg, M. Y. *Angew. Chem. Int. Ed.* **1999**, *38*, 3178.
64. Song, L.-C.; Yang, Z.-Y.; Bian, H.-Z.; Hu. Q.-M. *Organometallics* **2004**, *23*, 3082.
65. Song, L.-C.; Yang, Z.-Y.; Bian, H.-Z.; Liu, Y.; Wang, H.-T.; Liu, X.-F.; Hu. Q.-M. *Organometallics* **2004**, *24*, 6126.
66. Li, H.; Rauchfuss, T. B. *J. Am. Chem. Soc.* **2002**, *124*, 726.
67. Wang, Z.; Liu, J.-H.; He, C.-J.; Jiang, S.; Akermark, B.; Sun, L.-C. *J. Organomet. Chem.* **2007**, *692*, 5501.
68. Razavet, M.; Davies, S. C.; Hughes, D. L.; Pickett, C. J. *Chem. Commun.* **2001**, *9*, 847.

69. Boyke, C. A.; Rauchfuss, T. B.; Wilson, S. R.; Rohmer, M.-M.; Benard, M. *J. Am. Chem. Soc.* **2004**, *126*, 15151.
70. Tard, C.; Liu, X.; Ibrahim, S. K.; Bruschi, M.; De Gioia, L.; Davies, S. C.; Yang, X.; Wang, L.-S.; Sawers, G.; Pickett, C. J. *Nature* **2005**, *433*, 610.
71. Zeng, X. H.; Li, Z. M.; Xiao, Z. Y.; Wang, Y. W.; Liu, X. M. *Electrochem. Commun.* **2010**, *12*, 342.
72. Liu, Y.-C.; Yen, T.-H.; Tseng, Y.-J.; Hu, C.-H.; Lee, G.-H.; Chiang, M.-H. *Inorg. Chem.* **2012**, *51*, 5997.
73. Collin, J.-P.; Jouaiti, A.; Sauvage, J.-P. *Inorg. Chem.* **1988**, *27*, 1986.
74. Curtis, C. J.; Miedaner, A.; Ellis, W. W.; DuBois, D. L. *J. Am. Chem. Soc.* **2002**, *124*, 1918.
75. Curtis, C. J.; Miedaner, A.; Ciancanelli, R.; Ellis, W. W.; Noll, B. C.; DuBois, M. R.; DuBois, D. L. *Inorg. Chem.* **2003**, *42*, 216.
76. Wilson, A. D.; Newell, R. H.; McNevin, M. J.; Muckerman, J. T.; DuBois, M. R.; DuBois, D. L. *J. Am. Chem. Soc.* **2006**, *128*, 358.
77. DuBois, D. L. *Inorg. Chem.* **2014**, *53*, 3935.
78. Luca, O. R.; Blakemore, J. D.; Konezny, S. J.; Praetorius, J. M.; Schmeier, T. J.; Hunsinger, G. B.; Batista, V. S.; Brudvig, G. W.; Hazari, N.; Crabtree, R. H. *Inorg. Chem.* **2012**, *51*, 8704.
79. Jacques, P. A.; Artero, V.; Pecaut, J.; Fontecave, M. *Proc. Natl. Acad. Sci.* **2009**, *106*, 20627.
80. Shao, H.; Muduli, S. K.; Tran, P. D.; Soo, H. S. *Chem. Commun.* **2016**, *52*, 2948.
81. Zimmer, M.; Schulte, G.; Luo, X. L.; Crabtree, R. H. *Angew. Chem. Int. Ed. Engl.* **1991**, *30*, 193.
82. Sellmann, D.; Geipel, F.; Moll, M. *Angew. Chem. Int. Ed.* **2000**, *39*, 561.
83. Sellmann, D.; Prakash, R.; Heinemann, F. W. *Eur. J. Inorg. Chem.* **2004**, *9*, 1847.
84. Straistari, T.; Fize, J.; Shova, S.; Réglier, M.; Artero, V.; Orio M. *ChemCatChem.* **2017**, *9*, 2262.
85. Zarkadoulas, A.; Field, M. J.; Papatriantafyllopoulou, C.; Fize, J.; Artero, V.; Mitsopoulou, C. A. *Inorg. Chem.* **2016**, *55*, 432.
86. Fisher, B.; Eisenberg, R. *J. Am. Chem. Soc.* **1980**, *102*, 7361.
87. Razavet, M.; Artero, V.; Fontecave, M. *Inorg. Chem.* **2005**, *44*, 4786.
88. Baffert, C.; Artero, V.; Fontecave, M. *Inorg. Chem.* **2007**, *46*, 1817.
89. Marinescu, S. C.; Winkler, J. R.; Gray, H. B. *Proc. Natl. Acad. Sci.* **2012**, *109*, 15127.
90. Hu, X.; Cossairt, B. M.; Brunschwig, B. S.; Lewis, N. S.; Peters, J. C. *Chem. Commun.* **2005**, *37*, 4723.
91. Hu, X.; Brunschwig, B. S.; Peters, J. C. *J. Am. Chem. Soc.* **2007**, *129*, 8988.
92. Donck, S.; Fize, J.; Gravel, E.; Doris, E.; Artero V. *Chem. Commun.* **2016**, 52, 11783.
93. Wadsworth, B. L.; Beiler, A. M.; Khusnutdinova, D.; Jacob, S. I.; Moore, G. F. *ACS Catal.* **2016**, *6*, 8048.

94. Andreiadis, E. S.; Jacques, P.-A.; Tran, P. D.; Leyris, A.; Chavarot-Kerlidou, M.; Jousselme, B.; Matheron, M.; Pécaut, J.; Palacin, S.; Fontecave, M.; Artero, V. *Nat. Chem.* **2013**, *5*, 48.
95. Bigi, P.; Hanna, T. E.; Harman, W. H.; Chang, A.; Chang, C. J. *Chem. Commun.* **2010**, *46*, 958.
96. Sun, Y.; Bigi, J. P.; Piro, N. A.; Tang, M. L.; Long, J. R.; Chang, C. *J. Am. Chem. Soc.* **2011**, *133*, 9212.
97. Stubbert, B. D.; Peters, J. C.; Gray, H. B. *J. Am. Chem. Soc.* **2011**, *133*, 18070.
98. Mahammed, A.; Mondal, B.; Rana, A.; Dey, A.; Gross, Z. *Chem. Commun.* **2014**, *50*, 2725.
99. Sun, H.; Han, Y.; Lei, H.; Chena, M.; Cao, R. *Chem. Commun.* **2017**, *53*, 6195.
100. Jacobsen, G. M.; Yang, J. Y.; Twamley, B.; Wilson, A. D.; Bullock, R. M.; DuBois, M. R.; DuBois, D. L. *Energy Environ. Sci.* **2008**, *1*, 167.
101. Wiedner, E. S.; Yang, J. Y.; Dougherty, W. G.; Kassel, W. S.; Bullock, R. M.; DuBois, M. R.; DuBois, D. L. *Organometallics* **2010**, *29*, 5390.
102. Wiedner, E. S.; Roberts, J. A. S.; Dougherty, W. G.; Kassel, W. S.; DuBois, D. L.; Bullock, R. M. *Inorg. Chem.* **2013**, *52*, 9975.
103. Wiedner, E. S.; Appel, A. M.; DuBois, D. L.; Bullock, R. M. *Inorg. Chem.* **2013**, *52*, 14391.
104. Bhugun, I.; Lexa, D.; Savéant, J.-M. *J. Phys. Chem.* **1996**, *100*, 19981.
105. Bhugun I.; Lexa D.; Savéant J.-M. *J. Am. Chem. Soc.* **1996**, *118*, 3982.
106. Solis, B. H.; Maher, A. G.; Dogutan, D. K.; Nocera, D. G.; Hammes-Schiffer, S. *Proc. Natl. Acad. Sci.* **2016**, *113*, 485.
107. Lei, H.; Fang, H.; Han, Y.; Lai, W.; Fu, X.; Cao, R. *ACS Catal.* **2015**, *5*, 5145.
108. Fourmond, V.; Canaguier, S.; Golly, B.; Field, M. J.; Fontecave, M.; Artero, V. *Energy Environ. Sci.* **2011**, *4*, 2417.
109. Hou, K.; Poh, T. P.; Fan, W. Y. *Chem. Commun.* **2014**, *50*, 6630.
110. Zhang, P.; Wang, M.; Yang, Y.; Yao, T.; Sun, L.-C. *Angew. Chem. Int. Ed.* **2014**, *53*, 13803.
111. Liao, R.-Z.; Wang, M.; Sun, L.; Siegbahn, P. E. M. *Dalton Trans.* **2015**, *44*, 9736.

6 Bioinspired Models of Mo and W-Enzymes in CO_2 Reduction

Jagannath Bhuyan* and Sabyasachi Sarkar†,‡

*Department of Chemistry, North Eastern Regional Institute of Science and Technology (NERIST), Nirjuli, Itanagar, Arunachal Pradesh-791109, India

†Nanoscience and Synthetic Leaf Laboratory at Downing Hall, Center for Healthcare Science and Technology, Indian Institute of Engineering Science and Technology, Shibpur (IIESTS), Botanic Garden Pin-711103, West Bengal, India

‡abya@iitk.ac.in; sabby@chem.iiests.ac.in

Table of Contents

List of Abbreviations

Ad	adamantyl
AOR	aldehyde oxido-reductase
ATP	adenosine triphosphate
bdt	benzene-1,2-dithiolate
*Cm*FDH	FDH from *Candida methylica*
COD	carbon monoxide dehydrogenase
Cp	cyclopentadienyl (η^5-C_5H_5)
Cp*	1,2,3,4,5-pentamethylcyclopentadienyl (η^5-C_5Me_5)
*Ct*FDH	FDH from *Chaetomium thermophilum*
dcpe	bis(dicyclohexylphosphino)ethane ($Cy_2PCH_2CH_2PCy_2$)
dmpe	bis(dimethylphosphino)ethane ($Me_2PCH_2CH_2PMe_2$)
DMSOR	dimethyl sulfoxide reductase
dppe	bis(diphenylphosphino)ethane ($Ph_2PCH_2CH_2PPh_2$)
*Ec*FDH-H	FDH H from *Escherichia coli*
EPR	electron paramagnetic resonance
FDH	formate dehydrogenase
HBPin	4,4,5,5-tetramethyl-1,3,2-dioxaborolane

^{i}Pr	isopropyl
^{i}Prpz	3-isopropylpyrazol-1-yl
Me_2pz	3,5-dimethylpyrazol-1-yl
MoFDH	Mo-containing FDH
MV^{2+}	methyl viologen
NADH/NAD^+	nicotinamide adenine dinucleotide
NADPH/$NADP^+$	nicotinamide adenine dinucleotide phosphate
Nar	respiratory nitrate reductase
ndt	naphthalene-2,3-dithiolate
NMP	*N*-methyl-2-pyrrolidone
OTE	optical transparent conductive glass electrode
PDI	pyridine diimine
PGD	pyranopterin guanosine dinucleotide
PMI	pyridine monoimine
PSI	photosystem I
PSII	photosystem II
QOR	quinoline 2-oxidoreductase
R*c*FDH	FDH from *Rhodobacter capsulatus*
RuBisCO	ribulose-1,5-bisphosphate carboxylase/oxygenase
$S_2P^iPr_2$	diisopropyldithiophosphinate
S_2PPh_2	diphenyldithiophosphinate
SO	sulfite oxidase
TMAO	trimethylamine *N*-oxide
TMNOR	trimethylamine *N*-oxide reductase
TOF	turnover frequency
TON	turnover number
WFDH	W-containing FDH
XOR	xanthine oxidoreductase

I. Introduction

Increasing level of carbon dioxide in the atmosphere is a major cause of concern for the world climate due to its green house effects.[1] Ocean is a huge sink of carbon dioxide and therefore ocean acidification and warming is another concern due to increasing level of carbon dioxide in the atmosphere.[2] The carbon dioxide level in the atmosphere is increasing due to the energy demand and consumption of fossil fuels. However, the ancient carbon dioxide conversion to reduced organic compounds is the source of fossil fuels. Therefore, conversion of carbon dioxide to useful materials has been one of the most wanted end reaction over the years. It opens out a way of cleaning the environment and concurrently produces valuable fuels, which can serve as feedstock for industries and in domestic use. To mimic such reaction to reduce carbon dioxide, the activation of carbon dioxide is very difficult due to its kinetic and thermodynamic stability.[3] Nature fixes carbon dioxide in several ways using different enzymatic pathways where thermodynamically uphill reactions are coupled with favorable reactions to drive such reaction. The most common natural carbon dioxide fixation path is the plant photosynthesis, where three interrelated domains, photosystem I (PSI), photosystem II (PSII) and the Calvin cycle (dark reaction) work simultaneously to convert carbon dioxide to sugar molecules.[4] The PSI absorbs sunlight to provide energy to the Calvin cycle where the CO_2 on binding to the enzyme, ribulose-1,5-bisphosphate carboxylase/oxygenase, (RuBisCO) undergoes a large number of reactions to finally form sugar units. The electrons provided by PSI and PSII are used to form ATP and NADPH and which are finally used to reduce CO_2 in the dark reaction. The other two pathways of carbon dioxide fixation are the cleavage of C–O bond (to form carbon monoxide)[5] and the formation of C–H (to form formate) bonds.[6]

Nature's adaptation on Mg-chlorophyll in service to fix CO_2 follow the earlier reaction of Zn-bearing bacteriochlorophyll in photo reducing CO_2 to formate. Nature attempted similar reaction albeit without the use of any sunlight but involving Mo and W enzymes, which catalyze the reversible reaction of formate to $CO_{2.}$ The significant interest is in the reverse reaction that is in the reduction of CO_2 to formate due to the easier transport, storage of formate with respect to other hydrogen like fuels. In addition, formate is a stable source for methanol or methane.

II. Molybdenum and Tungsten Enzymes

The molybdenum containing mononuclear enzymes are classified into three major families. These are sulfite oxidase (SO), dimethyl sulfoxide reductase (DMSOR) and xanthine oxidoreducatse (XOR) families.[7] The SO family includes sulfite oxidases, which use molecular oxygen as an

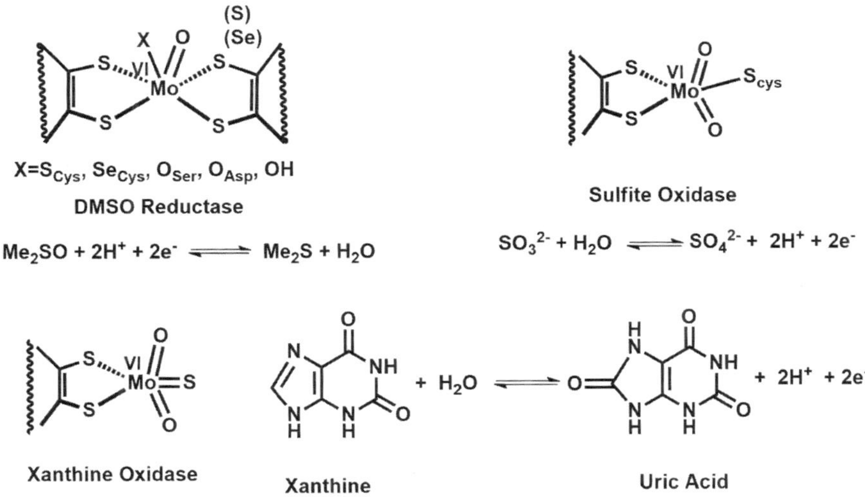

Figure 1. The major families of mono-nuclear Mo-enzymes with their catalytic reactions.[8] Cys, Asp and Ser means cysteine, aspartate and serine respectively.

electron acceptor, sulfite dehydrogenase and the assimilatory nitrate reductase. The dimethyl sulfoxide reductase (DMSOR), trimethylamine *N*-oxide reductase (TMNOR), periplasmic nitrate reductase (NAP) and formate dehydrogenases (FDH) are the most known enzymes of the DMSOR family.[8] The typical examples of the XOR family are xanthine oxidoreductase (XOR), aldehyde oxidoreductase (AOR), quinoline 2-oxidoreductase (QOR), carbon monoxide dehydrogenase (COD) etc. The hyperthermophilic archaea and bacteria with anaerobic metabolism uses tungsten enzymes. These tungsten enzymes mainly found as two families, known as, aldehyde oxidoreductases (AOR) and formate dehydrogenases (FDH).[9] The presence of Mo/W active centers in different organisms from the earliest hyperthermophilic archaea to higher organisms suggests the significance of Mo/W-containing enzymes in the evolutionary pathway.[10] In the early evolutionary period, when there was no oxygen in the earth, Mo- and W- were present as their sulfides MoS_2 and WS_2. Due to the higher solubility of WS_2 in water, tungsten was more available to organisms than molybdenum before the appearance of molecular oxygen on earth, which supports the presence of W-enzymes in anaerobic microorganisms.[11]

The Mo and W containing FDH enzymes mediate the reversible oxidation of formate to carbon dioxide and reduction of carbon dioxide to

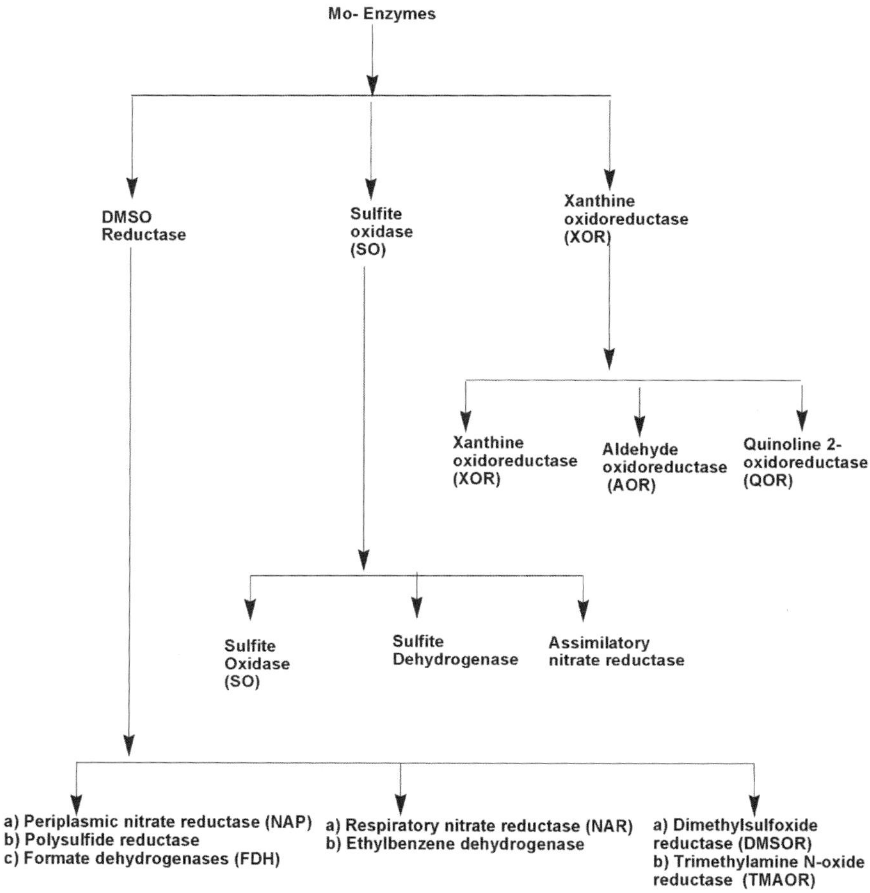

Figure 2. The different subfamilies of three main families (Hille classification) of mono-nuclear Mo-enzymes.

formate. Our interest would be on the conversion of carbon dioxide to formate by FDH enzymes and they belong to the family of DMSOR.

A. Dimethylsulfoxide Reductase (DMSOR) Family

The dimethylsulfoxide reductase (DMSOR) is the diverse group of prokaryotic enzymes widely distributed in bacteria and archaea.[12] Most enzymes of this family catalyze oxygen atom transfer reaction, oxidation/reduction, or hydroxylation/hydration reactions.[13] The enzymes of this family contain

Mo or W in the active site with coordinating to four sulfur ligands from the two dithiolene groups of two pyranopterin guanosine dinucleotide (PGD) molecules.[14] The fifth position is coordinated with oxygen, sulfur or selenium from amino acid side-chains and the sixth position is coordinated to oxygen (oxo, hydroxo) or sulfur (sulfide) atoms.[15] The members of DMSOR family have similar polypeptide fold in their catalytic subunits. Their dissimilarities in the structures are not only observed in the metal active site, but also with the neighboring amino acid chains.[12] The members of DMSOR family are classified into three important subfamilies:[12,14]

(a) Subfamily 1: Subfamily 1 consists of periplasmic nitrate reductases, polysulfide reductases and formate dehydrogenases.[14] In periplasmic nitrate reductases[16] and polysulfide reductases,[17] the active site is coordinated by a cysteine whereas in formate dehydrogenases it is a selenocysteine residue.[18]
(b) Subfamily 2: Respiratory nitrate reductase (Nar)[19] and ethylbenzene dehydrogenase[20] are considered in subfamily 2. In both enzymes the Mo ion is coordinated by one or two oxygen atoms (Figure 3).

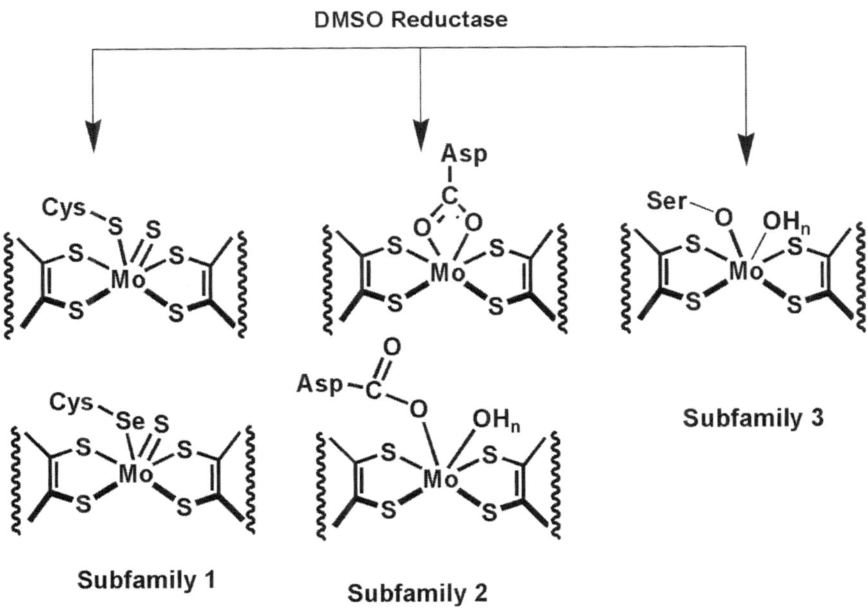

Figure 3. The main three subfamilies of enzyme DMSO reductase.[12,14]

(c) Subfamily 3: The DMSO reductase from *Rhodobacter capsulatus*[21] and trimethylamine N-oxide (TMAO) reductase from *Shewanella massilia*[22] belongs to this subfamily. In this subfamily, the fifth coordination site of the metal is bound through an oxygen atom of a serine side chain.

Based on these observations, a comprehensive discussion about bio-inspired models of enzyme FDH of DMSOR family with respect of carbon dioxide reduction is presented.

III. Formate Dehydrogenase

A large number of eukaryotic organisms, eubacteria and archaebacteria catalyze the reversible formation of formate from carbon dioxide in a variety of reactions.[23] The formate has significant interest as it is the first as well as stable intermediate of carbon dioxide reduction to methanol or other reduced product like methane. Therefore, formate has important applications in chemical industry as a viable energy source.[6] Moura and coworkers have been working on the enzyme formate dehydrogenase and reviewed this enzyme with an aim to inspire the design of a catalyst for carbon dioxide reduction.[24] FDHs catalyze the reversible oxidation of formate to carbon dioxide, which is a two-electron redox process according to the equation below.[25]

$$HCOO^- \leftrightharpoons CO_2 + 2e^- + H^+ \qquad E^0 = -420\ \text{mV} \qquad (1)$$

Based on the active site metal content, FDHs can be classified into two major categories. The category with no metal or redox co-factor is known as metal independent FDH while the category with a metal site or redox cofactors is known a metal dependent FDHs.[26] The metal dependent FDHs can again be classified into two sub-categories namely, Mo-containing FDHs (MoFDH) and W-containing FDHs (WFDH). The FDH enzymes from anaerobic bacteria contain metal Mo as active site whereas the archaea families contain either a metal Mo or W.[27] Although FDH is a member of the DMSO reductase family it favors the breaking of

C–H bond instead of catalyzing oxygen-atom transfer reactions similar to other DMSO reductase enzymes.

A. Metal Independent FDH

Metal independent FDHs are NAD^+-dependent enzymes and belong to the *D*-specific dehydrogenases of 2-oxyacids family.[6,28] These enzymes are widely found in aerobic bacteria, yeasts, fungi and plants.[6] Due to the absence of metal ions, metal independent FDHs oxidize formate to carbon dioxide through direct hydride transfer (Equation 2) from formate to NAD^+.[26]

$$HCOO^- + NAD^+ \rightarrow CO_2 + NADH \qquad (2)$$

The protein structure not only helps in substrate preorganization by locating the substrate formate and FDHs in close proximity but also in transition state stabilization with diminishing the activation barrier.[29]

B. Metal Dependent FDH

The metal dependent FDHs are Mo and W containing enzymes and are found in prokaryotes. Metal dependent FDHs have complex protein structures with redox metal Mo or W in the active site which participates in catalytic reversible formate oxidation.

The metal dependent FDHs are also NAD^+ dependent FDH but uses NAD^+ only as the terminal electron acceptor. These enzymes follow different enzymatic pathways for formate oxidation. A large number of metal dependent FDHs are known but their active site structure is well-conserved with a mononuclear Mo or W bound PGD moiety.[13] In the oxidized form of the active site, the metal is coordinated by one terminal sulfo group ($M = S$) and one S- or Se- atom from a cysteine or selenocysteine moiety besides the two dithiolene groups of pyranoterin cofactor (Figure 4) with distorted trigonal prismatic structure.[30] The structure in the reduced form is controversial.[13] There are reports which support the dissociation of the selenocysteine residue coordinated to the metal center during reduction. According to this, release of sulfo ligand is difficult but

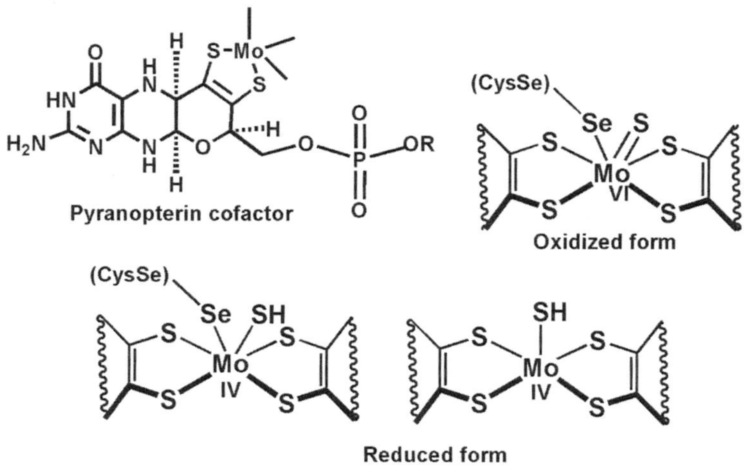

Figure 4. The pyranopterin cofactor and the structure of oxidized and reduced form of FDH active site. For the reduced structure both selenocystine bound and dissociated forms are shown.[6]

the selenocysteine/-cysteine dissociates to give rise to the observed square pyramidal geometry.[24]

C. Some Examples of Mo and W-containing Formate Dehydrogenase

The E. coli can encode three diverse MoFDHs:[13]

(i) Formate dehydrogenase H, the product of *fdhF* gene, but sometimes abbreviated in the literature as fdhf, FDH_H, and Fdh H.[18] We are referring to it as FDH H in this chapter.
(ii) Formate dehydrogenase N: It is a product of the *fdnGHI* operon and is composed of three subunits. In FDH N, the catalytic α subunit consists of Mo-metal coordinated two PGD units, a [4Fe–4S] cluster, and a SeCys residue. The β subunit contains one trans-membrane helix coordinated to four [4Fe–4S] clusters.[32] It is referred to in the literature as FdnGHI and $FDH_{N.}$
(iii) Formate dehydrogenase O: It is a product of *fdoDEGHI* operon and structure is similar to FDH N. FDH N and FDH O are heteromeric

membrane-bound respiratory enzyme that contain seven redox centers like Fe/S and hemes in addition to the Mo-site.[24]

The first crystal structure of FDH H was determined from *E. coli* Mo-FDH H, which showed coordination of four sulfur atoms of the two PGD moieties, one selenium atom from SeCys and a hydroxyl group.[18] After interpretation of W-FDH structure, which has a terminal sulfo-group attached to tungsten,[30] the structure of Mo FDH H was reinterpreted. It is now well-accepted the presence of metal sulfo moiety in the oxidized form of FDH (Figure 4).

Sulfate-reducing bacteria *Desulfovibrio* type contain different Mo-FDHs and W-FDHs. The Mo-FDH of *D. desulfuricans* is more complex heteromeric FDH which is comprised of several redox-active centers such as iron sulfur clusters and hemes in addition to the molybdenum part.[33] The kinetic and mechanistic study of Mo-FDH of *D. desulfuricans* was studied by Maia and co-workers. The enzyme not only oxidizes the formate but also reduces the CO_2. The overall reaction is thermodynamically and kinetically reversible.[6] Two FDHs from *D. alaskensis* bacterium were characterized, one has tungsten in the active site while the other can incorporate either Mo or W. The FDH *D. alaskensis* NCIMB 13491 is a heterodimeric protein containing either Mo or W in its active site and catalyzes oxidation of formate to CO_2. The temperature dependent EPR studies of the reduced *D. alaskensis* gives three rhombic signals in the temperature range 5–40 K and relaxation behavior typical of [4Fe–4S] clusters. It also displays two EPR signal with g-value less than two corresponds to mononuclear Mo^V and W^V.[34] The Mo containing FDH from *D. vulgaris* contains a molybdenum co-factor, Fe–S subunit and a c-type heme.[35]

A Mo-FDH from *Rhodobacter capsulatus* (RcFDH) facilitates the reversible conversion of formate to CO_2. In the oxidized form the Mo^{VI} is coordinated to a sulfido ligand (Mo = S) and a cysteine moiety besides the four sulfur groups of PGD moiety. Retention of the structure with Mo^V state was observed for the chemically reduced form.[31]

The W-FDH of *D. gigas* is a dimeric periplasmatic FDH with one W center and four Fe/S centers. The crystal structure of the tungsten formate dehydrogenase of *D. gigas* has been reported.[36] The authors studied structural data by multi-wavelength crystallography and found that W-atom in W-FDH is coordinated by two PGD molecules and a selenium atom of Se-Cys residue.[30,36]

IV. Metal Dependent FDH that Show the Activity of CO_2 Reduction

All metal dependent FDHs that can catalyze the reduction of carbon dioxide to formate is presented in Table 1 with the description of active site and subunits. The sulfate-reducing bacteria *D. desulfuricans* contain Mo-FDH which can reduce CO_2 with a higher k_{cat} of 50 s^{-1}.[6] The bacterium *D. vulgaris* Hildenborough express two soluble periplasmic enzymes FdhAB and FdhABC3 and one membrane bound periplasmic protein FdhM. The enzymes FdhAB and FdhABC3 are Mo dependent and can catalyze the reduction of CO_2 to formate. The FdhAB shows the catalytic activity in

Table 1. The formate dehydrogenases that show the carbon dioxide reduction activity.[24]

Enzyme FDH	Active site	Units	Ref.
Acetobacterium woodii FDH	Mo, SeCys, S	α (Mo, [4Fe–4S])	102
Desulfovibrio desulfuricans FDH	Mo, SeCys, S	$\alpha\beta\gamma$ (α: Mo, [4Fe–4S]; β: [4Fe–4S]; γ: 4 *c*-type hemes)	6, 33
Desulfovibrio vulgaris Hildenborough FDHABC3	Mo SeCys, S	$\alpha\beta\gamma$ (α: Mo, [4Fe–4S]; β: 3 [4Fe–4S] γ: 4 *c*-type hemes)	35
Rhodobacter capsulates FDH	Mo, Cys, S	$(\alpha\beta\gamma)_2$ (α; Mo, [2Fe–2S], 4[4Fe–4S]; β: [4Fe–4S], FMN; γ: [2Fe–2S])	31, 38
Escherichia coli FDH H	Mo, SeCys, S	α (Mo, [4Fe–4S])	18
Methylobacterium extorquens FDH	W, Cys, S	$\alpha\beta$ (α: W, ≥ 1Fe/S; β: [4Fe–4S], FMN)	43
Syntrophobacter fumaroxidans FDH1	W, SeCys, S	$(\alpha\beta\gamma)_2$ (W, Fe)	40
Clostridium carboxidivorans FDH	W, SeCys, S	α (W, [4Fe–4S])	97

the presence of W. During the growth of *D. vulgaris Hildenborough* with lactate and sulfate, the FDHs functions are not required, but in the absence of sulfate, these FDHs can reduce CO_2 to formate.[37] The *Rc*FDH localized in cytoplasm is an oxygen-tolerant protein with an $(\alpha\beta\gamma)_2$ subunits. This enzyme can catalyze the reduction of CO_2 to formate.[38] Molybdenum-containing formate dehydrogenase H from *E. coli* (*Ec*FDH-H) can also reduce carbon dioxide to formate reversibly. The *Ec*FDH H is an attractive model catalyst to study reversible CO_2 reduction.[39]

Two W containing FDH, namely WFDH-1 and WFDH-2 from the syntrophic propionate oxidizing bacterium *Syntrophobacter fumaroxidans* are known. These two enzymes can catalyze the reduction of CO_2 besides the oxidation of formate. The enzyme WFDH1 reduces CO_2 to formate at higher rate. In the presence of benzyl viologen as an artificial electron acceptor, the specific formate oxidation rate for enzyme FDH-1 was found as 700 U mg^{-1} and K_m for formate oxidation 0.04 mM. The FDH-2 oxidizes formate with a specific activity of 2700 U.mg^{-1} and K_m of 0.01 mM. When dithionite-reduced methyl viologen was used as the electron donor, FDH-1 and FDH-2 was found to reduce CO_2-to formate at rates of 900 Umg^{-1} and 89 Umg^{-1} respectively.[40] The purified FDH-1 is shown to activate an electrode for carbon dioxide reduction efficiently.[41] The WFDH from *Clostridium thermoaceticum* consists of two subunits with composition $\alpha_2\beta_2$ and the selenium resides in the two α subunits. This enzyme catalyzes CO_2 reduction with NADPH at neutral pH.[42] The W-containing NAD-dependent formate dehydrogenase from proteobacterium *Methylobacterium extorquens* AM1 was isolated under oxic conditions.[43]

V. Mechanism of Formate Dehydrogenase

It was believed that the molybdenum enzymes undergoes two-electron redox reactions at molybdenum center coupled with an oxygen atom transfer from water to the substrate through molybdenum or the reverse.[7,44] The mechanistic studies of metal dependent FDH revealed that formate oxidation is not an oxygen atom transfer reaction which forms bicarbonate (Equation 3).[45] The oxidation of formate catalyzed by FDH follows either by abstracting one proton and two electrons (equation 1) or by abstracting one hydride from the substrate (Equation 4).[24,46] The generation of $^{13}C^{16}O_2$ instead of $^{13}C^{18}O_2$ in the oxidation of ^{13}C-labeled formate in $H_2{}^{18}O$ corroborates the results.[45,47] The mechanism for both Mo-FDH and W-FDH is similar.

$$HCOO^- + H_2O \rightarrow HOCOO^- + 2e^- + 2H^+ \quad \text{Does not take place} \quad (3)$$

$$HCOO^- \leftrightharpoons CO_2 + H^- \quad (4)$$

The first mechanism of the FDH catalyzed oxidation of formate to carbon dioxide was proposed on the basis of first crystal structure of FDH-H (*E. coli*). The release of the hydroxyl group (later proved as sulfo-group) was proposed to form a penta-coordinated Mo center to create a vacant site for formate binding.[18] The binding of molybdenum center by formate molecule is followed by reduction of Mo^{VI} to Mo^{IV} and the breaking of the C-H bond of formate. The re-interpretation of the crystallographic data of the *E. coli* FDH H revealed that in reduced form selenocysteine is not bound to Mo center.[48]

An extensive density functional calculations (DFT)[49] results the sulfur shift mechanism which is based on the following stages:

(i) First formate approaches towards the Mo center and pushes the sulfo group (Mo = S) by oxygen of the formate towards the position which was initially occupied by the selenocysteine residue, and yield a structure (Figure 5) with CysSe–S–Mo bond, concurrently Mo metal is reduced to Mo^{IV} creating a vacant coordination position for formate binding.[50]

(ii) In the second stage, the CysSe–S–Mo bond is cleaved to form the non-coordinating selenocysteine residue–Cys–Se^-. The non-coordinating residue is stabilized by a hydrogen bond from the histidine molecule of side chains and abstracts the proton of formate to yield-CysSeH. The carbon dioxide species bound to the Mo metal (Mo–S–CO_2) structure is generated, which is eventually released and form the penta-coordinated $M^{IV} = S$ structure.[48]

(iii) In the final stage, oxidation of M(IV) to M(VI) occurs through intramolecular electron transfer to other redox species to complete the catalytic cycle by the deprotonation of the selenocysteine residue.

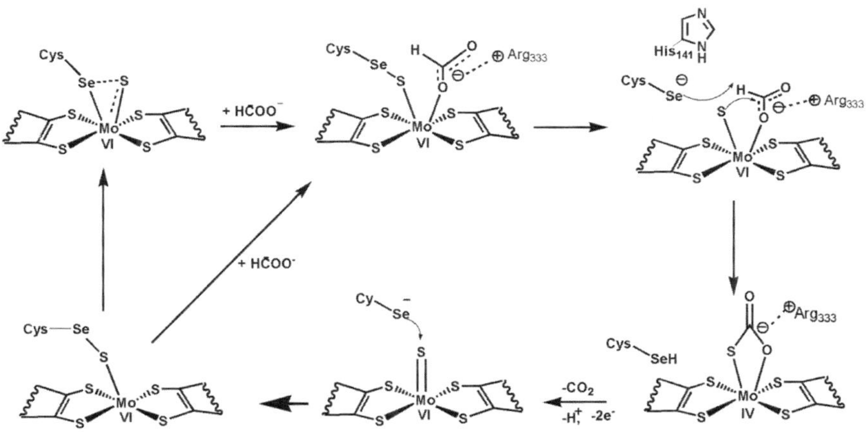

Figure 5. The sulfur shift mechanism for formate oxidation catalyzed by metal dependent FDH.[14]

Tiberti *et al.*[51] proposed a different mechanism where the oxidation of formate occurs *via* the hydride transfer from the formate to the metal through a Metal–H (M–H) intermediate followed by the transfer of the proton from the metal to the selenocysteine. According to this mechanism the metal mediates the hydride transfer, while in the sulfur shift mechanism formate hydrogen atom is abstracted by the selenocysteine moiety directly in the form of a proton. Another hydride transfer mechanism has been proposed for both carbon dioxide reduction and formate oxidation, where the reduced active site M^{IV}-SH acts as a hydride donor and oxidized M^{VI}=S acts as a hydride acceptor[6,24] (Figure 6). In the carbon dioxide reduction reaction, as the carbon dioxide molecule approaches to reduced active site the conserved arginine helps to align the carbon atom of CO_2 towards the protonated sulfo ligand through hydrogen bonds to occur easier hydride transfer from the protonated sulfo group of the reduced active site to the carbon of carbon dioxide molecule (Figure 6).[24] Then reduction of metal M^{VI} to M^{IV} happens through intramolecular electron transfer from other redox centers concurrently release of the formate molecule.[6,24] The reaction mechanism of FDH have been studied by computational methods and found that the sulfdo ligand abstracts a hydride ion from the formate to form a M(IV)–SH state and a thiocarbonate ion attached to Cys. The CO_2 is released from thiocarbonate ion when the metal center is oxidized back

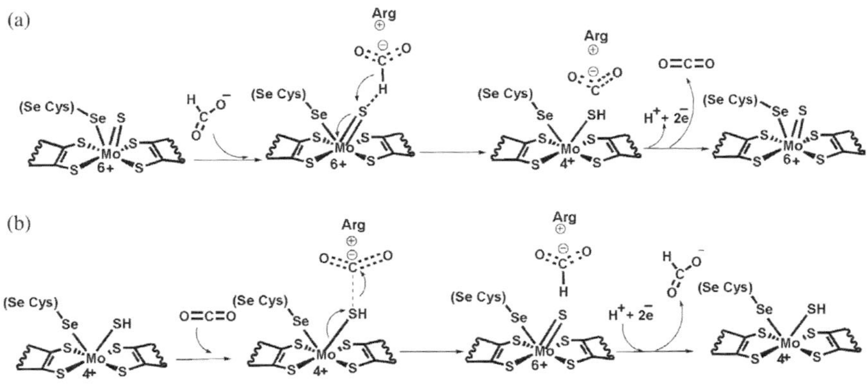

Figure 6. The proposed reaction mechanism (*via* hydride transfer) for the FDH-catalyzed (a) formate oxidation and (b) carbon dioxide reduction to formate.[24]

to the M(VI) state.[52] The main drawback of hydride transfer mechanism is formation of intermediate M–H complex, which is rare with Mo and W redox-active chemistry. The Mo and W prefer formation of oxides and sulfides rather than Metal-H complexes. Recently, Hille and coworkers studied the spectroscopy and kinetic property of NAD^+-dependent formate dehydrogenase from *Ralstonia eutropha*. They found an EPR spectrum of Mo^V and proposed the direct transfer of the substrate C_α hydrogen to the molybdenum-sulfur group of metal center.[52,53]

An alternative way for reduction of CO_2 has been reported where FDHs from *Candida methylica* (*Cm*FDH) and *Chaetomium thermophilum* (*Ct*FDH) reduces CO_2 in the form of hydrogen carbonate (HCO_3^-) to formate in a NADH-dependent reaction.[54]

VI. Models of the Enzyme FDHs for CO_2 Reduction

A. Models of Mo-FDHs for CO_2 Reduction

The synthesis of the structural units [Mo^{VI}(dithiolene)$_2$S(Se-cysteine)] and [Mo^{IV}(dithiolene)$_2$SH] is needed for modeling the oxidized and reduced form of active site of FDH enzymes respectively. The synthetic model of Mo-FDH which mimics the structural model of enzyme Mo-FDH is rare in the literature. A dithiolene complex $[Mo^{IV}(2Ad\text{-}Se)(S_2C_2Me_2)_2]^{1-}$

1, (2-Ad = 2-adamantyl) was synthesized and structurally characterized as model of reduced site of enzyme FDH.[55] With non-diothiolene ligands two series of mononuclear hexa-coordinated Mo(IV) complexes:*

(1) [{HB(Me$_2$pz)$_3$}MoS(S$_2$CNR$_2$)] **2** with R = Me, Et, ^{n}Pr, ^{n}Bu and
(2) [HB(3-^{i}Prpz)$_2$(5-^{i}Prpz)MoS(S$_2$PR$_2$)] **3** with R = ^{i}Pr and Ph

have been reported.[56–58] Here the tridentate non-diothiolene ligands were used to avoid the hydrolysis of Mo^{IV}S complexes. There is a report of synthesis of mononuclear five coordinate complex [MoS(L)$_2$]$^{2-}$ **4** (L = cyclohexene-1,2-dithiolate) by reaction of [Mo(CO)$_2$(L)$_2$] with sodium sulfide.[59] The treatment with ferrocenium cation with complex **4** produces the corresponding mononuclear Mo^V intermediate species [MoS(L)$_2$]$^{1-}$ **5** which then undergoes further oxidation to a sulfide bridged dimeric Mo^V complex (Figure 7).

B. Models of W-FDH for CO_2 Reduction

Similar to Mo-FDH for modeling the active site of W-FDH enzymes, synthesis of complexes with structure [W^{VI}(dithiolene)$_2$S(Se-Cysteine)] or [W^{IV}(dithiolene)$_2$SH] is required.[60] Several dithiolene tungsten complexes with ligand, benzene-1,2-dithiolate (bdt) and naphthalene-2,3-dithiolate (ndt)

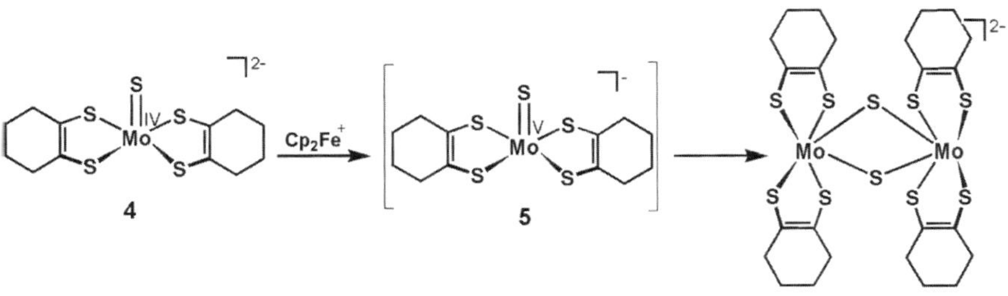

Figure 7. Conversion of a Mo^{IV}S (**4**) species to a Mo^{V}S (**5**) species and its subsequent decomposition.[59]

* HB(Me$_2$pz)$_3$ = hydrotris(3,5-dimethylpyrazol-1-yl)borate; HB(3-^{i}Prpz)$_2$(5-^{i}Prpz) = hydrobis(3-isopropylpyrazol-1-yl)(5-isopropylpyrazol-1-yl)borate; S$_2$P^{i}Pr$_2$ = diisopropyldithiophosphinate and S$_2$PPh$_2$ = diphenyldithiophosphinate)

such as, $[W^{V}O(bdt)_2]^{1-}$,$[W^{IV}O(bdt)_2]^{2-}$, $[W^{VI}O_2(bdt)_2]^{2-}$ have been reported.[61] Davies and coworkers reported the synthesis of several complexes like $[W^{IV}O(S_2C_2R_2)_2]^{2-}$ (R = phenyl, pyridin-2-yl, pyridin-3-yl, pyridin-4-yl or quinoxalin-2-yl) using the reaction of $[WO_2(CN)_4]^{4-}$ with a series of dimethylamino or thione-protected asymmetric dithiolenes (ene-1,2-dithiolates).[62] Some other W^{IV} and W^{VI} dithiolene complexes are presented in a review article.[63] Some of the reported W^{IV} complexes undergo conversion to the corresponding W^{VI} complexes, but no FDH reaction was reported for the complexes. Holm and coworkers first reported the detailed investigation and structural studies of bis(dithiolene) tungsten selenolate and analogous thiolate complexes with relevance to FDHs.[64] They synthesized several complexes of the types $[W^{IV}(QR)(S_2C_2Me_2)_2]^{-}$, $[W^{VI}O(QR)(S_2C_2Me_2)_2]^{-}$, and $[W^{VI}S(QR)(S_2C_2Me_2)_2]^{-}$ (Q = S, Se; R = tert-butyl, 1-adamantyl) and the desoxo species $[W^{VI}(SR)(OSiR'_3)(S_2C_2Me_2)_2]$ (R′ = Me, Ph).

The compound $[W^{VI}S(Se\text{-}1\text{-}Ad)(S_2C_2Me_2)_2]^{1-}$ **6** is the representative structural model for the oxidized active site of W-FDH and was synthesized by reaction of $[W(Se\text{-}1\text{-}Ad)(S_2C_2Me_2)_2]^{1-}$ with Ph_3SbS at −30°C (Figure 8) The single crystal structure of this model compound was not achieved but the single crystal structure of model compound $[W^{IV}(Se\text{-}1\text{-}Ad)(S_2C_2Me_2)_2]^{1-}$ **7** and similar analogs of the reduced active site structure have been reported.[64]

C. Structural-Functional Models of W-FDH

A compound $(NEt_4)_2[W^{IV}O(S_2C_2(CN)_2)_2]$ **8** was synthesized by reaction of Na_2WO_4, and $Na_2S_2C_2(CN)_2$ (Na_2mnt) in acidified (pH 5.5) aqueous

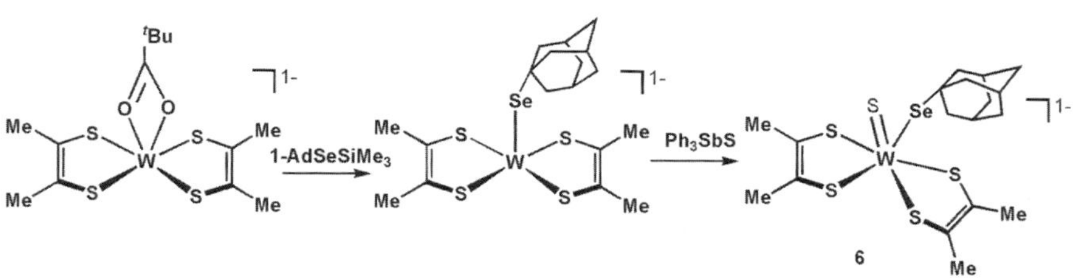

Figure 8. Synthesis of complex **6** for a synthetic model for W-FDH.[64]

medium in the presence of excess of sodium dithionite and Et_4NBr.[65,66] The compound, $(NEt_4)_2[W^{IV}O(S_2C_2(CN)_2)_2]$ **8** although, is not similar to the active site structure of W-FDH, can catalyze the reduction of CO_2/ HCO_3^- (pH=7.5) to formate along with the formation of $(NEt_4)_2[W^{VI}O_2(S_2C_2(CN)_2)_2]$ (Figure 9). This complex was the first synthesized tungsten compound functionally similar to W-FDH in relation to the reduction of CO_2.[65,66]

The authors also reported the displacement of W by Mo from the complex **8** using Na_2MoO_4 similar to a reaction used for the reconstitution assay of W-FDH.[67]

Holm and coworkers synthesized the complexes $[W^{IV}(CO)(HCO_2)(S_2C_2Me_2)_2]^{1-}$ **9** and $[W^{VI}O(O_2CH)(S_2C_2Me_2)_2]^{1-}$ **10** from $[W(CO)_2(S_2C_2Me_2)_2]$ whose carbonyls are labile for substitution. In the presence of formate the complex $[W(CO)_2(S_2C_2Me_2)_2]$ in acetonitrile/THF medium produces complex **9** whereas in Me_2SO medium produces compound **10** (Figure 10). Complex **9** and **10** are the first structurally characterized model complexes of W-FDH which shows the formate binding to mononuclear W^{IV} and W^{VI}.[68] In these two complexes the selenocysteinate ligand found in the enzyme active site was absent and the attempts to synthesize the corresponding selenolate complexes were not successful. There is a recent report of a reduced W-FDH model complex

$$\underset{\mathbf{8}}{\left[W^{IV}O(mnt)_2\right]^{2-}} + HCO_3^- \longrightarrow \left[W^{VI}O_2(mnt)_2\right]^{2-} + HCOO^-$$

$$\underset{\mathbf{8}}{\left[W^{IV}O(mnt)_2\right]^{2-}} + MoO_4^{2-} \longrightarrow \left[Mo^{IV}O(mnt)_2\right]^{2-} + WO_4^{2-}$$

Figure 9. The FDHs activity of complex **8**.[66]

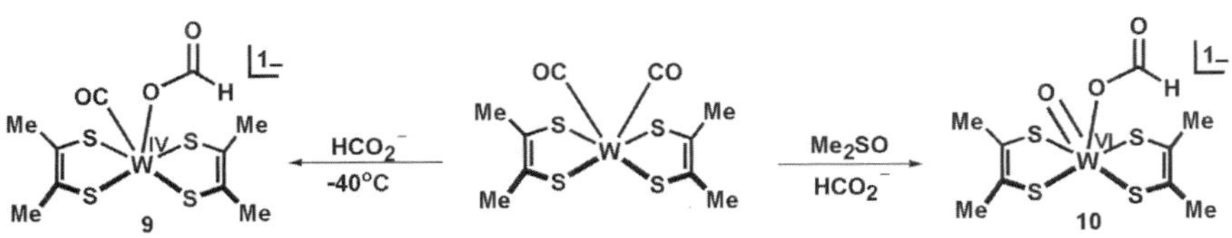

Figure 10. Synthesis of compound **9** and **10** in the presence of formate.[68]

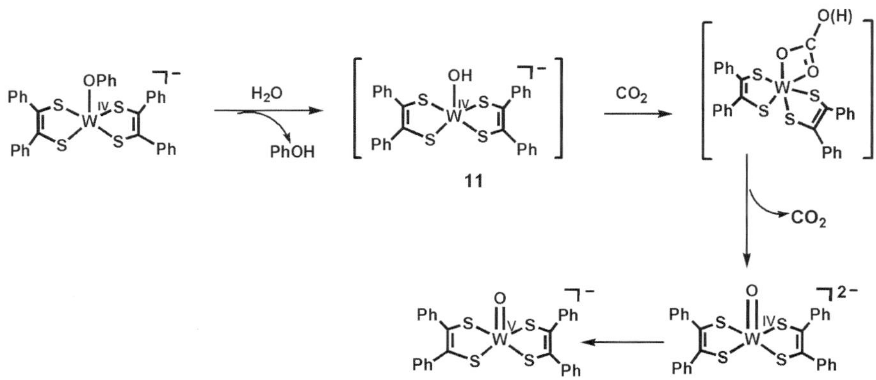

Figure 11. Synthesis of compound **11** and its reactivity with CO_2.[69]

$[W^{IV}(OH)(S_2C_2Ph_2)_2]^-$ **11**. The complex **11** was synthesized through hydrolysis of complex $[W^{IV}(OPh)(S_2C_2Ph_2)_2]^-$ and the reactivity of **11** with CO_2 was studied.[69] The reaction between **11** and CO_2 at room temperature results the formation of $[W^{IV}(O)(S_2C_2Ph_2)_2]^{2-}$, which slowly oxidizes to $[W^{V}(O)(S_2C_2Ph_2)_2]^-$ (Figure 11). The isotopic labeling experiments indicate that the oxygen atom in $[W^{IV}(O)(S_2C_2Ph_2)_2]^{2-}$ was incorporated from CO_2 molecule.[69]

VII. Other Ligands (Not Related to FDHs) that Respond to CO_2 Reduction

A. Mo-Complexes

The first cyclopentadienyl-Mo complex was synthesized by reaction between molybdenum hexacarbonyl and cyclopentadienyl ligand (Cp) in 1954.[70] By using molybdocene complex $[Cp_2Mo(PhC{\equiv}CPh)]$, which has a weakly bonded ligand, carbon dioxide bound molybdocene **12** was synthesized by simply passing CO_2 gas (Figure 12). The CO_2 unit was found as coordinated to the metal in a side on η^2–C,O pattern (Figure 12) with bending the Cp ligand to provide cavity for CO_2 unit.[71]

The molybdocene derivative Cp_2MoH_2 shows photochemical disproportion of CO_2 to CO and CO_3^{2-}.[72] Although CO_2 is a neutral molecule, the charge present in CO_2 unit of Mo–CO_2 is negative similar to other

$$Cp_2Mo(PhC{\equiv}CPh) + CO_2 \longrightarrow [Cp_2Mo(CO_2)] + PhC{\equiv}CPh$$

Figure 12. Synthesis of carbon dioxide bound molybdocene **12**.[71]

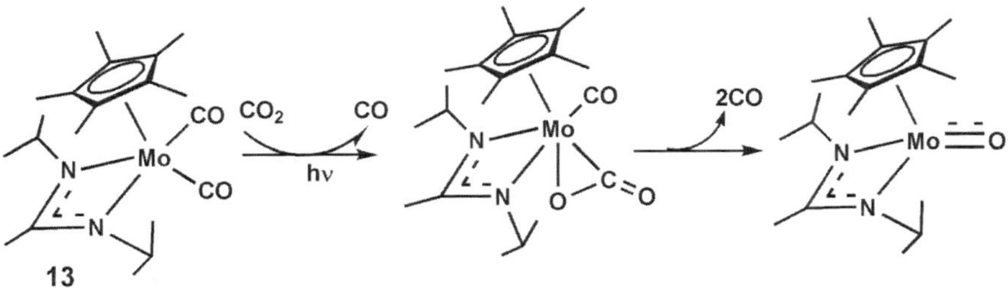

13

Figure 13. Photocatalytic reaction of complex **13** in the presence of CO_2.[75]

transition metal-CO_2 complexes, which indicates the charge transfer from metal to the ligand CO_2.[73] The neutral $CpMo(CO)_3$ does not interact with CO_2 molecule,[74] but some complexes like [Cp*Mo{N(*i*Pr)C(Me)N(*i*Pr)}$(CO)_2$] **13**, Cp* = η^5-C_5Me_5 shows photocatalytic reactions with CO_2 to form M = O species and CO (Figure 13).[75]

The pentamethylcyclopentadienyl molybdenum trimethylphosphine complex $Cp^*Mo(PMe_3)_3H$ **14** undergoes CO abstraction reaction from CO_2 resulting in $Cp^*Mo(PMe_3)_2(CO)H$ and Me_3PO (Figure 14). The $Cp^*Mo(PMe_3)_2$ (CO)H can be used as a catalyst for the decarboxylation of HCO_2H to CO_2 and H_2.[76]

There are a few remarkable examples of catalytic reductions of CO_2 with molybdenum complexes (not as enzymatic FDH). There is a report of hydrogenation of carbon dioxide using $MoCl_3$ combined with ligand phosphines such as dcpe ($Cy_2PCH_2CH_2PCy_2$) or dppe ($Ph_2PCH_2CH_2PPh_2$). The pre-catalysts has TON up to 63 and TOF of 8.4 per hour.[77]

A low-valent molybdenum complex, $(PN^{Me}P)Mo(C_2H_4)_2$, (pincer ligand $PN^{Me}P$ = $MeN(CH_2CH_2PPh_2)_2$) was prepared and characterized. There is an agostic interaction present between the metal Mo and the *N*-methyl substituent in the complex $(PN^{Me}P)Mo(C_2H_4)_2$. Upon exposure to CO_2, the β-agostic C–H bond is cleaved by molybdenum and produce a cyclometalated molybdenum formate complex, $(\kappa^4$-$PN^{Me}P)$-$Mo(C_2H_4)$ $(\kappa^2$-$O_2CH)$ **15** (Figure 15). This species mediates CO_2 hydrogenation to formate under basic conditions.[78]

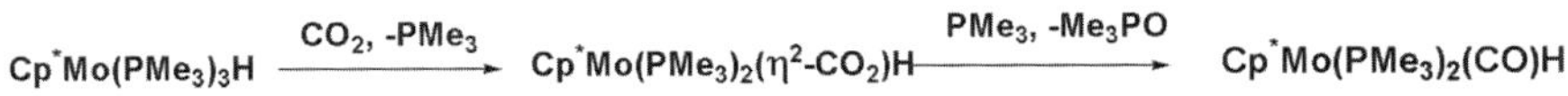

14

Figure 14. The CO abstraction reaction from CO_2 by complex **14**.[76]

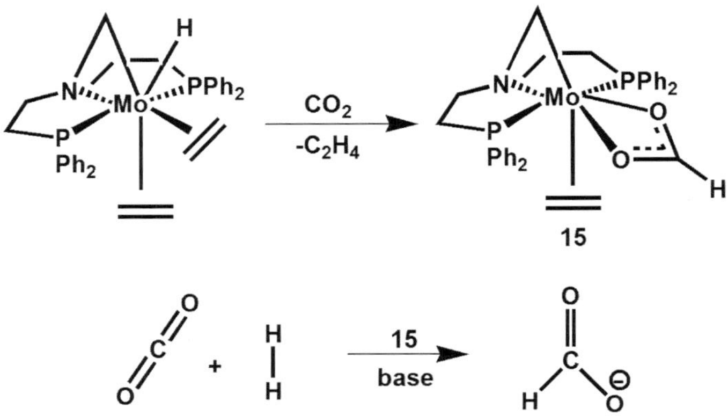

Figure 15. Synthesis of complex **15** and its catalytic activity for hydrogenation of carbon dioxide to formate.[78]

The catalytic properties of a pyridine diimine (PDI) molybdenum catalyst (κ^6-*P,N,N,N,C,P*-^{Ph2PPr}PDI)MoH **16** for the reduction of CO_2 has been reported recently. The catalyst was synthesized from (^{Ph2PPr}PDI) Mo(CO) following several steps. The treatment of CO_2 with **16** yields the η^1-formate complex, (κ^6-P,N,N,N,C,P-^{Ph2PPr}PDI)Mo(OCOH) due to the facile insertion of CO_2. The use of complex **16** as a CO_2 reduction catalyst in the presence of 4,4,5,5-tetramethyl-1,3,2-dioxaborolane, HBPin produces the selective formation of $H_3COBPin$ and $O(BPin)_2$.[79] Bipyridine coordinated molybdenum carbonyl complex **17** and **18** work as electrocatalyst for carbon dioxide reduction through a two electron reduction pathway.[80]

The species $[Mo(CO)_3(bpy)]^{2-}$ was found as the active catalyst in the electrochemical reduction of CO_2. The catalytic conversion of CO_2 was found more efficient in solvent *N*-methyl-2-pyrrolidone (NMP) than tetrahydrofuran.[81]

The use of pyridine monoimine (PMI) molybdenum tetracarbonyl complex $[(^{iPr_2Ph}PMI)Mo(CO)_4]$ **19** for electrochemical and stoichiometric

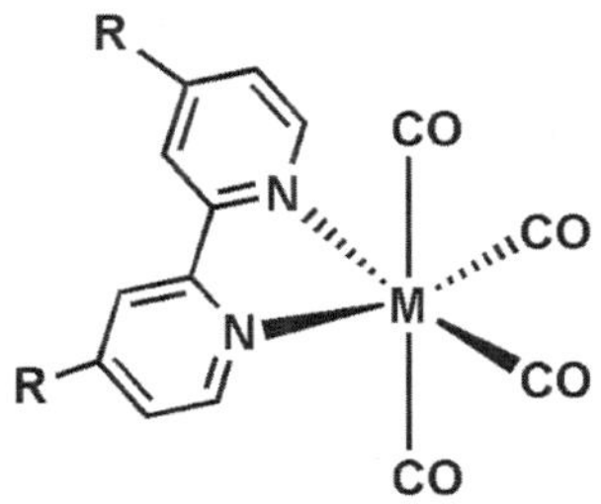

17, M = Mo, R = H

18, M = Mo, R = ^{t}Bu

Figure 16. Structure of some Mo-complexes that responds to CO_2 reduction.[80]

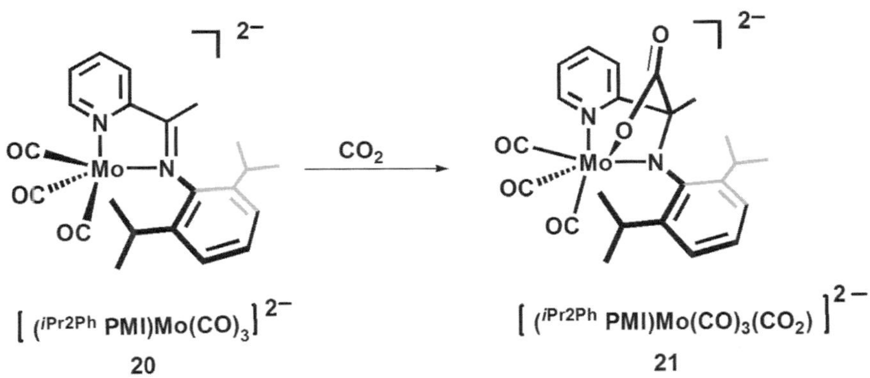

Figure 17. Proposed reaction path for the reaction of the dianion [($^{iPr}2^{Ph}$PMI) $Mo(CO)_3]^{2-}$ with CO_2.[82]

CO_2 reduction is known. The reduced product $[(^{iPr_2Ph}PMI)Mo(CO)_3]^{2-}$ **20** reacts with carbon dioxide to form a complex $[(^{iPr_2Ph}PMI)Mo(CO)_3(CO_2)]^{2-}$, **21**, where carbon atom of CO_2 binds on the ligand and the oxygen atom on the metal center (Figure 17).[82] An easily available metal oxide, MoO_2, can act as an electro-catalyst for CO_2 reduction in organic solvents. The efficiency depends on the temperature and water content.[83]

There is a report of synthesis of unusual CO_2-ligated bimetallic compound, $(C_3H_5N_2)_3(C_3H_4N_2)[PMo_{11}CoO_{38}(CO_2)]\cdot4H_2O$ **22** and $(C_3H_5N_2)_4[SiMo_{11}CoO_{38}(CO_2)]4H_2O$ **23** which demonstrates easy CO_2

fixation by polyoxoanions. The compound **23** was synthesized from an aqueous solution containing $[H_2PMo_{11}CoO_{40}]^{5-}$ polyoxoanion with bubbled CO_2. These two compounds, represent the CO_2 coordination by inorganic polyoxoanions in aqueous solution with typical the μ-η^1, η^1-OCO linear coordination mode. Although CO_2 reduction was not observed the linear way of the CO_2 binding may be considered as structural and functional model for catalytic reduction of CO_2 into useful chemicals.[84] The molybdate dianion, $[MoO_4]^{2-}$ easily binds with CO_2 and forms monocarbonate complex $[MoO_3(k^2\text{-}CO_3)]^{2-}$ **24** and in the presence of excess CO_2 forms dioxo dicarbonate $[MoO_2(k^2\text{-}CO_3)_2]^{2-}$ **25** species. The monocarbonate **24** is shown to react with triethylsilane in acetonitrile under an atmosphere of CO_2 and gives formate along with some silylated molybdates.[85]

B. W-Complexes

Most of the tungsten analogs of above discussed bipyridine coordinated molybdenum carbonyl complexes also work as electro-catalysts for carbon dioxide reduction. Theoretical calculations using DFT was performed for the reduced species of a W complex, W(bpy-*t*Bu)$(CO)_4$ **26** to understand the species formed upon electrochemical reduction.[80] To enhance the catalytic activity for CO_2 reduction tetracarbonyl complexes of nonclassical diimine ligands $W(CO)_4$(4,6-diphenyl-2,2′-bipyridine) **27**, $W(CO)_4$(6-(2,6-dimethoxyphenyl)-4-phenyl-2,2′-bipyridine) **28** and $W(CO)_4$(2,2′-dipyridylamine) **29** were synthesized and found to have electrochemical catalytic activity to reduce carbon dioxide to carbon monoxide.[86]

Several other tungsten complexes are known which reacts with carbon dioxide in a typical way. Photochemical excitation (irradiation λ = > 420 nm) of dimer $W_2(CO)_{10}{}^{2-}$ results the formation $W(CO)_5{}^-$. In the presence of PR_3 (PPh_3 and PBu_3), the monomer $W(CO)_5{}^-$ is converted to $W(CO)_5PR_3{}^-$, **30** which is an efficient catalyst for reduction of carbon dioxide to formate and CO.[87]

Tungsten carbyne hydride complex *trans*-W(CMes)$(dmpe)_2$H **31**, (dmpe = $Me_2PCH_2CH_2PMe_2$) reacts with carbon dioxide to form formato-O-complex *trans*-W(CMes)$(dmpe)_2$(OCHO) **32** due to facile insertion of C = O unit in the W–H bond.[88] There is a report of deoxygenation of carbon dioxide using transient terminal phosphinidene tungsten com-

$$W_2(CO)_{10}^{2-} \overset{h\nu}{\rightleftharpoons} 2W(CO)_5^-$$

$$W(CO)_5^- + PR_3 \overset{h\nu}{\rightleftharpoons} \underset{\mathbf{30}}{W(CO)_5PR_3^-}$$

$$W(CO)_5PR_3^- + CO_2 \longrightarrow W(CO)_5PR_3 + CO_2^-$$

Figure 18. Synthesis of complex **30** from photochemical excitation of dimer $W_2(CO)_{10}^{2-}$ followed by reaction with PR_3 and its reactions.[87]

plex.[89] Phosphine ligated complexes like $WC1_2(PMePh_2)_4$ **33** reacts with carbon dioxide and forms tungsten(IV) oxo-carbonyl complex, W(O) $CI_2(CO)(PMePh_2)_2$ **34** by breaking the carbon dioxide molecule.[90]

VIII. Application of FDHs for Carbon Dioxide Reduction

There is a growing interest to design homogeneous photochemical CO_2 reduction system based on combination of the process of photo-reduction of NAD^+ or MV^{2+} (as an electron relay) using photosensitization of dye molecules and CO_2 reduction with FDH.[91,92] Amao and coworkers have developed several systems for carbon dioxide reduction to formic acid using formate dehydrogenase and reduced form of methyl viologen ($MV^{\bullet+}$). The design of such as system using TEOA (triethanolamine), $ZnTMePyP^{4+}$ (zinc tetrakis(4-methylpyridyl)porphyrin), MV^{2+} and FDH is shown in Figure 19.

A system has been developed for hydrogen and formic acid production coupling the glucose oxidation (with glucose dehydrogenase) and hydrogen and formic acid generation (with platinum nano-particle and FDH) using the photo-reduction of methyl viologen.[93] A visible-light operated CO_2-glucose biofuel cell is known. The cell is comprised of an anode (a chlorin-e_6 immobilized on TiO_2 thin layer film onto optical transparent conductive glass electrode (OTE)), a cathode (FDH and viologen with long alkyl chain co-immobilized OTE) and the solution having glucose, glucose dehydrogenase and NAD^+.[94]

The photo-reduction of CO_2 to formic acid was observed in nano-porous reactor using a photosensitizer, $Ru(bpy)_3^{2+}$, methyl viologen

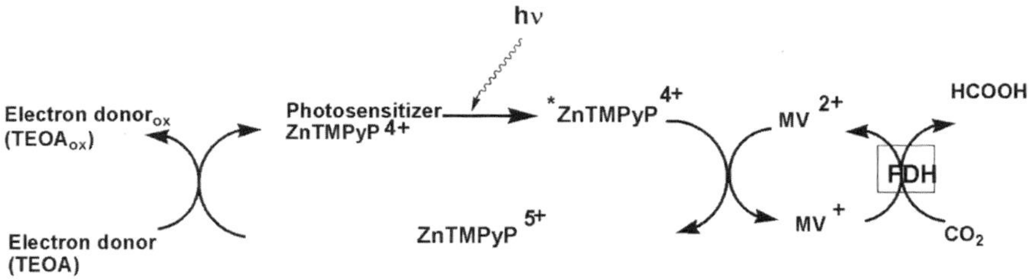

Figure 19. Photochemical and enzymatic carbon dioxide conversion to formic acid using the system consisting of TEOA, $ZnTMePyP^{4+}$, MV^{2+} and FDH.[92]

(MV^{2+}) and formate dehydrogenase (FDH) under ambient conditions.[95] There is a report of electro-enzymatic cell using *Candida boidinii* formate dehydrogenase for the conversion of CO_2 to formate.[96]

The formate dehydrogenase from the acetogen *Clostridium carboxidivorans* strain $P7^T$ expressed in *E. coli* displays NADH-dependent catalytic activity of carbon dioxide reduction.[97] FDH from *Pyrococcus furiosus and Methanobacterium thermoformicicum* when expressed in *E. coli* JM109(DE3) produces formate through hydrogenation of CO_2.[98] There is a recent report of conversion of CO_2 to formate in a hybrid biofuel cell system using fabricated enzymatic cathode with FDH.[99] A novel bioprocess for reduction of CO_2 to formate using a whole cell catalysis was developed using three different *Desulfovibrio* species namely, *D. vulgaris, D. alaskensis* G20, and *D. desulfuricans* ATCC 27774. Among these species *D. desulfuricans* showed the highest catalytic activity.[100] A gas-diffusion-type biocathode for CO_2 reduction with W-containing FDH from *Methylobacterium* extorquens AM1 is known.[101] The bacterial hydrogen-dependent carbon dioxide reductase from *Acetobacterium woodii* catalyzes the reaction of hydrogen dependent carbon dioxide reduction. Whole-cells of the enzyme can be used as catalyst for formation of formate from dihydrogen (H_2) and CO_2.[102]

IX. Conclusions

There is no doubt that the activation of carbon dioxide is a very difficult challenge. The plant photosynthesis which uses the enzyme RuBbisCO to reduce carbon dioxide to sugar units is a slow process. The other pathway of carbon dioxide fixation by the cleavage of C–O bond to carbon monoxide is not an environment friendly process. Therefore, FDHs are very interesting systems for designing assemblies of carbon dioxide fixation or formate production. The use of biomimetic catalyst, FDH over the use of other chemical catalysts has several advantages. These enzymes have the advantage of specificity in the formation of only product formate and the capability of working as homogeneous catalyst. In addition, these systems have the advantage in consuming of atmospheric carbon dioxide which concurrently helps in cleaning the environment. The major challenges for construction of such a system are the oxygen sensitivity of FDHs and protein stability. Another factor is higher affinities for the reverse formate oxidation reaction.[24]

Although the model complexes of Mo and W containing FDH enzymes are limited to date, the complexes that have been synthesized exhibit a remarkable diversity. The synthesis of model complexes along with the photo-driven system where FDS act as a catalyst will grow significantly for their use in carbon dioxide reduction purposes in coming years. The catalytic rates of carbon dioxide reduction need to be enhanced for the use in industrial scale. The knowledge gained from natural enzymes should be used to develop bio-inspired, stable efficient catalysts in carbon dioxide reduction.

X. References

1. Thomas, C. D.; Cameron, A.; Green, R. E.; Bakkenes, M.; Beaumont, L. J.; Collingham, Y. C.; Erasmus, B. F. N.; de Siqueira, M. F.; Grainger, A.; Hannah, L.; Hughes, L.; Huntley, B.; van Jaarsveld, A. S.; Midgley, G. F.; Miles, L.; Ortega-Huerta, M. A.; Peterson, A. T.; Phillips O. L.; Williams, S. E. *Nature* **2004**, *427*, 145.
2. Heinze, C.; Meyer, S.; Goris, N.; Anderson, L.; Steinfeldt, R.; Chang, N.; Quéré, C. L.; Bakker D. C. E. *Earth Sys. Dynam.* **2015**, *6*, 327.
3. Sakakura, T.; Choi, J. C.; Yasuda, H. *Chem. Rev.* **2007**, *107*, 2365.
4. Bhuyan, J.; Sarkar, R.; Sarkar, S. *Angew. Chem. Int. Ed.* **2011**, *50*, 10603.
5. Fontecilla-Camps, J. C.; Amara, P.; Cavazza, C.; Nicolet, Y.; Volbeda, A. *Nature* **2009**, *460*, 814.
6. Maia, L. B.; Fonseca, L.; Moura, J. J. G.; Moura, I. *J. Am. Chem. Soc.* **2016**, *138*, 8834.

7. Hille, R. *Chem. Rev.* **1996**, *96*, 2757.
8. Majumdar, A. *Dalton Trans.* **2014**, *43*, 8990.
9. Johnson, M. K.; Rees, D. C.; Adams M. W. W. *Chem. Rev.* **1996**, *96*, 2817.
10. Bevers, L. E.; Hagedoorn, P. L.; Hagen, W. R. *Coord. Chem. Rev.* **2009**, *253*, 269.
11. L'vov, N. P.; Nosikov, A. N.; Antipov, A. N. *Biochemistry (Moscow)* **2002**, *67*, 196.
12. McDevitt, C. A. ; Hugenholtz, P.; Hanson, G. R.; McEwan, A. G. *Mol. Microbiol.* **2002**, *44*, 1575
13. Hille, R.; Hall, J.; Basu, P. *Chem. Rev.* **2014**, *114*, 3963.
14. Gonzalez, P. J.; Rivas, M. G.; Mota, C. S.; Brondino, C. D.; Moura, I.; Moura, J. J. G. *Coord. Chem. Rev.* **2013**, *275*, 315.
15. Hille, R. *Trends Biochem. Sci.* **2002**, *27*, 360.
16. Dias, J. M.; Than, M. E.; Humm, A.; Huber, R.; Bourenkov, G. P.; Bartunik, H. D.; Bursakov, S.; Calvete, J.; Caldeira, J.; Carneiro, C.; Moura, J. G.; Moura, I.; Romão, M. J. *Structure* **1999**, *7*, 65.
17. Jormakka, M.; Yokoyama, K.; Yano, T.; Tamakoshi, M.; Akimoto, S.; Shimamura, T.; Curmi, P.; Iwata, S. *Nat. Struct. Mol. Biol.* **2008**, *15*, 730.
18. Boyington, J. C.; Gladyshev, V. N.; Khangulov, S. V.; Stadtman, T. C.; Sun, P. D. *Science* **1997**, *275*, 1305.
19. Bertero, M. G.; Rothery, R. A.; Palak, M.; Hou, C.; Lim, D.; Blasco, F.; Weiner, J. H.; Strynadka, N. C. J. *Nat. Struc. Biol.* **2003**, *10*, 681.
20. Kloer, D. P.; Hagel, C.; Heider, J.; Schulz, G. E. *Structure* **2006**, *14*, 1377.
21. Schneider, F.; Lowe, J.; Huber, R.; Schindelin, H.; Kisker, C.; Knablein, J. *J. Mol. Biol.* **1996**, *263*, 53.
22. Zhang, L.; Nelson, K. J.; Rajagopalan, K. V.; George, G. N. *Inorg. Chem.* **2008**, *47*, 1074.
23. Dobbek, H. *Coord. Chem. Rev.* **2011**, *255*, 1104.
24. Maia, L. B.; Moura, I.; Moura, J. J. G. *Inorg. Chim. Acta.* **2017**, *455*, 350.
25. Romao, M. J. *Dalton Trans.* **2009**, 4053.
26. Maia, L. B.; Moura, J. J. G.; Moura I. *J. Biol. Inorg. Chem.* **2015**, *20*, 287.
27. Ferry, J. G. *FEMS Microbiol. Rev.* **1990**, *87*, 377.
28. Vinals, C.; Depiereux, E.; Feytmans, E. *Biochem. Biophys. Res. Commun.* **1993**, *192*, 182.
29. Castillo, R.; Oliva, M.; Marti, S.; Moliner, V. *J. Phys. Chem. B* **2008**, *112*, 10012.
30. Raaijmakers, H.; Macieira, S.; Dias, J. M.; Teixeira, S.; Bursakov, S.; Huber, R.; Moura, J. J. G.; Moura, I.; Romao, M. J. *Structure* **2002**, *10*, 1261.
31. Schrapers, P.; Hartmann, T.; Kositzki, R.; Dau, H.; Reschke, S.; Schulzke, C.; Leimkühler, S.; Haumann, M. *Inorg. Chem.* **2015**, *54*, 3260.
32. Jormakka, M.; Byrne, B.; Iwata, S. *Curr. Opin. Struct. Biol.* **2003**, *13*, 418.
33. Costa, C.; Teixeira, M.; LeGall, J.; Moura J. J. G.; Moura, I. *J. Biol. Inorg. Chem.* **1997**, *2*, 198.
34. Brondino, C. D.; Passeggi, M. C. G.; Caldeira, J.; Almendra, M. J.; Feio, M. J.; Moura, J. J. G.; Moura, I. *J. Biol. Inorg. Chem.* **2004**, *9*, 145.

35. Sebban, C.; Blanchard, L.; Bruschi, M.; Guerlesquin, F. *FEMS Microbiology Letters* **1995**, *133*, 143.
36. Raaijmakers, H.; Teixeira, S.; Dias, J. M.; Almendra, M. J.; Brondino, C. D.; Moura, I.; Moura, J. J. G.; Romao, M. J. *J. Biol. Inorg. Chem.* **2001**, *6*, 398.
37. da Silva, S. M.; Voordouw, J.; Leitao, C.; Martins, M.; Voordouw, G.; Pereira, I. A. C. *Microbiology* **2013**, *159*, 1760.
38. Hartmann, T.; Leimkuhler, S. *FEBS J.* **2013**, *280*, 6083.
39. Bassegoda, A.; Madden, C.; Wakerley, D. W.; Reisner, E.; Hirst J. *J. Am. Chem. Soc.* **2014**, *136*, 15473.
40. de Bok, F. A.; Hagedoorn, P. L.; Silva, P. J.; Hagen, W. R.; Schiltz, E.; Fritsche, K.; Stams, A. J. M. *Eur. J. Biochem.* **2003**, *270*, 2476.
41. Reda, T.; Plugge, C. M.; Abram, N. J.; Hirst, J. *Proc. Natl. Acad. Sci. U.S.A.* **2008**, *105*, 10654.
42. Yamamoto, I.; Saikit, T.; Liu, S. M.; Ljungdahl, L. G. *J. Biol. Chem.* **1983**, *258*, 1826.
43. Laukel, M.; Chistoserdova, L.; Lidstrom, M. E.; Vorholt, J. A. *Eur. J. Biochem.* **2003**, *270*, 325.
44. George, G. N.; Costa, C.; Moura, J. J. G.; Moura, I. *J. Am. Chem. Soc.* **1999**, *121*, 2625.
45. Khangulov, S. V.; Gladyshev, V. N.; Dismukes, G. C.; Stadtman, T. C. *Biochemistry* **1998**, *37*, 3518.
46. Thauer, R. K.; Kaufer, B.; Fuchs, G. *Eur. J. Biochem.* **1975**, *55*, 111.
47. Shiraishi, T.; Fukusaki, E.; Kajiyama, S.; Kobayashi, A. *J. Chromatogr. A.* **1999**, *855*, 337.
48. Raaijmakers, H. C.; Romao, M. J. *J. Biol. Inorg. Chem.* **2006**, *11*, 849.
49. Mota, C. S.; Rivas, M. G.; Brondino, C. D.; Moura, I.; Moura, J. J. G.; Gonzalez, P. J.; Cerqueira, N. M. F. S. A. *J. Biol. Inorg. Chem.* **2011**, *16*, 1255.
50. Cerqueira, N. M. F. S. A.; Gonzalez, P. J.; Fernandes, P. A.; Moura, J. J. G.; Romas, M. J. *Inorg. Chem.* **2013**, *52*, 10766.
51. Tiberti, M.; Papaleo, E.; Russo, N.; Gioia, L. D.; Zampella, G. *Inorg. Chem.* **2012**, *51*, 8331.
52. Dong, G.; Ryde, U. J. *Biol. Inorg. Chem.* **2018** (accepted).
53. Niks, D.; Duvvuru, J.; Escalona, M.; Hille, R. *J. Biol. Chem.* **2016**, *291*, 1162.
54. Aslan, A. S.; Valjakka, J.; Ruupunen, J.; Yildirim, D.; Turner, N. J.; Turunen, O.; Binay, B. *Protein. Eng. Des. Sel.* **2016**, *30*, 47.
55. Lim, B. S.; Holm, R. H. *J. Am. Chem. Soc.* **2001**, *123*, 1920.
56. Young, C. G.; Roberts, S. A.; Ortega, R. B.; Enemark, J. H. *J. Am. Chem. Soc.* **1987**, *109*, 2938.
57. Young C. G.; Laughlin, L. J.; Colmanet S.; Scrofani S. D. B. *Inorg. Chem.* **1996**, *35*, 5368.
58. Young, C. G.; Gable, R. W.; Hill, J. P.; George, G. N. *Eur. J. Inorg. Chem.* **2001**, 2227.

59. Sugimoto, H.; Sakurai, T.; Miyake, H.; Tanaka, K.; Tsukube, H. *Inorg. Chem.* **2005**, *44*, 6927.
60. Majumdar, A.; Sarkar, S. *Coord. Chem. Rev.* **2011**, *255*, 1039.
61. Oku, H.; Ueyama, N.; Nakamura, A. *Bull. Chem. Soc. Jpn.* **1996**, *69*, 3139.
62. Davies, E. S.; Aston, G. M.; Beddoes, R. L.; Collison, D.; Dinsmore, A.; Docrat, A.; Joule, J. A.; Wilson, C. R.; Garner, C. D. *J. Chem. Soc.; Dalton Trans.* **1998**, 3647.
63. Enemark, J. H.; Cooney, J. J. A.; Wang, J. J.; Holm, R. H. *Chem. Rev.* **2004**, *104*, 1175.
64. Groysman, S.; Holm, R. H. *Inorg. Chem.* **2007**, *46*, 4090.
65. Das, S. K.; Biswas, D.; Maiti, R.; Sarkar, S. *J. Am. Chem. Soc.* **1996**, *118*, 1387.
66. Sarkar, S.; Das, S. K. Proc. *Indian. Acad. Sci. (Chem. Sci.)* **1992**, *104*, 533.
67. Deaton, J. C.; Solomon, E. I.; Durfor, C. N.; Wetherbee, P. J.; Burgess, B. K.; Jacobs, D. B. *Biochem. Biophys. Res. Commun.* **1984**, *121*, 1042.
68. Jiang, J.; Holm, R. H. *Inorg. Chem.* **2004**, *43*, 1302.
69. Seo, J.; Kim, E. *Inorg. Chem.* **2012**, *51*, 7951.
70. Wilkinson, G. *J. Am. Chem. Soc.* **1954**, *76*, 209.
71. Gambarotta, S.; Floriani, C.; Villa, A. C.; Guastin, C. *J. Am. Chem. Soc.* **1985**, *107*, 2986.
72. Belmore, K. A.; Vanderpool, R. A.; Tsai, J. C.; Khan, M. A.; Nicholas, K. M. *J. Am. Chem. Soc.* **1988**, *110*, 2004.
73. Liu, C. G. *Mol. Phys.* **2013**, *111*, 259.
74. Sun, X. Z.; Nikiforov, S. M.; Dedieu, A.; George, M. W. *Organometallics* **2001**, *20*, 1515.
75. Yonke, B. L.; Reeds, J. P. Zavalij, P. Y.; Sita, L. R. *Angew. Chem. Int. Ed.* **2011**, *50*, 12342.
76. Shin, J. H.; Churchill, D. G.; Parkin, G. *J. Organomet. Chem.* **2002**, *642*, 9.
77. Tai, C. C.; Chang, T.; Roller, B.; Jessop, P. G. *Inorg. Chem.* **2003**, *42*, 7340.
78. Zhang, Y.; Williard, P. G.; Bernskoetter, W. H. *Organometallics* **2016**, *35*, 860.
79. Pal, R. Groy, T. L.; Trovitch, R. J. *Inorg. Chem.* **2015**, *54*, 7506.
80. Clark, M. L.; Grice, K. A.; Moore, C. E.; Rheingold, A. L.; Kubiak, C. P. *Chem. Sci.* **2014**, *5*, 1894.
81. Tory, J.; Setterfield-Price, B.; Dryfe, R. A. W.; Hartl, F. *ChemElectroChem* **2015**, *2*, 213.
82. Sieh, D.; Lacy, D. C.; Peters, J. C.; Kubiak, C. P. *Chem. Eur. J.* **2015**, *21*, 8497.
83. Oh, Y.; Vrubel, H.; Guidoux, S.; Hu, X. *Chem. Commun.* **2014**, *50*, 3878.
84. Gao, G.; Li, F.; Xu, L.; Liu, X.; Yang, Y. *J. Am. Chem. Soc.* **2008**, *130*, 10838.
85. Knopf, I.; Ono, T.; Temprado, M.; Tofan, D.; Cummins, C. C. *Chem. Sci.* **2014**, *5*, 1772.
86. Franco, F.; Cometto, C.; Sordello, F.; Minero, C.; Nencini, L.; Fiedler, J.; Gobetto, R.; Nervi, C. *ChemElectroChem* **2015**, *2*, 1372.
87. Silavwe, N. D.; Goldman, A. S.; Ritter R.; Tyler, D. R. *Inorg. Chem.* **1989**, *28*, 1231.

88. Furno, F.; Fox, T.; Schmalle, H. W.; Berke, H. *Organometallics* **2000**, *19*, 3620.
89. Schulten, C.; von Frantzius, G.; Schnakenburg, G.; Espinosab, A.; Streubel, R. *Chem. Sci.* **2012**, *3*, 3526.
90. Bryan, J. C.; Geib, S. J.; Rheingold, A. L.; Mayer, J. M. *J. Am. Chem. Soc.* **1987**, *109*, 2826.
91. Tsujisho, I.; Toyoda, M.; Amao, Y. *Catal. Commun.* **2006**, *7*, 173.
92. Amao, Y.; *ChemCatChem* **2011**, *3*, 458.
93. Amao, Y.; Takahara, S.; Sakai, Y. *Int. J. Hydrogen Energy* **2014**, *39*, 20771.
94. Amao, Y.; Shuto, N. *J. Porphyr. Phthalocyanines* **2015**, *19*, 459.
95. Noji, T.; Jin, T.; Nango, M.; Kamiya, N.; Amao, Y. *Appl. Mater. Interfaces.* **2017**, *9*, 3260.
96. Kim, S.; Kim, M. K.; Lee, S. H.; Yoon, S.; Jung, K. D. *J. Mol. Catal. B: Enzym.* **2014**, *102*, 9.
97. Alissandratos, A.; Kim, H. K.; Matthews, H.; Hennessy, J. E.; Philbrook, A.; Easton, C. J. *Appl. Environ. Microbiol.* **2013**, *79*, 741.
98. Alissandratos, A.; Kim, H. K.; Easton, C. J. *Bioresour. Technol.* **2014**, *164*, 7.
99. Zhang, L.; Ong, J.; Liu, J.; Li, S. F. Y. *Renew. Energy* **2017**, *108*, 581.
100. Mourato, C.; Martins, M.; da Silva, S. M.; Pereira, I. A. C. *Bioresour. Technol.* **2017**, *235*, 149.
101. Sakai, K.; Kitazumi, Y.; Shirai, O.;Takagi, K; Kano, K. *Electrochem. Commun.* **2016**, *73*, 85.
102. Schuchmann, K.; Müller, V. *Science* **2013**, *342*, 1382.

7 Tyrosinases: Enzymes, Models and Related Applications

Aleksandar Bijelic*, Annette Rompel*,‡ and Catherine Belle†,§

*Universität Wien, Fakultät für Chemie,
Institut für Biophysikalische Chemie, Althanstraße 14,
1090 Wien, Austria
†Université Grenoble-Alpes, CNRS, DCM, 38000 Grenoble, France
‡annette.rompel@univie.ac.at
§catherine.belle@univ-grenoble-alpes.fr

Table of Contents

List of Abbreviations

CuA	copper A
CuB	copper B
DBED	*N,N′*-di-tert-butyl-ethylene diamine
EPR	electron paramagnetic resonance
H	histidine
HBRPz3	tris-pyrazolylborate
H6M4h	1,2-bis[2-di-(6-methyl-2-pyridyl)methyl-6-pyridyl]ethane
L55	α, α′-bis[[bis(1-methyl-2-benzimidazolyl)methyl]amino]-*m*-xylene
L55Bu4	(*S*)-*N*,*N*-[1,3-phenylenebis(methylene)]bis(1-{1-[(*S*)-2-methylbutyl]-1*H*-benzimidazol-2-yl}-*N*-[(1-{(*S*)-2-methylbutyl}-1*H*-benzimidazol-2-yl)methyl]methanamine)
L66	α, α′-bis{bis[2-(1′-methyl-2′-benzimidazolyl)ethyl]amino}-m-xylene
L-DOPA	*L*-3,4-dihydroxyphenylalanine
^{C}P	*cis* peroxo
^{S}P	μ-η^2-η^2 peroxo
^{T}P	*trans* peroxo
PPO	polyphenol oxidase
QM/MM	quantum mechanics/molecular mechanics

O	bis μ-oxo
SDS	sodium dodecyl sulfate
S-PzPHI	(*S*)-2-piperazinemethanamine,1,4-bis[2-((*N*-(1-acetoxy-3-(1-methyl-1*H*-imidazol-4-yl))-2-(*S*)-propyl)-(*N*-(1-methyl-1*H*-imidazol-2-ylmethyl)))ethyl]-*N*-(phenylmethyl)-*N*-(acetoxy),
TACN	1,4,7-triazacyclononane
TON	turnover number
TYR	tyrosinase
TYRP	tyrosinase related protein
UV-Vis	ultraviolet-visible
XAS	X-ray absorption spectroscopy

I. Tyrosinases: Enzymes

A. Introduction

Tyrosinases (TYRs) are type-III-copper enzymes and belong to the class of polyphenol oxidases (PPOs). They are widely distributed in bacteria, fungi, invertebrates, higher plants and mammals.[1–3] TYR is the rate-limiting enzyme in the biosynthesis of the pigment melanin as it provides the starting material for this pathway by catalyzing the oxygenation of the monophenol *L*-tyrosine.[4] This TYR-catalyzed oxygenation consists of two linked reactions, namely the *ortho*-hydroxylation of monophenols to *o*-diphenols (monophenolase activity, EC 1.14.18.1) and the subsequent two-electron oxidation of these *o*-diphenols to the corresponding *o*-quinones (diphenolase activity, EC 1.10.3.1), whereby molecular oxygen is reduced to water (Figure 1).[5]

The *o*-diphenols generated by the *o*-hydroxylation step remain bound to the enzyme and are directly oxidized, leading to the sole release of the final *o*-quinones.[6]

These quinones are highly reactive and undergo numerous spontaneous, non-enzymatic reactions like polymerization to high-molecular-weight melanins.[4] Melanins are heterogeneous polyphenolic polymers that are widely distributed in nature. Thus, TYRs are responsible for the browning of wounded tissue, which represents a huge agricultural problem, the browning of fruits and vegetables. Apart from providing

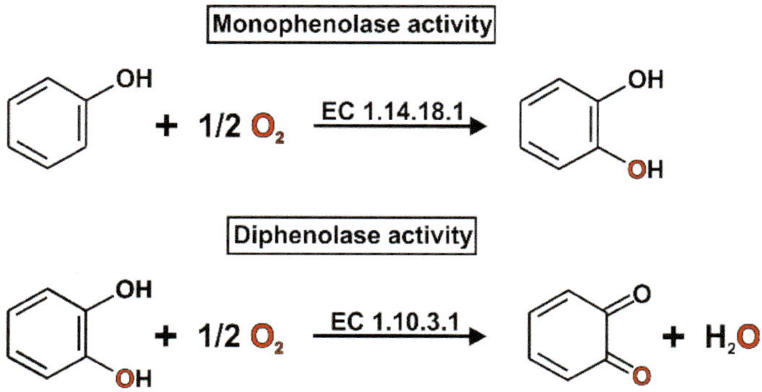

Figure 1. Reactions catalyzed by TYR.

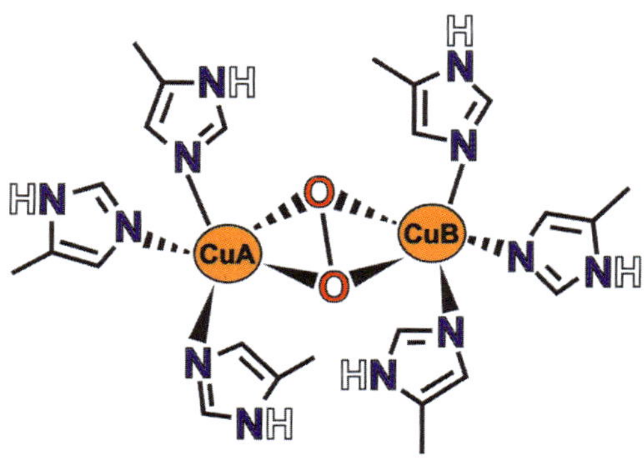

Figure 2. Active site of TYR in its *oxy*-form. Oxygen is bound as peroxide in the μ-η^2:η^2 mode and both copper ions exhibit the oxidation state II.

pigmentation in fungi, plants, invertebrates, mammals and humans, melanins are involved in defense mechanism (immunity) and wound healing in many organisms,[7–10] morphogenesis in fungi[11] and sclerotization after molting in arthropods.[12,13]

TYR possesses a coupled binuclear type-III-copper active site, where each copper ion (CuA and CuB) is coordinated by three histidyl side chains (H_{A1-3} and H_{B1-3}) (Figure 2). The active site can bind molecular oxygen in the specific peroxo μ-η^2:η^2 bridging mode yielding the catalytically active *oxy*-form of the enzyme.[14]

B. Catalytic Mechanism of Tyrosinases

The active center of TYR can exist in three different catalytically relevant states, namely in the *deoxy*-, *met*- and the above described *oxy*-state. The states differ in the oxidation state of the two copper ions (+I in the *deoxy*-form and +II in the *met*- and *oxy*-form), the distance between them (ranging from 2.8 to 4.6 Å) and their respective role in the catalytic cycle of TYR (Figure 3).[15]

The *deoxy*-form of TYR represents the inactive form of the enzyme with the copper ions, $Cu(I)_2$, being 4.1 to 4.6 Å apart making this state the one with the longest Cu–Cu distance. After binding of molecular oxygen, the active site turns into its *oxy*-state, $Cu(II)_2(\mu\text{-}\eta^2\text{:}\eta^2\text{-}O_2)$, thus becoming active and able to bind and convert both mono- and diphenols to the respective quinones.[16] The Cu–Cu distance of the *oxy*-state is 2.8 to 3.2 Å and thus the shortest among all states.[17] Monophenols bound to the *oxy*-form are first *ortho*-hydroxylated in an electrophilic aromatic substitution

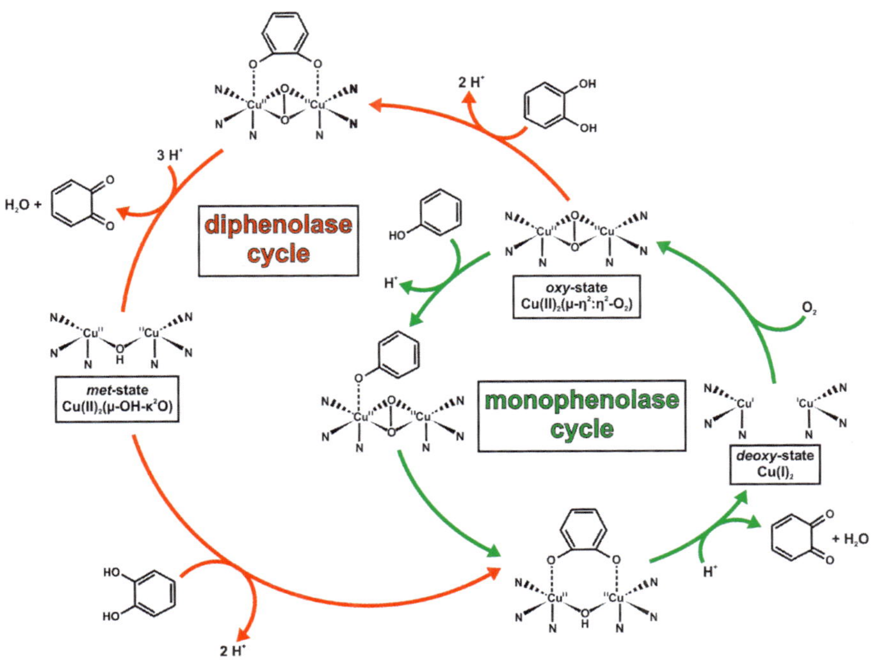

Figure 3. Mono- and diphenolase cycle of TYR according to Ref. [12].

reaction and then oxidized to provide the corresponding *o*-quinones (monophenolase cycle) with the hydroxylation being the rate-determining step.[18] Upon the release of the product and water, the *deoxy*-form is regenerated and the enzyme becomes ready for the next cycle. When an external diphenol is bound by the *oxy*-state, the substrate is oxidized to the respective *o*-quinone leaving the enzyme in its resting *met*-form. In the *met*-form, $Cu(II)_2(\mu\text{-}OH\text{-}\kappa^2O)$, the copper ions are bridged by a hydroxide or water molecule and exhibit a Cu–Cu distance of 3.2 to 4.0 Å.[19] Recent QM/MM calculations strongly suggest that water represents the copper bridging moiety as it was energetically more favorable than the hydroxide ion.[20] The *met*-form is able to bind both mono- and diphenols, however, only the latter is converted to the corresponding *o*-quinone by oxidation, which transforms the active center back into its *deoxy*-state (diphenolase cycle).[21]

C. Domain Organization of Tyrosinases

TYRs are very diverse regarding their structural properties, functions, tissue distributions and cellular location as there is no common TYR that occurs across all species.

TYRs normally consist of three main domains, namely an N-terminal domain, a central domain harboring the active site and a C-terminal domain.[22] The occurrence and function of the N- and C-terminal domain can vary among organisms as only TYRs from higher eukaryotes and plants possess all three domains with the main domain identifying the protein as TYR (Figure 4).

The C-terminal domain occurs in animal, plant and fungal TYRs and shields the active site of TYR, thereby preventing substrate binding. As long as the C-terminal domain stays attached to the enzyme, TYRs exhibit either very weak or no enzymatic activity at all and are thus referred to as latent TYRs.[15] Therefore, one function of the C-terminal extension is to keep the enzyme inactive inside the cell. However, the size (and complexity) of the C-terminal domain is too large to just ensure the latency of the enzyme and therefore this segment most likely has further important but yet unknown functions, especially when considering that the C-terminus of some animal TYRs contains a helical

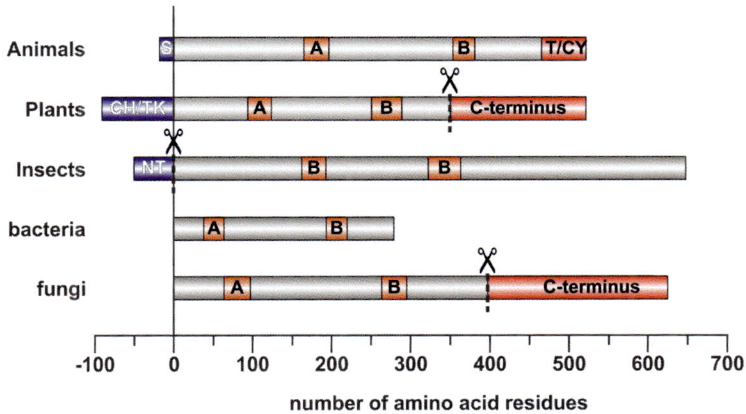

Figure 4. Domain organization of TYRs. Numbering of amino acids starts at the main domain (silver). N- and C-terminal domain are colored in blue and red, respectively, whereas the copper coordinating regions within the central domain are illustrated in brown (with A and B indicating CuA and CuB, respectively). S = signal peptide, which in the case of plant TYRs consists of two signal domains, namely a chloroplast transit peptide (CH) followed by a thylakoid transfer domain (TK). NT = N-terminal domain, which in the case of insects fulfils the role of shielding the active site (similar to the function of the C-terminal domain in other organisms). Furthermore, in insects, both copper-coordinating regions are of the CuB-type. T = transmembrane helix and CY = cytosolic space as the C-terminal domain of higher animals is membrane-associated with a tail protruding into the cytosolic space. The cleavage sites for TYR activation are marked by scissors.

transmembrane region indicating their membrane-association.[15,23] For example, the C-terminal domain of TYR from *Pholiota nameko* was shown to be required for the correct folding of the catalytically active main domain, which is also true for a series of plant and fungal TYRs, however, this fact is not reflected in the literature.[24] Fungal TYRs possess a characteristic CXXC (C = cysteine and X = any amino acid) motif within their C-terminus, which is believed to be involved in the copper uptake of the enzyme as it was also identified in copper chaperones.[25] In contrast, two structurally known bacterial TYRs, those from *Bacillus megaterium* and *Streptomyces castaneoglobisporus*, lack a C-terminal domain.[26,27] The former TYR is directly expressed in its active form, whereas the latter forms a heterodimer with a so-called caddie protein, which covers the enzyme's active site instead of a C-terminal domain.

Insect TYRs are also not equipped with a C-terminal domain but the shielding function is taken over by its N-terminal domain.[28]

The N-terminal domain is present in most plant and animal TYRs and is mainly responsible for the subcellular targeting of the enzyme after its expression.[15] In plants the N-terminal region usually consists of a chloroplast transit peptide and a thylakoid transfer domain, which determine the destination of the TYR, mostly chloroplasts.[2] The lack of this domain in fungal TYRs can largely be explained by the fact that they are generally suggested to be cytoplasmic.[29] Bacterial TYRs are often indicated as extracellular enzymes and therefore their secretion is suggested to be assisted by a second protein harboring the required signal sequences.[3,30,31] The main domain contains the two copper regions, where each copper ion, CuA and CuB, is coordinated by three fully conserved histidines. The two regions consist of the motifs H-X_n-H-X_8-H (around CuA) and H-X_3-H-X_n-H (around CuB) with H = histidine, X = any amino acid and n = variable integer number ranging from < 10 to > 20.[32]

D. Activation Process of Tyrosinases

Most TYRs are expressed as latent pro-enzymes *in vivo*, that is, in the non-active form, which needs to be activated by the removal of the active site shielding C-terminal or N-terminal (insect TYRs) domain, respectively. However, for most of the TYRs, the activation mechanism *in vivo* is unknown, although it is supposed to be proteolytic activation. This assumption arises from sequence differences between the genes of some fungal and walnut TYRs and the experimentally determined sequences of the respective isolated enzymes as the latter mainly represent the active form lacking the C-terminal domain.[33–38]

The activation mechanism has only been determined for three insect TYRs, namely those from *Manduca sexta*, *Bombyx mori* and *Holotrichia diomphalia*, revealing the involvement of diverse proteins and a cascade of proteases in the regulation of PPO activation.[28] The activation process differs among the three insects but the major activation steps can be briefly described as follows: In response to pathogen infections, different peptidoglycan- or pattern-recognition proteins bind to the surface polysaccharides of the invading microorganism and thus trigger the activation of initiator protease(s). The initiator protease(s) in turn activate(s) a protease cascade, which includes species-specific terminal serine proteases that cleave and activate TYR (this final step differs the most among the three insects).

However, recently, it was found that PPO from apricot underwent spontaneous activation during storage at 4°C suggesting that some PPOs and thus TYRs might be able to exhibit some kind of self-activation without the need of external proteases.[39]

Latent TYRs can also be activated *in vitro* by applying acidic pHs,[33,40–42] detergents like SDS,[38,40,42–44] lipids,[45] alcohols[46] or limited proteolysis.[38,40,42,47] While the mechanism of activation is clear for the last method (proteolytic cleavage of the C-terminal domain), it is rather unclear for the other approaches, however, it is assumed that they in general induce structural and conformational rearrangements at the active site region, most likely at the interface between the active site and the C-terminal domain, which facilitate the influx of external substrates into the active center.

E. Structural Features of Tyrosinases

In 2006, the first TYR structure from the bacterium *Streptomyces castaneoglobisporus* was published.[26] This structure was later followed by those of TYR from the insect *Manduca sexta,*[48] the bacterium *Bacillus megaterium,*[27] the fungi *Agaricus bisporus*[35,36,49] and *Aspergillus oryzae,*[50] the prawn *Masupenaeus japonicas*[51] and the walnut *Juglans regia*[52] yielding in total eight TYR structures. Thus, the structural knowledge of TYRs is mainly based on these eight structures lacking a TYR structure from mammals. However, recently, the crystal structure of the human TYR related protein 1 (TYRP1) was solved which surprisingly contains zinc instead of copper in the active site (no oxidase activity) and represents the first structure of a mammalian melanogenic enzyme.[53] Despite the current lack of knowledge about TYRP1's function, its structure represents a breakthrough in this field and will hopefully provide important information on the mechanism of human and other mammal TYRs.[54,55] Interestingly, despite originating from different sources and thus having different sequences, functions and overall structures, all TYRs possess nearly identical active site architectures.

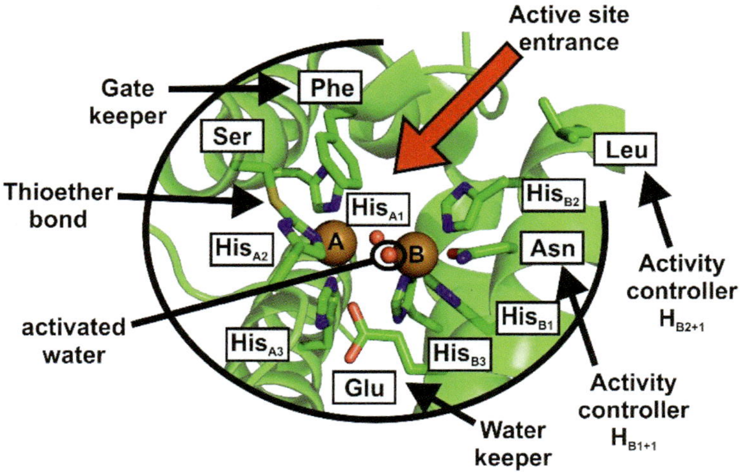

Figure 5. Active site of TYR. The active site structure of walnut TYR is shown as an example highlighting all in the text discussed structural features. The only unmarked atom is a hydroxyl (red sphere), which is located between CuA and CuB, as the enzyme was crystallized in its *met*-form. The enzyme is shown as green cartoon with crucial amino acid residues being illustrated in ball-and-stick representation. Color code: green, carbon; blue, nitrogen; red, oxygen; yellow, sulfur and brown, copper.

The active site consists of a bundle of four α-helices that surrounds the catalytic binuclear copper center (Figure 5). Each copper ion is coordinated by three histidines, whereby five of them (His_{A1}, His_{A3} and His_{B1-3}) are typically provided by the α-helices forming the bundle, whereas one, the second CuA coordinating histidine (His_{A2}), is located on a flexible loop. However, in TYR from *Manduca sexta* and *Masupenaeus japonicas* all six histidines are located on α-helices. The second histidine His_{A2} of all structurally known fungal and plant TYRs forms an unusual thioether bond with an adjacent cysteine residue that has an yet unknown function.[56] It was suggested that this thioether bridge stabilizes the position of the second histidine and thus the entire binuclear copper site in order to facilitate the electron transfer during the catalytic reaction as it depends on the correct positioning of the copper ions.[57] Another role for the thioether bond was proposed during the crystallographic studies of the apo- and holo-form (copper-lacking and copper-containing form) of pro-TYR from *Aspergillus oryzae* suggesting that the cysteine residue together with the cysteines from the CXXC motif might be involved in the copper incorporation process as the bond is only formed upon the addition of copper.[50]

There is a glutamic acid located at the entrance of the active site, which is highly conserved among all structurally investigated TYRs and therefore it was defined as a critical residue for TYR activity.[58,59] This glutamic acid (water-keeper) is assumed to stabilize and together with another amino acid (mostly asparagine) activate a highly conserved water molecule. The activated water molecule is supposed to deprotonate monophenolic substrates during the catalytic cycle of TYRs and is thus putatively responsible for the enzyme's monophenolase activity.[60] This theory is supported by a number of dicopper model systems, which showed that the μ-η^2:η^2-peroxo site only exhibits *ortho*-hydroxylation on phenolate instead of phenol.[61]

Another important residue within the active site is the so-called gate-keeper or blocker residue, which is located atop of CuA. The terms were coined as it was initially thought that a bulky residue, more precisely phenylalanine, at this position blocks monophenols from binding to CuA, thus being the reason for the lack of monophenolase activity in catechol oxidases, a related type-III-copper enzyme only able to exhibit diphenolase activity.[62] This idea was back then additionally supported as, in comparison to the structurally known catechol oxidases, the first TYR structures from bacteria, fungi and insects contained only small residues at this position. According to this theory, monophenolic substrates do only bind to CuA and diphenols only to CuB to initiate mono- and diphenolase activity, respectively.[60,63] This controversial theory was finally refuted by the crystal structure of walnut TYR which, despite exhibiting both mono- and diphenolase activity, has a bulky phenylalanine at the blocker residue position.[52] Based on the structure and catalytic behavior of walnut TYR, it was suggested that the phenylalanine residue does not block phenolic substrates from accessing CuA but rather stabilizes them via π-π-interactions. This was later supported by molecular dynamic studies of aurone synthase from *Coreopsis grandiflora*, a catechol oxidase with hydroxylase activity, as simulations with different substrates did indeed reveal π–π-interactions between the aromatic rings of the substrates and the enzyme's gate-keeper (phenylalanine).[64]

Two other interesting amino acid positions are those located next to the first two CuB coordinating histidines (His_{B1+1} and His_{B2+1}). The first residue, His_{B1+1}, was suggested to assist the above described conserved glutamic acid in the stabilization and activation of the putatively substrate deprotonating water.[65] This residue is in all structurally known TYRs an asparagine (with the exception of PPO4 from *Agaricus bisporus* which possesses an aspartic acid instead) and is therefore believed to be essential for monophenolase activity, especially as most catechol oxidases possess other amino acids at this position. However, sequences of other TYRs, for example, two TYRs from apple disprove the necessity of asparagine at this position as the two apple TYRs contain an alanine and a glycine at this position, respectively.[42]

The second residue, His_{B2+1}, was described to be a critical factor for the stabilization of the correct orientation of some substrates within the active site and thus crucial for substrate specificty.[27,64,66] Polar or charged residues at this position are able to interact with substrates containing a polar group at their tail (tail of the binding substrate is located adjacent to position His_{B2+1}), whereas, substrates lacking a polar tail would rather be stabilized by hydrophobic groups at this position. Very recently it was suggested that the monophenolase/diphenolase activity ratio of TYRs can be controlled by an appropriate combination of these two residues as, depending on the combination, the polarity of the active site is changed and therefore also the preference for mono- *versus* diphenolic substrates.[42,67]

Thus, the term "activity controllers" was coined for these residues (H_{B1+1} and H_{B2+1}).[42] All above described structural features and amino acids are summarized in Table 1 for all structurally characterized TYRs.

F. Applications of Tyrosinases

Due to their high reactivity and ability to both hydroxylate and oxidize various phenolic compounds, TYRs exhibit a wide range of application potential.

The use of TYR for the synthesis of *o*-diphenols like *L*-DOPA (*L*-3,4-dihydroxyphenylalanine) is very promising for the pharmaceutical field as *L*-DOPA has been widely used as a drug to treat the Parkinson dis-

Table 1. Overview of common structural features of TYRs.

Organism (PPO)	PDB Entry[a]	Thioether bridge	Gate keeper residue	Activity controller	Reference
Streptomyces castaneoglobisporus	1WX2	No	Gly	Asn/Val	26
Bacillus megaterium	3NM8	No	Val	Asn/Arg	27, 68
Aspergillus oryzae	3W6W	Yes	Val	Asn/Asn	50
Agaricus bisporus (PPO3)	2Y9W	Yes	Val	Asn/Phe	49
Agaricus bisporus (PPO4)	4OUA	Yes	Ala	Asp/Gly	35, 38, 69
Manduca sexta (PPO1)	3HHS	No	Glu	Asn/Ser	48
Manduca sexta (PPO2)	3HHS	No	Ser	Asn/Val	48
Marsupenaeus japonicas	3WKY	No	Val	Asn/Asp	51
Juglans regia	5CE9	Yes	Phe	Asn/Leu	52

[a] Note that there are more deposited structures in the PDB for some TYRs, however, here only one representative PDB entry is shown. For more details, see provided references.

ease.[70] The commercial process for *L*-DOPA synthesis by asymmetric hydrogenation of vanillin using organometallic catalysis includes several elaborate reaction steps and is hampered by poor conversion rates and low enantioselectivity.[71,72] TYR-mediated production of *L*-DOPA *via* enzymatic conversion of *L*-tyrosine requires only one single step under mild operational conditions (neutral pH and around room temperature) exhibiting outstanding enantioselectivity at low costs.[73] However, the enzymatic approach also suffers from critical limitations as the resulting *L*-DOPA product is immediately converted to the corresponding DOPA-quinone by the enzyme's diphenolase activity. To obtain *L*-DOPA, despite the consecutive reaction, the antioxidant ascorbic acid can be used which reduces the *o*-quinone back to the diphenolic *L*-DOPA.[74] Unfortunately, it is known that ascorbic acid partially inactivates TYRs.[75] One strategy to tackle all above mentioned shortcomings is the immobilization of TYR as it facilitates the separation of the enzyme from the reaction media and thus enhances the stability and recoverability of the system and also the productivity.[76]

One major industrial concern is the toxicity of phenols which are mainly found in the wastewater of industries such as coal conversion, pharmacy, plastics, textiles, dyes, iron/steel, pulp/paper and petroleum refineries posing severe risks for the environment and human health. Therefore, researchers have studied the use of enzymes in wastewater treatment to solve this problem and thereby found PPOs like TYRs to be important and effective tools for removing these contaminants.[77] The commercially available TYR from *Agaricus bisporus* is the best studied TYR concerning TYR-mediated bioremediation. Besides the removal of phenols from industrial wastewater, TYRs can also be used as biosensors to detect phenols at low concentration, which is important for the analysis of phenolic compounds or their detection in food as their presence there is an indication for lack of freshness. Since TYR-catalyzed reactions are electron-transfer reactions, a chemical signal is produced during the catalytic cycle, which in turn is transformed into an electric signal (detection) by the transducer surface on which the enzyme is attached.[78]

TYRs can induce protein-protein cross-linking by hydroxylating and oxidizing tyrosyl residues located on the protein's surface and thus forming protein-attached *o*-quinones, which in turn can react further with a second proteinogenic quinone (*via* phenol coupling), the free thiol group (*via* Michael addition) or free amino group (*via* condensation reaction) of another protein.[79] In addition, these quinones can react with the nucleophilic amino acid side chains of lysine, tyrosine, cysteine and histidine.[80] In this way, TYRs are able to cross-link and modify food biopolymers, which is of great interest for the food industry.[81] For example, TYRs were shown to improve the functionalization of dairy products by biopolymer cross-linking, including cross-linking of diverse whey proteins, which resulted in the prevention of syneresis, firming of soft textures and improvement of the antioxidant properties.[79,82–84]

II. Tyrosinases: Synthetic Analogs

Despite long-standing researches, the X-ray structural data from TYRs has been available for only a decade. During this time, chemists have produced a significant variety of compounds to reproduce the spectroscopic data and investigate the properties of the enzyme but also to reproduce the high reactivity and ability of TYRs to hydroxylate and oxidize various phenolic compounds, using O_2, a reaction not easy to achieve from an industrial point of view. Several recent reviews describing advancements in the field are available.[85–90]

A. Model Complexes and Reactive Intermediates

The *ortho*-hydroxylation reactions of phenol substrates catalyzed by TYRs include the formation of a copper-oxygen core (*oxy*-tyrosinase form) as the key intermediate. With the aim of fundamental knowledge, low-molecular-weight copper complexes have been developed to help identify and understand the reaction mechanisms involved. They are particularly inspired by this *oxy*-tyrosinase active form. Model chemistry, upon oxygenation of reduced copper(I) complexes, has highlighted different possible Cu_2/O_2 species (Figure 6) where the nature of the supporting ligands employed controls the final structure of the Cu_2/O_2 species.

The pioneering work of Kitajima[14] was an important break-through in the field. From the sterically hindered tris-pyrazolylborate ligands (HB^RPz_3, Figure 7), he obtained a dicopper(II) complex $[HB^RPz_3Cu(II)]_2(\mu\text{-}O_2)$ with dioxygen bonded as a peroxide anion in a μ-η^2:η^2 fashion (**^{S}P**) (Figure 6). To reproduce the biological function of TYRs, Karlin and coworkers starting from a dicopper(I) complex supported by a dinuclear ligand with a

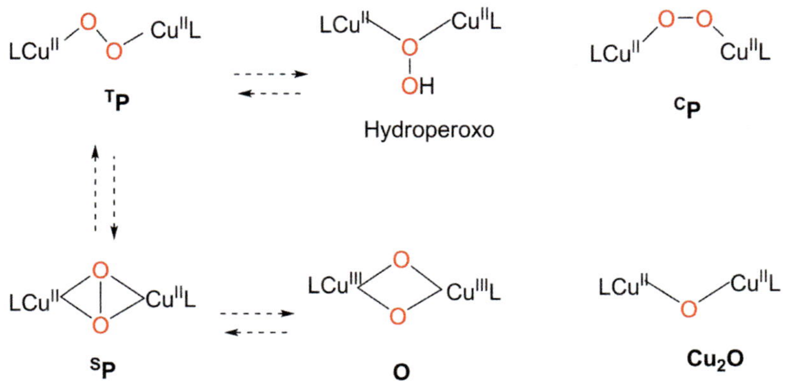

Figure 6. Cu_2/O_2 adducts characterized by X-ray from model complexes and their possible interconversions.

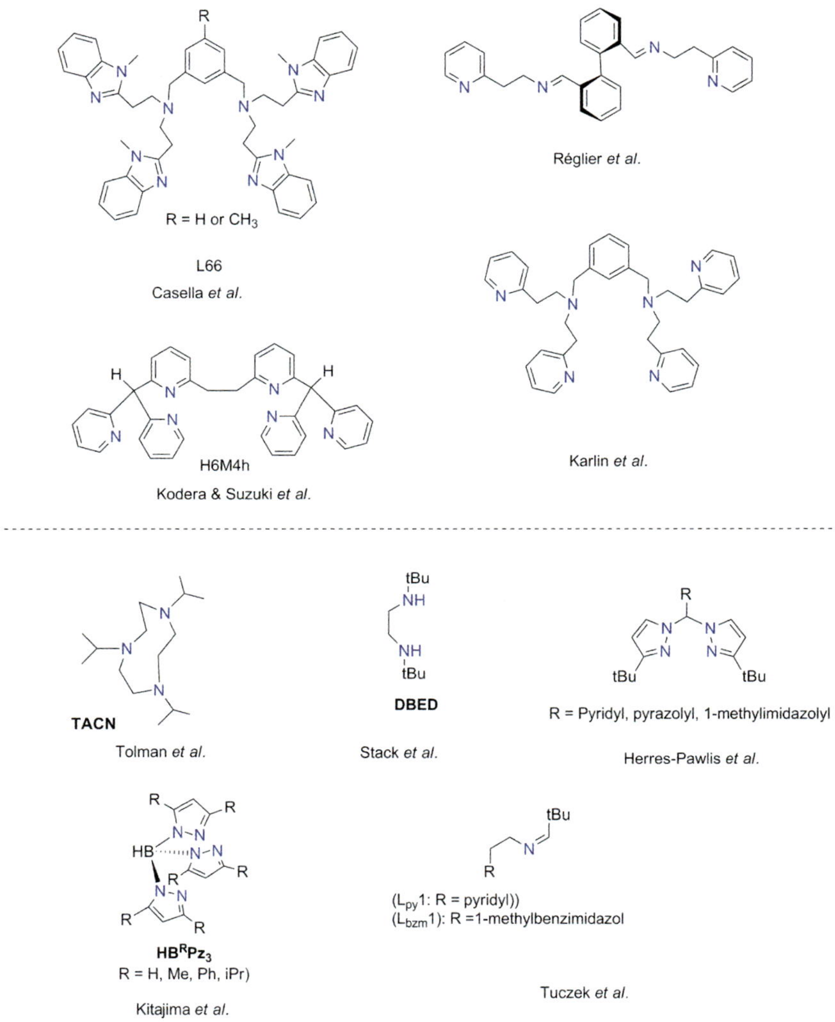

Figure 7. Examples of dinuclear (top) and mononuclear (down) ligands used to synthesize Cu_2/O_2 species proposed as reactive intermediates mediating hydroxylation of phenolic substrates.

m-xylene spacer (Figure 7), obtained hydroxylation in the 2-position after the binding of dioxygen and formation of a μ-η^2:η^2 peroxodicopper(II) intermediate.[91] Further works[92–94] supported the formation of an the $^{\mathbf{S}}\mathbf{P}$ species as the hydroxylating species of phenolic compounds by an electrophilic aromatic substitution and the subsequent two-electron oxidation to *o*-quinones in TYRs. Later, existence of a possible equilibrium between $^{\mathbf{S}}\mathbf{P}$ unit and a bis(μ-oxo)dicopper(III) (**O**) species, described by Tolman,[95] brings **O** as an alternative intermediate for the aromatic hydroxylation by TYRs.[96] Using DBED (*N,N′*-di-tert-butyl-ethylene diamine) as the ligand (Figure 7), Stack and coworkers demonstrated that after coordination of a phenolate substrate on the corresponding μ-η^2:η^2 peroxodicopper, a phenolate-bound bis(μ-oxo) dicopper(III) complex is observed leading to the hydroxylation and the subsequent oxidation into the corresponding quinone.[97] More recently, possible interconversion between $^{\mathbf{T}}\mathbf{P}$ and **O** species has been observed[98] as well as between hydroperoxo and $^{\mathbf{S}}\mathbf{P}$ species.[99] A few assemblies with the *cis*-1,2-peroxo ($^{\mathbf{C}}\mathbf{P}$)[100] or the μ-oxo (Cu_2O)[101] core have been characterized and along with $^{\mathbf{T}}\mathbf{P}$ or hydroperoxo species, they have no direct relevance to tyrosinase enzymes. In fact, the μ-η^2:η^2 peroxodicopper $^{\mathbf{S}}\mathbf{P}$ is the only intermediate identified by spectroscopy and X-Ray crystallography on TYRs.[26,102] Nevertheless the question about the exact nature of active species in TYRs remains under discussion as the facile equilibrium observed between μ-η^2:η^2 peroxo dicopper(II) and bis(μ-oxo) dicopper(III) species in synthetic complexes[95] could occurred similarly in the enzyme. Furthermore from model studies, the **O** species has appeared to be competent for performing phenolate hydroxylation.[97,103]

Everything considered, from mono or dinuclear systems, results from different research groups evidence the importance of ligand design on the reactivity: the presence of substituents are required for control of the bulkiness and the electronic properties (donor/acceptor groups) as well as flexibility of the ligand. The majority of bidentate ligands have been used to generate **O** species or equilibrium mixtures of $^{\mathbf{S}}\mathbf{P}$ and **O** species. Tridentate ligands (aromatic, aliphatic or mixed aromatic/aliphatic) provide the stabilization of the $^{\mathbf{S}}\mathbf{P}$ unit.[88] Spectroscopic distinctions of the nature of O_2 binding are mainly based on a combination of UV-Vis, resonance Raman and X-ray absorption (XAS) spectroscopies, at low temperature. Important structural and spectroscopic features for each type of Cu_2/O_2 species are reported in Table 2.

Table 2. Range of metric and spectroscopic parameters of crystallographically characterized **O**, ^{S}**P**, ^{T}**P**, ^{C}**P** species.[86,88,100,104]

Species	**O**	^{S}**P**	^{T}**P**	^{C}**P**
Cu...Cu (Å)	2.7–2.8	3.3–3.6	4.3–4.6	3.8
O–O (Å)	NA	1.4–1.6	1.3–1.5	1.5
λ_{max}, nm (ε, M^{-1} cm^{-1})	313–500 (10,000–28,000) 408–494 (330–28,000)	333–400 (18,000–25,000) 500–550 (900–1500)	515–550 (10,000–15,000)	500 (3000)
ν_{Cu2-O2} (cm^{-1})	590–653	~300	539–586	437
ν_{O-O} (cm^{-1})	NA	721–773	820–839	799

B. Models Complexes and External Catalytic Phenolic Hydroxylation

Contrary to the successful development of copper complexes that mediate the oxidation of organic compounds, synthetic catalysts based on copper complexes acting as mimics of tyrosinase are limited.[105] Two different types of model system have been explored.

1. *Dinuclear Copper Complex Approach*

Regarding the complexes derived from dinuclear ligands, besides copper complexes able to perform stoichiometric hydroxylation, few are able to perform catalytic hydroxylation of phenolic derivatives. Starting from a dicopper(I) complex based on benzimidazole groups (L66, Figure 7), in the presence of dioxygen, Casella performed phenol hydroxylation with a turnover number (TON) of 1.2.[106] Significant catalytic hydroxylation was reported later by Réglier and coworkers using a dicopper(I) complex based on a binaphtyl ligand, with a TON of 16 (Figure 7).[107]

2. *Mononuclear Copper Complex Approach*

The reproduction of tyrosinase activity starting from mononuclear copper(I) complexes to generate an active Cu_2/O_2 species after exposure to O_2 has been more recently developed. The first system was reported by Tuczek starting from a $^{S}\mathbf{P}$ species supported by a bidendate ligand containing a pyridyl arm and an imine function (Figure 7).[108] Significant hydroxylation of phenolic compound was observed with a TON of 22. Modifications of the heterocyclic group led to serie of bidentate mononucleating ligands[109,110] and the most effective catalyst with TON of 31 was obtained from a ligand with a methylbenzimidazol arm (Figure 7).[110] Later, Stack and Herres-Pawlis from a room-temperature stable $^{S}\mathbf{P}$ species generated from a (pyridyl)bis(pyrazolyl)methane ligand (Figure 7) performed catalytic oxygenation and oxidation of phenolic compounds in quinones (TON = 10–15).[111]

C. Models Complexes and Diversification of Phenol Hydroxylation Reactions

1. *Alternative Substrates*

Catalytic oxidations using molecular oxygen are processes of considerable interest compared to the methods based on peroxides as oxidant. In this context, the O_2-insertion ability of TYRs is of great interest and was expanded to alternative substrates such as substituted phenols, *N*-acetyltyrosine methyl ester,[112] estrogenic compound[113] or sulfoxide as thioanisole.[114] Synthetic systems models of TYRs have been successfully applied to these alternative substrates. For instance, the complex based on L66 ligand (Figure 7) catalytically oxidized thioanisole in methylphenylsulfoxide in the presence of dioxygen and with hydroxylamine as sacrificial reductant through a dinuclear peroxodicopper(II) intermediate, with a turnover number of 420.[115] Furthermore, the catalytic hydroxylation abilities toward non biological substrates and beyond that possible in enzymes have been demonstrated on 8-hydroxyquinoline by μ-η^2:η^2 peroxodicopper complexes developed from substituted pyridyl or imidazole bis(pyrazol) methane ligands by Stack and Herres-Pawlis (Figure 7) in a catalytic mode (TONs between 8 and 21).[111,116] When the ligand displays a substituted pyridyl moiety a high velocity in the hydroxylation is observed.

2. *Stereoselective Reactions*

An important aspect for synthetic applications is the development of systems capable of catalytic stereoselective reactions. This feature has been particularly studied using copper salts in combination with chiral adducts or simple copper complexes applied to a wide range of asymmetric transformations.[105] Surprisingly, stereoselective reactions based on tyrosinase models are restricted to the contributions of Casella's group.[117] Several attempts to introduce a group for chiral discrimination into the arms of the dinucleating ligand or in the linker (Figure 8) have been described.[118–122] Presence of a chiral alkyl substituent on the benzymidazole ligand (L55Bu_4*, Figure 8) instead of the achiral group of the parent ligand led to a corresponding copper complex used for stereoselective reactions of enantiomeric mixtures of biologicaly relevant catechols (*L*-DOPA or thioanisoles).[121]

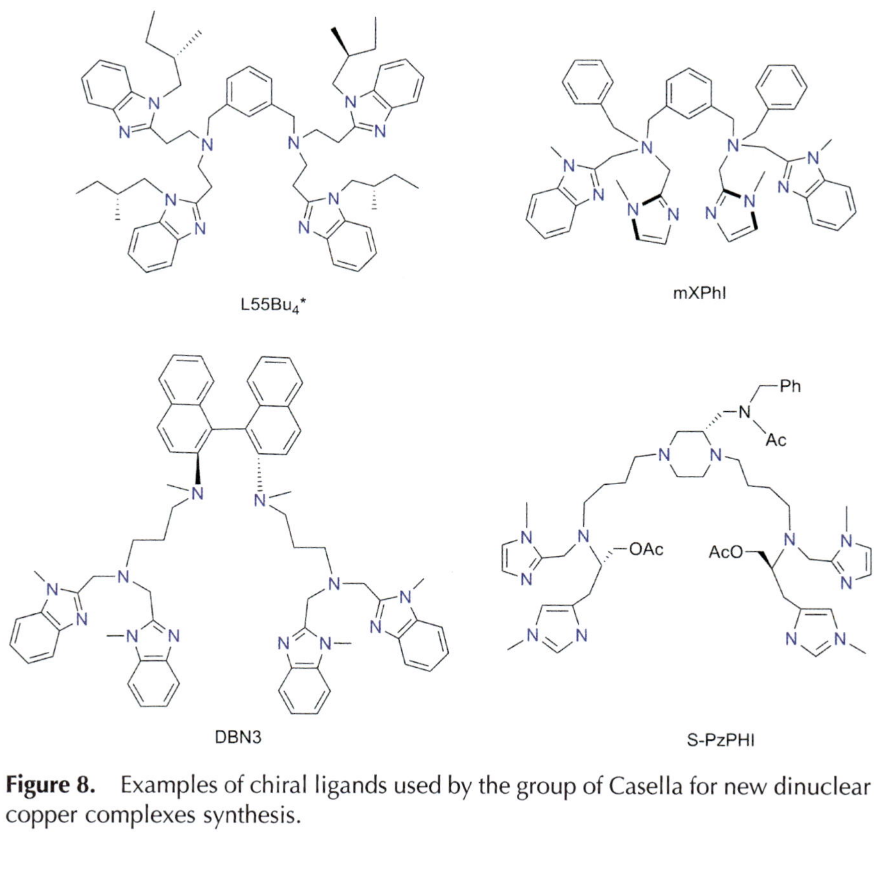

Figure 8. Examples of chiral ligands used by the group of Casella for new dinuclear copper complexes synthesis.

To date the chiral copper(II) complex $[Cu_2(L55Bu_4*)]^{4+}$ displays a low enantioselectivity in oxidation of sulfide[121] compared to complex $[Cu_2(L66)]^{4+}$ (Figure 7) but a marked increase is observed using complex $[Cu_2(mXPhI)]^{4+}$ (Figure 8) with more 40% of e.e with thioanisole or *p*-tolyl methyl sulfide.[117]

3. *Hydroxylation and Oxidative Coupling*

Among the numerous attempts to mimic TYRs activity, in some case catalytic *ortho*-hydroxylation of phenolic compounds is accompanied by oxidative coupling *via* a phenoxyl radical species, leading to corresponding biphenols (Figure 9). Possible control of these competitive pathways has been demonstrated providing catalyst selectivity between *ortho*-hydroxylation and oxidative coupling.[123] Choice of ligand, Cu salt, reaction temperature and presence of desiccant are parameters to consider in such reactions extended to various functionalizations of phenolic compounds.[123,124] Furthermore the formed *ortho*-quinones can be involved in successive reactions as cross-coupling with phenols for synthesis of aryl ethers (Figure 9).[125]

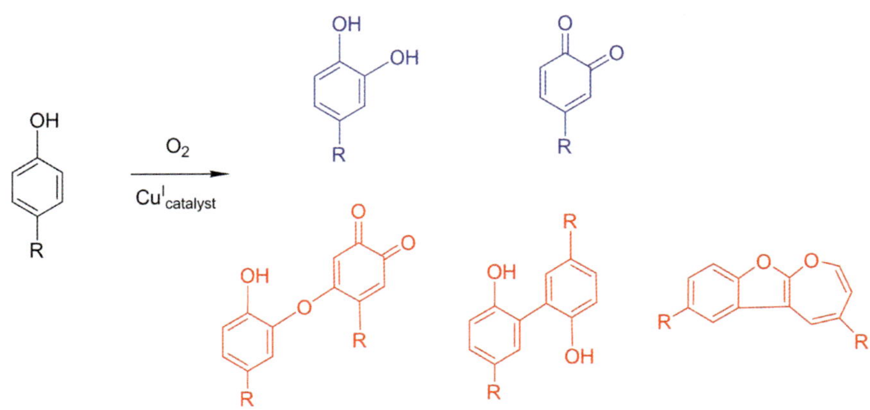

Figure 9. Examples of resulting products from 'tyrosinase like activity' (blue) and oxidative coupling (red) for *para*-substituted phenols.

4. *Oxidation for Phenol Pollutants Degradation*

Degradation of phenol pollutants such as halophenols by a wide variety of polyphenol oxidases and particularly tyrosinases (See part 1.F) have been extensively investigated.[126–129] Extensions of the aerobic oxygenation/oxidative coupling properties of phenols by copper(I) bio-inspired models complexes/O_2 is an interesting alternative avoiding the limitations, of enzyme uses in terms of stability (temperature, narrow range of conditions. etc.). Recently Lumb and co-workers have demonstrated the viability of this approach using a copper(I) complex reported previously supported by DBED as the ligand (Figure 7) for the degradation of various halophenols.[130]

D. Model Complexes as an Approach to Identify Inhibitors of Melanins Biosynthesis

Due to the historical lack of structure data, model complexes have been especially useful in a first approach to understanding the ability and the binding mode of inactivated substrates[131–133] or inhibitors of TYRs. Ty inhibition is a well-known approach against increased production and accumulation of melanins (see part 1.A) and the binuclear copper site represents a well-known target to achieve efficient and selective TYRs inhibition. In this regard, the detailed studies of small molecules binding on model complexes in relation with the inhibition behavior are important to establish binding property/inhibition activity relationships.

For example, interaction studies of known inhibitors such as kojic acid[134] (5-hydroxy-2(hydroxymethyl)-4-pyrone) or substituted pyridine-*N*-oxide[135–137] with TYRs *met*-form models have been reported (Figure 10) and have allowed new advances in: (1) elucidation of their possible binding mode on the dicopper(II) center; (2) knowing how coordination environment of the dicopper center affects the binding of the inhibitor and (3) understanding the topological modifications induced by the binding. Structural and spectroscopic studies associated with theoretical modeling has led to propose new scaffolds,[138,139] which better target the TYRs active site for more selective and potent inhibitor developments.

(a) **Binding mode of 2-hydroxypyridine-N-oxide inhibitor on models of TYRs.**

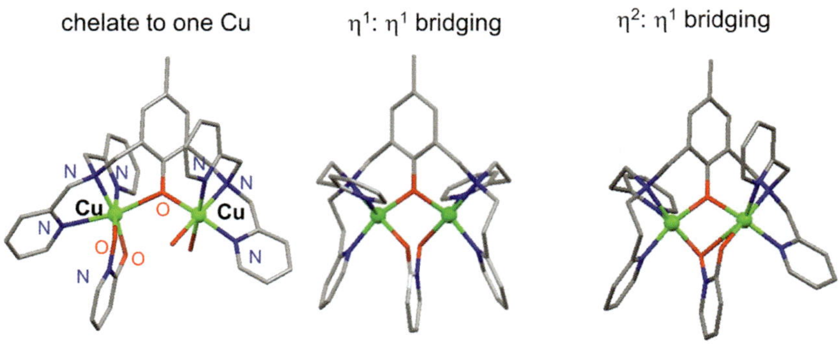

(b) **Main structural and spectroscopic features**

d Cu..Cu = 3.927 Å	d Cu..Cu = 3.271 Å	d Cu..Cu = 3.229 Å
S = 1, EPR active Cu(II) centers in octahedral geometry	*S* = 0, no EPR spectrum Cu(II) centers in square based pyramidal geometry	*S* = 0, no EPR spectrum one Cu(II) in octahedral geometry and one Cu(II) in square based pyramidal geometry

Figure 10. Structural and functional models of TYRs active site in interaction with the inhibitor 2-hydroxypyridine-*N*-oxide.

III. Acknowledgments

This work is supported by the Austrian Science Fund (FWF) P25217 and P29144 and by the French National Research Agency in the framework of the Labex ARCANE (ANR-11-LABX-0003-01) and of the "Investissements d'avenir" program Cosmethics (ANR-15-IDEX-02).

IV. References

1. Körner, A.; Pawelek, J. *Science* **1982**, *217*, 1163.
2. Mayer, A. M. *Phytochemistry* **2006**, *67*, 2318.
3. Claus, H.; Decker, H. *Syst. Appl. Microbiol.* **2006**, *29*, 3.
4. Mason, H. S. *J. Biol. Chem.* **1948**, *172*, 83.
5. Rodríguez-López, J. N.; Tudela, J.; Varón, R.; García-Carmona, F.; García-Cánovas, F. *J. Biol. Chem.* **1992**, *267*, 3801.

6. Ramsden, C. A.; Riley, P. A. *Bioorg. Med. Chem.* **2014**, *22*, 2388.
7. Felton, G. W.; Donato, K.; Vecchio, R. J. D.; Duffey, S. S. *J. Chem. Ecol.* **1989**, *15*, 2667.
8. Thipyapong, P.; Hunt, M. D.; Steffens, J. C. *Planta* **2004**, *220*, 105.
9. Anderson, J. V.; Fuerst, E. P.; Tedrow, T.; Hulke, B.; Kennedy, A. C. *J. Agric. Food Chem.* **2010**, *58*, 10597.
10. Fuerst, E. P.; Okubara, P. A.; Anderson, J. V.; Morris, C. F. *Front. Plant Sci.* **2014**, *5*, 689.
11. Largeteau, M. L.; Latapy, C.; Minvielle, N.; Regnault-Roger, C.; Savoie, J.-M. *Appl. Microbiol. Biotechnol.* **2010**, *85*, 1499.
12. Sánchez-Ferrer, A.; Rodríguez-López, J. N.; García-Cánovas, F.; García-Carmona, F. *Biochim. Biophys. Acta* **1995**, *1247*, 1.
13. Sugumaran, M., *Pigment Cell Res.* **2002**, *15*, 2.
14. Kitajima, N.; Fujisawa, K.; Morooka, Y.; Toriumi, K. *J. Am. Chem. Soc.* **1989**, *111*, 8975.
15. Pretzler, M.; Bijelic, A.; Rompel, A. In *Reference Module in Chemistry, Molecular Sciences and Chemical Engineering*; Elsevier **2015**, doi:10.1016/B978-0-12-409547-2.11521-5.
16. Makino, N.; Mason, H. S. *J. Biol. Chem.* **1973**, *248*, 5731.
17. Rompel, A.; Fischer, H.; Büldt-Karentzopoulos, K.; Meiwes, D.; Zippel, F.; Nolting, H.-F.; Hermes, C.; Krebs, B.; Witzel, H. *J. Inorg. Biochem.* **1995**, *59*, 715.
18. Yamazaki, S.; Itoh, S. *J. Am. Chem. Soc.* **2003**, *125*, 13034.
19. Eickman, N. C.; Himmelwright, R. S.; Solomon, E. I. *Proc. Natl. Acad. Sci. U. S. A.* **1979**, *76*, 2094.
20. Zou, C.; Huang, W.; Zhao, G.; Wan, X.; Hu, X.; Jin, Y.; Li, J.; Liu, J. *Mol. Basel Switz.* **2017**, *22,* 1836.
21. Ros, J.; Rodríguez-López, J.; García-Cánovas, F. *Biochim. Biophys. Acta BBA — Protein Struct. Mol. Enzymol.* **1994**, *1204*, 33.
22. van Gelder, C. W.; Flurkey, W. H.; Wichers, H. J. *Phytochemistry* **1997**, *45*, 1309.
23. Pretzler, M.; Rompel, A. *Inorg. Chim. Acta* **2018**, *481*, 25.
24. Moe, L. L.; Maekawa, S.; Kawamura-Konishi, Y. *Appl. Microbiol. Biotechnol.* **2015**, *99*, 5499.
25. Fujieda, N.; Murata, M.; Yabuta, S.; Ikeda, T.; Shimokawa, C.; Nakamura, Y.; Hata, Y.; Itoh, S. *J. Biol. Inorg. Chem.* **2013**, *18*, 19.
26. Matoba, Y.; Kumagai, T.; Yamamoto, A.; Yoshitsu, H.; Sugiyama, M. *J. Biol. Chem.* **2006**, *281*, 8981.
27. Sendovski, M.; Kanteev, M.; Ben-Yosef, V. S.; Adir, N.; Fishman, A. *J. Mol. Biol.* **2011**, *405*, 227.
28. Lu, A.; Zhang, Q.; Zhang, J.; Yang, B.; Wu, K.; Xie, W.; Luan, Y.-X.; Ling, E. *Front. Physiol.* **2014**, *5*, 252.
29. Rast, D. M.; Baumgartner, D.; Mayer, C.; Hollenstein, G. O. *Phytochemistry* **2003**, *64*, 339.

30. Leu, W. M.; Chen, L. Y.; Liaw, L. L.; Lee, Y. H. *J. Biol. Chem.* **1992**, *267*, 20108.
31. Tsai, T.-Y.; Lee, Y.-H. W. *J. Biol. Chem.* **1998**, *273*, 19243.
32. García-Borrón, J. C.; Solano, F. *Pigment Cell Res.* **2002**, *15*, 162.
33. Espín, J. C.; van Leeuwen, J.; Wichers, H. J. *J. Agric. Food Chem.* **1999**, *47*, 3509.
34. Halaouli, S.; Asther, M.; Sigoillot, J.-C.; Hamdi, M.; Lomascolo, A. *J. Appl. Microbiol.* **2006**, *100*, 219.
35. Mauracher, S. G.; Molitor, C.; Al-Oweini, R.; Kortz, U.; Rompel, A. *Acta Crystallogr. D Biol. Crystallogr.* **2014**, *70*, 2301.
36. Mauracher, S. G.; Molitor, C.; Michael, C.; Kragl, M.; Rizzi, A.; Rompel, A. *Phytochemistry* **2014**, *99*, 14.
37. Zekiri, F.; Molitor, C.; Mauracher, S. G.; Michael, C.; Mayer, R. L.; Gerner, C.; Rompel, A. *Phytochemistry* **2014**, *101*, 5.
38. Pretzler, M.; Bijelic, A.; Rompel, A. *Sci. Rep.* **2017**, *7*, 1810.
39. Derardja, A. E.; Pretzler, M.; Kampatsikas, I.; Barkat, M.; Rompel, A. *J. Agric. Food Chem.* **2017**, *65*, 8203.
40. Seo, S.-Y.; Sharma, V. K.; Sharma, N. *J. Agric. Food Chem.* **2003**, *51*, 2837.
41. Selinheimo, E.; NiEidhin, D.; Steffensen, C.; Nielsen, J.; Lomascolo, A.; Halaouli, S.; Record, E.; O'Beirne, D.; Buchert, J.; Kruus, K. *J. Biotechnol.* **2007**, *130*, 471.
42. Kampatsikas, I.; Bijelic, A.; Pretzler, M.; Rompel, A. *Sci. Rep.* **2017**, *7*, 8860.
43. Fujita, Y.; Uraga, Y.; Ichisima, E. *Biochim. Biophys. Acta* **1995**, *1261*, 151.
44. Ichishima, E.; Maeba, H.; Amikura, T.; Sakata, H. *Biochim. Biophys. Acta* **1984**, *786*, 25.
45. Jiménez-Cervantes, C.; García-Borrón, J. C.; Lozano, J. A.; Solano, F. *Biochim. Biophys. Acta* **1995**, *1243*, 421.
46. Okot-Kotber, M.; Liavoga, A.; Yong, K.-J.; Bagorogoza, K. *J. Agric. Food Chem.* **2002**, *50*, 2410.
47. Espín, J. C.; Wichers, H. J. *J. Agric. Food Chem.* **1999**, *47*, 3518.
48. Li, Y.; Wang, Y.; Jiang, H.; Deng, J. *Proc. Natl. Acad. Sci. U. S. A.* **2009**, *106*, 17002.
49. Ismaya, W. T.; Rozeboom, H. J.; Weijn, A.; Mes, J. J.; Fusetti, F.; Wichers, H. J.; Dijkstra, B. W. *Biochemistry* **2011**, *50*, 5477.
50. Fujieda, N.; Yabuta, S.; Ikeda, T.; Oyama, T.; Muraki, N.; Kurisu, G.; Itoh, S., *J. Biol. Chem.* **2013**, *288*, 22128.
51. Masuda, T.; Momoji, K.; Hirata, T.; Mikami, B. *FEBS J.* **2014**, *281*, 2659.
52. Bijelic, A.; Pretzler, M.; Molitor, C.; Zekiri, F.; Rompel, A. *Angew. Chem. Int. Ed.* **2015**, *54*, 14677.
53. Lai, X.; Wichers, H. J.; Soler-Lopez, M.; Dijkstra, B. W. *Angew. Chem. Int. Ed. Engl.* **2017**, *56*, 9812.
54. Decker, H.; Tuczek, F. *Angew. Chem. Int. Ed.* **2017**, *56*, 14352.
55. Lai, X.; Wichers, H. J.; Soler-Lopez, M.; Dijkstra, B. W. *Chem. — Eur. J.* **2018**, *24*, 47.
56. Lerch, K. *J. Biol. Chem.* **1982**, *257*, 6414.

57. Klabunde, T.; Eicken, C.; Sacchettini, J. C.; Krebs, B. *Nat. Struct. Biol.* **1998**, *5*, 1084.
58. Solem, E.; Tuczek, F.; Decker, H. *Angew. Chem. Int. Ed.* **2016**, *55*, 2884.
59. Decker, H.; Solem, E.; Tuczek, F. *Inorg. Chim. Acta* **2018**, *481*, 32.
60. Goldfeder, M.; Kanteev, M.; Isaschar-Ovdat, S.; Adir, N.; Fishman, A. *Nat. Commun.* **2014**, *5*, 4505.
61. Quist, D. A.; Diaz, D. E.; Liu, J. J.; Karlin, K. D. *J. Biol. Inorg. Chem.* **2017**, *22*, 253.
62. Magnus, K. A.; Hazes, B.; Ton-That, H.; Bonaventura, C.; Bonaventura, J.; Hol, W. G. J. *Proteins Struct. Funct. Bioinforma.* **1994**, *19*, 302.
63. Olivares, C.; García-Borrón, J. C.; Solano, F. *Biochemistry* **2002**, *41*, 679.
64. Molitor, C.; Mauracher, S. G.; Rompel, A. *Proc. Natl. Acad. Sci. U. S. A.* **2016**, *113*, E1806.
65. Kanteev, M.; Goldfeder, M.; Fishman, A. *Protein Sci. Publ. Protein Soc.* **2015**, *24*, 1360.
66. Prexler, S. M.; Singh, R.; Moerschbacher, B. M.; Dirks-Hofmeister, M. E. *Plant Mol. Biol.* **2018**, *96*, 151.
67. Kampatsikas, I.; Bijelic, A.; Pretzler, M.; Rompel, A. *Acta Crystallogr. Sect. F Struct. Biol. Commun.* **2017**, *73*, 491.
68. Deri, B.; Kanteev, M.; Goldfeder, M.; Lecina, D.; Guallar, V.; Adir, N.; Fishman, A. *Sci. Rep.* **2016**, *6*, 34993.
69. Mauracher, S. G.; Molitor, C.; Al-Oweini, R.; Kortz, U.; Rompel, A. *Acta Crystallogr. Sect. F Struct. Biol. Commun.* **2014**, *70*, 263.
70. Gelb, D. J.; Oliver, E.; Gilman, S. *Arch. Neurol.* **1999**, *56*, 33.
71. Knowles, W. S. *Acc. Chem. Res.* **1983**, *16*, 106.
72. Knowles, W. S. *J. Chem. Educ.* **1986**, *63*, 222.
73. Chávez-Béjar, M. I.; Báez-Viveros, J. L.; Martínez, A.; Bolívar, F.; Gosset, G. *Process Biochem.* **2012**, *47*, 1017.
74. Satô, M. *Phytochemistry* **1969**, *8*, 353.
75. Munoz-Munoz, J. L.; Garcia-Molina, F.; García-Ruiz, P. A.; Varon, R.; Tudela, J.; García-Cánovas, F.; Rodriguez-Lopez, J. N. *Biochim. Biophys. Acta BBA — Proteins Proteomics* **2009**, *1794*, 244.
76. Mateo, C.; Palomo, J. M.; Fernandez-Lorente, G.; Guisan, J. M.; Fernandez-Lafuente, R. *Enzyme Microb. Technol.* **2007**, *40*, 1451.
77. Ikehata, K.; Nicell, J. A. *Biotechnol. Prog.* **2000**, *16*, 533.
78. Sassolas, A.; Blum, L. J.; Leca-Bouvier, B. D. *Biotechnol. Adv.* **2012**, *30*, 489.
79. Ito, S.; Kato, T.; Shinpo, K.; Fujita, K. *Biochem. J.* **1984**, *222*, 407.
80. Bittner, S. *Amino Acids* **2006**, *30*, 205.
81. Dunnewind, B.; Van, V.; Orsel, R. *J. Cereal Sci.* **2002**, *36*, 357.
82. Thalmann, C. R.; Lötzbeyer, T. *Eur. Food Res. Technol.* **2002**, *214*, 276.
83. Halaouli, S.; Asther, M.; Kruus, K.; Guo, L.; Hamdi, M.; Sigoillot, J.-C.; Asther, M.; Lomascolo, A. *J. Appl. Microbiol.* **2005**, *98*, 332.
84. Zaidi, K. U.; Ali, A. S.; Ali, S. A.; Naaz, I. *Biochem. Res. Int.* **2014**, *2014*, 1.

85. Solomon, E. I.; Heppner, D. E.; Johnston, E. M.; Ginsbach, J. W.; Cirera, J.; Qayyum, M.; Kieber-Emmons, M. T.; Kjaergaard, C. H.; Hadt, R. G.; Tian, L. *Chem. Rev.* **2014**, *114*, 3659.
86. Elwell, C. E.; Gagnon, N. L.; Neisen, B. D.; Dhar, D.; Spaeth, A. D.; Yee, G. M.; Tolman, W. B. *Chem. Rev.* **2017**, *117*, 2059.
87. Rolff, M.; Schottenheim, J.; Decker, H.; Tuczek, F. *Chem. Soc. Rev.* **2011**, *40*, 4077.
88. Mirica, L. M.; Ottenwaelder, X.; Stack, T. D. P. *Chem. Rev.* **2004**, *104*, 1013.
89. Itoh, S.; Fukuzumi, S. *Acc. Chem. Res.* **2007**, *40*, 592.
90. Hamann, J. N.; Herzigkeit, B.; Jurgeleit, R.; Tuczek, F. *Coord. Chem. Rev.* **2017**, *334*, 54.
91. Pidcock, E.; Obias, H. V.; Zhang, C. X.; Karlin, K. D.; Solomon, E. I. *J. Am. Chem. Soc.* **1998**, *120*, 7841.
92. Itoh, S.; Kumei, H.; Taki, M.; Nagatomo, S.; Kitagawa, T.; Fukuzumi, S. *J. Am. Chem. Soc.* **2001**, *123*, 6708.
93. Kodera, M.; Katayama, K.; Tachi, Y.; Kano, K.; Hirota, S.; Fujinami, S.; Suzuki, M. *J. Am. Chem. Soc.* **1999**, *121*, 11006.
94. Mandal, S.; Mukherjee, J.; Lloret, F.; Mukherjee, R. *Inorg. Chem.* **2012**, *51*, 13148.
95. Halfen, J. A.; Mahapatra, S.; Wilkinson, E. C.; Kaderli, S.; Young, V. G.; Que, L.; Zuberbühler, A. D.; Tolman, W. B. *Science* **1996**, *271*, 1397.
96. Holland, P. L.; Rodgers, K. R.; Tolman, W. B. *Angew. Chem. Int. Ed Engl.* **1999**, *38*, 1139.
97. Mirica, L. M.; Vance, M.; Rudd, D. J.; Hedman, B.; Hodgson, K. O.; Solomon, E. I.; Stack, T. D. P. *Science* **2005**, *308*, 1890.
98. Kieber-Emmons, M. T.; Ginsbach, J. W.; Wick, P. K.; Lucas, H. R.; Helton, M. E.; Lucchese, B.; Suzuki, M.; Zuberbühler, A. D.; Karlin, K. D.; Solomon, E. I. *Angew. Chem. Int. Ed Engl.* **2014**, *53*, 4935.
99. Kindermann, N.; Dechert, S.; Demeshko, S.; Meyer, F. *J. Am. Chem. Soc.* **2015**, *137*, 8002.
100. Dalle, K. E.; Gruene, T.; Dechert, S.; Demeshko, S.; Meyer, F. *J. Am. Chem. Soc.* **2014**, *136*, 7428.
101. Haack, P.; Kärgel, A.; Greco, C.; Dokic, J.; Braun, B.; Pfaff, F. F.; Mebs, S.; Ray, K.; Limberg, C. *J. Am. Chem. Soc.* **2013**, *135*, 16148.
102. Solomon, E. I.; Sundaram, U. M.; Machonkin, T. E. *Chem. Rev.* **1996**, *96*, 2563.
103. Company, A.; Palavicini, S.; Garcia-Bosch, I.; Mas-Ballesté, R.; Que, L.; Rybak-Akimova, E. V.; Casella, L.; Ribas, X.; Costas, M. *Chem. — Eur. J.* **2008**, *14*, 3535.
104. Goswami, V. E.; Walli, A.; Förster, M.; Dechert, S.; Demeshko, S.; C. Holthausen, M.; Meyer, F. *Chem. Sci.* **2017**, *8*, 3031.
105. Allen, S. E.; Walvoord, R. R.; Padilla-Salinas, R.; Kozlowski, M. C. *Chem. Rev.* **2013**, *113*, 6234.
106. Casella, L.; Gullotti, M.; Radaelli, R.; Di Gennaro, P. *J. Chem. Soc. Chem. Commun.* **1991**, 22, 1611.
107. Réglier, M.; Jorand, C.; Waegell, B. *J. Chem. Soc. Chem. Commun.* **1990**, 24, 1752.

108. Rolff, M.; Schottenheim, J.; Peters, G.; Tuczek, F. *Angew. Chem. Int. Ed.* **2010**, *49*, 6438.
109. Hamann, J. N.; Tuczek, F. *Chem. Commun.* **2014**, *50*, 2298.
110. Schottenheim, J.; Fateeva, N.; Thimm, W.; Krahmer, J.; Tuczek, F. *Z. Anorg. Allg. Chem.* **2013**, *639*, 1491.
111. Hoffmann, A.; Citek, C.; Binder, S.; Goos, A.; Rübhausen, M.; Troeppner, O.; Ivanović-Burmazović, I.; Wasinger, E. C.; Stack, T. D. P.; Herres-Pawlis, S. *Angew. Chem. Int. Ed.* **2013**, *52*, 5398.
112. Kahn, V.; Ben-Shalom, N. *Pigment Cell Res.* **1998**, *11*, 24.
113. Westerfeld, W. W. *Biochem. J.* **1940**, *34*, 51.
114. Pievo, R.; Gullotti, M.; Monzani, E.; Casella, L. *Biochemistry* **2008**, *47*, 3493.
115. Gamba, I.; Palavicini, S.; Monzani, E.; Casella, L. *Chem. — Eur. J.* **2009**, *15*, 12932.
116. Liebhäuser, P.; Keisers, K.; Hoffmann, A.; Schnappinger, T.; Sommer, I.; Thoma, A.; Wilfer, C.; Schoch, R.; Stührenberg, K.; Bauer, M., *et al. Chem. — Eur. J.* **2017**, *23*, 12171.
117. Presti, E. L.; Monzani, E.; Santagostini, L.; Casella, L. *Inorg. Chim. Acta* **2018**, *481*, 47.
118. Gullotti, M.; Santagostini, L.; Pagliarin, R.; Granata, A.; Casella, L. *J. Mol. Catal. Chem.* **2005**, *235*, 271.
119. Mutti, F. G.; Gullotti, M.; Casella, L.; Santagostini, L.; Pagliarin, R.; Andersson, K. K.; Iozzi, M. F.; Zoppellaro, G. *Dalton Trans.* **2011**, *40*, 5436.
120. Mimmi, C. M.; Gullotti, M.; Santagostini, L.; Battaini, G.; Monzani, E.; Pagliarin, R.; Zoppellaro, G.; Casella, L. *Dalton Trans.* **2004**, 2192.
121. Perrone, M. L.; Lo Presti, E.; Dell'Acqua, S.; Monzani, E.; Santagostini, L.; Casella, L. *Eur. J. Inorg. Chem.* **2015**, *2015*, 3493.
122. Perrone, M. L.; Salvadeo, E.; Lo Presti, E.; Pasotti, L.; Monzani, E.; Santagostini, L.; Casella, L. *Dalton Trans.* **2017**, *46*, 4018.
123. Esguerra, K. V. N.; Fall, Y.; Petitjean, L.; Lumb, J.-P. *J. Am. Chem. Soc.* **2014**, *136*, 7662.
124. Askari, M. S.; Esguerra, K. V. N.; Lumb, J.-P.; Ottenwaelder, X. *Inorg. Chem.* **2015**, *54*, 8665.
125. Huang, Z.; Lumb, J.-P. *Angew. Chem. Int.* **2016**, *55*, 11543.
126. Marino, S. M.; Fogal, S.; Bisaglia, M.; Moro, S.; Scartabelli, G.; De Gioia, L.; Spada, A.; Monzani, E.; Casella, L.; Mammi, S., *et al. Arch. Biochem. Biophys.* **2011**, *505*, 67.
127. Wada, S.; Ichikawa, H.; Tastsumi, K. *Biotechnol. Bioeng.* **1995**, *45*, 304.
128. Wada, S.; Ichikawa, H.; Tatsumi, K. *Biotechnol. Bioeng.* **1993**, *42*, 854.
129. Ba, S.; Vinoth, V. K. *Crit. Rev. Biotechnol.* **2017**, *37*, 819–832.
130. Esguerra, K. V. N.; Fall, Y.; Lumb, J.-P. *Inorg. Chim. Acta* **2018**, *481*, 197.
131. Ackermann, J.; Meyer, F.; Kaifer, E.; Pritzkow, H. *Chem. — Eur. J.* **2002**, *8*, 247.
132. Karlin, K. D.; Gultneh, Y.; Nicholson, T.; Zubieta, J. *Inorg. Chem.* **1985**, *24*, 3725.

133. Battaini, G.; Monzani, E.; Casella, L.; Santagostini, L.; Pagliarin, R. *J. Biol. Inorg. Chem.* **2000**, *5*, 262.
134. Bochot, C.; Gouron, A.; Bubacco, L.; Milet, A.; Philouze, C.; Réglier, M.; Serratrice, G.; Jamet, H.; Belle, C. *Chem. Commun.* **2014**, *50*, 308.
135. Peyroux, E.; Ghattas, W.; Hardré, R.; Giorgi, M.; Faure, B.; Simaan, A. J.; Belle, C.; Réglier, M. *Inorg. Chem.* **2009**, *48*, 10874.
136. Orio, M.; Bochot, C.; Dubois, C.; Gellon, G.; Hardré, R.; Jamet, H.; Luneau, D.; Philouze, C.; Réglier, M.; Serratrice, G., *et al. Chem. — Eur. J.* **2011**, *17*, 13482.
137. Bochot, C.; Favre, E.; Dubois, C.; Baptiste, B.; Bubacco, L.; Carrupt, P.-A.; Gellon, G.; Hardré, R.; Luneau, D.; Moreau, Y., *et al. Chem. — Eur. J.* **2013**, *19*, 3655.
138. Dubois, C.; Haudecoeur, R.; Orio, M.; Belle, C.; Bochot, C.; Boumendjel, A.; Hardré, R.; Jamet, H.; Réglier, M. *Chem. Bio. Chem.* **2012**, *13*, 559.
139. Haudecoeur, R.; Carotti, M.; Gouron, A.; Maresca, M.; Buitrago, E.; Hardré, R.; Bergantino, E.; Jamet, H.; Belle, C.; Réglier, M., *et al. ACS Med. Chem. Lett.* **2017**, *8*, 55.

8 Modeling the Mononuclear Copper Monooxygenase Active Site

Alda Lisa Concia, Alessia Munzone, Maria-Chrysanthi Kafentzi, Amélie Kochem, Marius Réglier, Christophe Decroos and A. Jalila Simaan*

Aix Marseille Univ, CNRS, Centrale Marseille, iSm2, Marseille, France
*jalila.simaan@univ-amu.fr

Table of Contents

List of Abbreviations

4-ampy	4-aminopyridine
A	alanine
AA	auxiliary activities
ACN	acetonitrile
$AcrH_2$	*N*-methyl-9,10-dihydroacridine
An	*Aspergillus nidulans*
Ao	*Aspergillus oryzae*
BNAH	1-benzyl-1,4-dihydronicotinamide
BP	bispidine
BPPA	bis(6-pivalamide-2-pyridylmethyl)(2-pyridylmethyl)amine ligand
BzImH	1,3-dimethyl-2,3-dihydrobenzimidazole
CAZy	carbohydrate-active enzymes
CBM	carbohydrate-binding module
CDH	cellobiose dehydrogenase
CT	charge transfer
DBE	bond dissociation energy
DBED	*N,N′*-di-tert-butyl-ethylene diamine
DCM	dichloromethane
DFT	density functional theory

DHA	dihydroanthracene
DβM	dopamine β-monooxygenase
EPR	electron paramagnetic resonance
ESI	electrospray ionization
ET	electron transfer
EXAFS	X-ray absorption fine structure
GH	glycoside hydrolase
GMC	glucose-methanol-choline
GOx	glucose oxidase
H	histidine
HAA	H-atom abstraction
HAT	H-atom transfer
HYSCORE	hyperfine sub-level correlation
IYT	*N*-α-acetyl-3,5-dioiodotyrosyl-D-threonine
Jd	*Jonesia denitrificans*
KIE	kinetic isotope effect
LPMO	lytic polysaccharide monooxygenase
LRET	long-range electron transfer
Ls	*Lentinus similis*
Mes	mesityl
MOX	monooxygenase X
MPPA	bis(pyrid-2-ylmethyl){[6-(pivalamido)pyrid-2-yl]meth-yl} amine
MS	mass spectroscopy
Mt	*Myceliophthora thermophila*
N	asparagine
PAL	peptidyl-α-hydroxyglycine α-amidating lyase
PAM	peptidylglycine α-amidating monooxygenase
Pc	*Pycnoporus coccineus*
PCET	proton-coupled electron transfer
PDB	protein data bank
PDH	pyranose dehydrogenase
Ph	phenyl
phen	phenanthroline
PHM	peptidylglycine α-hydroxylating monooxygenase
PMO	polysaccharide monooxygenase

PT	proton transfer
QM/MM	quantum mechanical/molecular mechanical
R_3tren	tris(2-aminoethyl)amine
rR	resonance Raman
S	serine
Sav	streptavidin
Sm	*Serratia marcescens*
Td	*Thermobia domestica*
TEMPO-H	2,2,6,6-tetramethylpiperidine-*N*-hydroxide
THF	tetrahydrofuran
TMG	tetramethyl guanidine
TMPA	tris(2-pyridylmethyl)amine
tren	tris(2-aminoethyl)amine
TβM	tyramine β-monooxygenase
UV-vis.	ultraviolet-visible
XAS	X-ray absorption spectroscopy
YVG	NH_2-tyrosine-valine-glycine-COOH

I. Introduction

Facing the 21st century environmental issues, the need to develop chemical processes respectful of our environment becomes urgent. In accordance with the principles of green chemistry,[1] effort of chemists should particularly focus on reactions having energy efficiencies (mild conditions of temperature and pressure) and using catalysis in safer solvents (water).[2] Because living systems perform chemical transformation in mild conditions, inspiration by nature and its enzymatic toolkit must guide chemists in the search of sustainable chemical transformations. In this context, the catalyzed-activation of dioxygen for the functionalization of C–H bonds to form alcohols represents one of the major challenges in chemistry.[3,4] Monooxygenases are enzymes that catalyze the insertion of a single O-atom from dioxygen into an organic substrate.[5,6] To perform this transformation, monooxygenases can use organic (flavin, pterin, etc.) or inorganic cofactors (iron, copper, etc.) to activate dioxygen and overcome its spin-forbidden reaction with organic substrates. Because, monooxygenases function/work in mild conditions and are highly chemo-, regio-, and/or enantioselective, they deserve a peculiar attention from chemists in the development of modern catalysts. Among monooxygenases that contain an inorganic cofactor, those containing copper have been less studied than their iron-containing homologs. Depending on the nature of the substrate, living organisms deployed different strategies and used different copper-oxygen species to oxidize organic substrates.[7] As an example, tyrosinases that hydroxylate phenols into catechols/*o*-quinones involve in the catalytic cycle the formation of a transient (μ-η^2:η^2-peroxo)Cu_2(II) species (Chapter 7).[8] Derived from the (μ-η^2:η^2-peroxo)Cu_2(II) core, some reactive Cu_2/O complexes are now well-identified such as the (μ-oxo)Cu_2(III)[9] or the mixed valent (μ-oxo)Cu(II)Cu(III) cores.[10–14] Despite a controversy about the binuclear *versus* mononuclear copper structure of the particulate methane monooxygenases (pMMO),[15] it was proposed that these species could be involved in the transformation of methane into methanol.[16–18]

Mononuclear copper-oxygen species are involved in other copper-containing monooxygenase such as dopamine/tyramine β-monooxygenases (DβM/TβM), peptidylglycine α-hydroxylating monooxygenases (PHM) and lytic polysaccharide monooxygenases (LPMOs). For these enzymes, several Cu/O species have been postulated. Thanks to the use of a bioinorganic approach, some of these species have been prepared and characterized both in enzyme and in model complexes. We will describe these achievements in the following chapter after the description of the corresponding enzymatic systems. Given the controversy about the nature of pMMO's active site, this enzymatic system will not be discussed in the present chapter.

II. Dopamine/Tyramine β-Monooxygenases (DβM/TβM) and Peptidylglycine α-Hydroxylating Monooxygenase (PHM), Non-Coupled Binuclear Copper Containing Monooxygenases

Dopamine β-monooxygenase (DβM), tyramine β-monooxygenase (TβM) and peptidylglycine α-hydroxylating monooxygenase (PHM) constitute a unique class of metalloenzymes present in eukaryotes where they catalyze important reactions in neurotransmitter and hormone biosynthetic pathways.[19,20] A fourth family member, monooxygenase X (MOX) was identified from human and chicken genome analysis,[21] but its catalytic activity and function are still unknown to date.[22]

Whereas DβM[23] is active as a tetrameric glycoprotein of 75 kDa/monomer, TβM[24] and PHM[25] are active as monomer of 69 kD and 35 kD, respectively. Despite these large differences in macroscopic features, the three enzymes exhibit strong similarities. They: (i) catalyze the cleavage of dioxygen to form hydroxylated products and water (monooxygenase activity); (ii) require the presence of two independent T2 copper sites[26] per subunit for full activity and (iii) use ascorbate as a two-electron donor. Even though they possess two T2 copper sites, these enzymes are considered as mononuclear because the two copper centers have two distinct roles; one center is involved in electron transfer (ET) and the second one in dioxygen activation and hydroxylation reaction (*vide infra*).

A. Biological Functions of the Non-Coupled Binuclear Copper Containing Monooxygenases

Dopamine β-monooxygenase (DβM) is a mammalian enzyme involved in catecholamine biosynthetic pathway where it transforms dopamine into norepinephrine (Scheme 1a, X = OH).[27] Its insect homolog, Tyramine β-monooxygenase (TβM) transforms tyramine into octopamine in a similar way (Scheme 1a, X = H).[28] Peptidylglycine α-hydroxylating monooxygenase (PHM) participates in the biosynthesis of α-amidated peptide hormones in mammals and insects. PHM is a catalytic domain of peptidylglycine α-amidating monooxygenase (PAM), a bifunctional enzyme involved in the post-translational amidation of C-terminal glycine-extended peptides,[29] a key step in the activation of several neuropeptides and peptide hormones (substance P, oxytocin, neuropeptide Y, adrenomedullin, etc.). PAM contains two independent enzymatic domains: (i) a monooxygenase domain (PHM) that catalyzes the stereospecific hydroxylation of the C-terminal glycine residue into hydroxyglycine and (ii) a lyase domain (peptidyl-α-hydroxyglycine α-amidating lyase, PAL)[30] which in turn transforms hydroxyglycine into amidopeptide and glyoxylic acid (Scheme 1b). The active site of the PAL domain contains a Zn(II) ion coordinated by three histidine residues to which the substrate binds with its α-hydroxyl group.[31] The two isolated domains, either separated through endoproteolytic cleavage or through independent expression, retain their enzymatic activities. Both enzymatic activities have

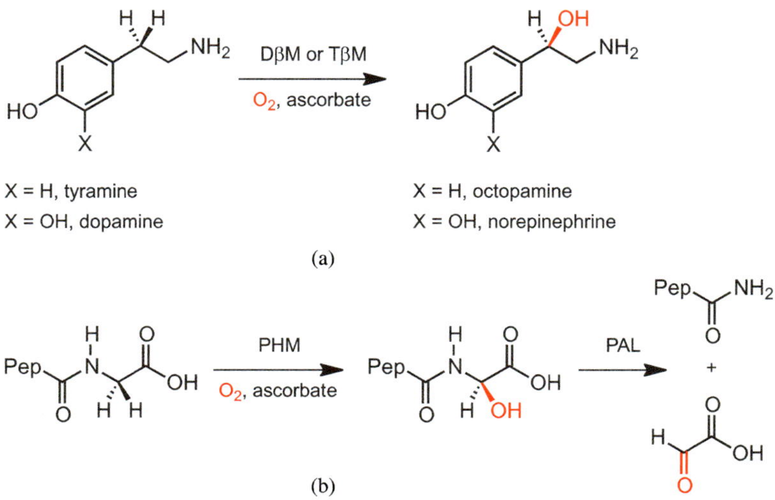

Scheme 1. Reactions catalyzed by mononuclear copper monooxygenases: (a) dopamine β-monooxygenase (DβM) and tyramine β-monooxygenase (TβM); (b) peptidylglycine α-hydroxylating monooxygenase (PHM).

broad substrate scope, since more than 20 α-amidated peptides were characterized coming from the same PAM enzyme.[32]

B. Structures of the Non-Coupled Binuclear Copper Containing Monooxygenases

DβM, TβM and PHM share around 28% sequence identity extending through a common catalytic domain of 270 residues, which includes the conserved two copper ligands (Figure 1).[19,33]

Numerous structural and functional data have been collected for DβM that allow having a quite precise description of its functioning at the molecular level.[19,34] Extended X-ray absorption fine structure (EXAFS) has been used to characterize the ligand environment of the two Cu-ions in both oxidized and reduced forms.[35,36] Without any evidence of back scattering between the two copper sites, the distance between the two copper centers was estimated to exceed 4 Å. The absence of any spin coupling in the electron paramagnetic resonance (EPR) confirmed this feature.[37,38] These spectroscopic data have provided early evidence

Cu_H (Cu_A) domain

```
DβM HIIMYEAIVTEGNEALVHHMEVFQCTN-ESEAFPMFNGPCDSKMKPDRLN 296
TβM HIVQFEPLIR--TPGIVHHMEVFHCEAGEHEEIPLYNG--DCEQLPRAK  342
PHM FVIDFKPRAS---MDTVHHMLLFGCNMPSSTGSYWFCDEGTCTDKAN--- 136
                    ****  * *

DβM YCRHVLAAWALGAKAFYYPEEAGVPLGSSGSSRFLRLEVHYHNPRNIQG- 345
TβM ICSKVMVLWAMGAGTFTYPPEAGLPIGGPGFNPYVRLEVHFNNPEKQSG- 391
PHM ----ILYAWARNAPPTRLPKGVGFRVGGETGSKYFVLQVHYGDISAFRDN 182
            **  *     *   *   *         * **
```

Cu_M (Cu_B) domain

```
DβM RCTQMALPKSGIRIFASQLHTHLTGRKVITVLARDGQQREVVNRDNHYSP 445
TβM DCTRAALPATGIIIFGSQLHTHLRGVRVLTRHFRGEQELREVNRDDYYSN 541
PHM SCQYKMYP---MHVFAYRVHTHHLGKVVSGYRVRNGQWTLIGRQNPQLP- 271
     *     *      *    ***  *  *     *  *

DβM HFQEIRMLKNAVTVHQGDVLITSCTYNTENRTMATVGGFGILEEMCVNYV 495
TβM HFQEMRTLHYKPRVLPGDALVTTCYYNTKDDKTAALGGFSISDEMCVNYI 541
PHM --QAFYPVEHPVDVTFGDILAARCVFTGEGRTEATHIGGTSSDEMCNLYI 319
      *          *  ** *   *         *   *     ***  *
```

Figure 1. Multiple sequence alignment of amino acid sequences. The alignment was generated with Clustal Omega. Sequences for DβM (NP_037290.2) and PHM (NP_037132.2) are from *Rattus norvegicus* and TβM (NP_001284996.1) from *Drosophila melanogaster*. The histidine residues involved in the coordination of copper are highlighted in yellow and methionine residue in blue. The conserved residues are marked with an *asterisk* (*).

against a reactive binuclear center and, instead, have suggested separate functions for the two copper centers that have been confirmed later. On one side, the Cu_A (Cu_H in PHM), surrounded by three histidine residues and a water molecule, has been assigned as the ET site where ascorbate binds and delivers two electrons. On the other side, Cu_B (Cu_M in PHM), coordinated by two *N*-atoms from histidine residues and a water molecule together with an elongated bond to the sulfur atom from a methionine residue has been assigned as the substrate binding and hydroxylation site.[34] These features allow to classify these copper-containing enzymes as non-coupled binuclear copper containing monooxygenases with the respect to their structure but mononuclear with the respect to the mechanism.

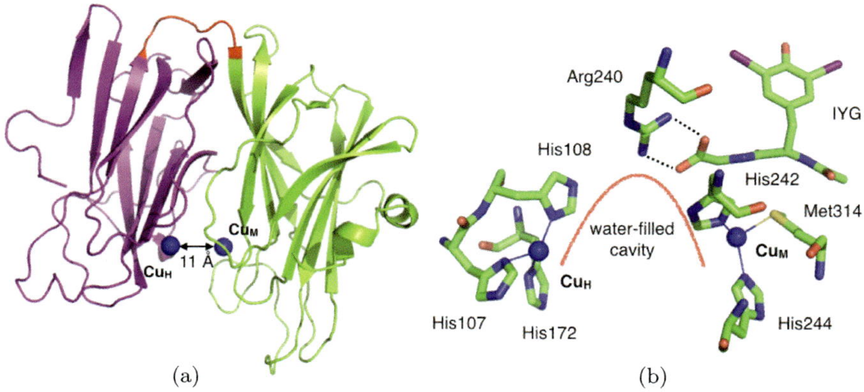

Figure 2. (a) Face view of the PHMcc from Protein Data Bank (PDB code 1OMP) showing the two domains connected by a loop (in red) in which each domain binds a single copper atom (Cu_M domain in green and Cu_H domain in purple) and (b) the active site with bound substrate *N*-α-acetyl-3,5-dioiodotyrosyl-glycine (IYG). The two copper ions are represented as blue spheres.

1. *Crystal Structure of PHM*

The crystal structures of oxidized and reduced PHM forms with or without a bound peptide substrate (*N*-α-acetyl-3,5-dioiodotyrosyl-glycine, IYG) (Figure 2) have confirmed the conclusions of the earlier spectroscopic data.[39] The important features determined from the crystal structures of the oxidized or reduced forms of the PHM catalytic core (PHMcc) are: (i) a two-domain structure connected by a loop in which each domain binds a single copper atom, (ii) a distance of 10.59 Å between the two copper sites, (iii) the absence of closure in the copper binding domains in all enzyme forms studied, and (iv) the identification of a water-filled cavity that is located at the solvent interface and connect the two copper binding domains. Several structures of oxidized and reduced forms of PHMcc with bound carbon monoxide, azide and nitrite on Cu_M center are now available and confirm a differential reactivity between the two copper sites.[40]

2. *Crystal Structure of DβM*

From the accumulated functional and structural data on DβM and by analogy with the PHM structures, a three-dimensional model of human DβM

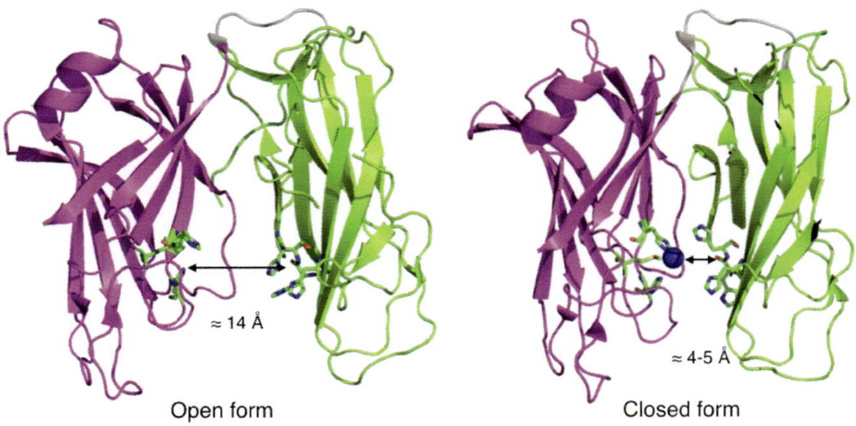

(a) (b)

Figure 3. Structure of the human DβM (PDB code 4ZEL). The Cu_A (Cu_H) domain is displayed in green and the Cu_B (Cu_M) domain in magenta; (a) the open and (c) the closed forms. Only the Cu_B (Cu_M) domain of the closed form contains a Cu-atom.

was reported in 2011.[41] It is only very recently that the crystal structure of the dimeric form of the human DβM at 2.9 Å resolution was published.[42] The overall three-dimensional structure of dimeric human DβM reveals that both chains fold into four domains: (i) the DOMON domain (DβM *N*-terminal domain); (ii) the catalytic Cu_A (Cu_H) and Cu_B (Cu_M) domains and (iii) the C-terminal dimerization domain (Figure 3). In this structure that validates in part the model of Kapoor *el al.*,[41] the catalytic core with the copper-binding domains has the same topology as described for PHM.[39] However, this structure reveals intriguing features. First of all, only one in four copper centers of the catalytic copper domains, the Cu_B (Cu_M), is occupied, the three other sites are free of any Cu-atoms. Secondly, the catalytic domains present both an open active site similar to that of PHM with a copper binding site distance estimated to 14 Å and a closed form with a copper binding site distance estimated to 4–5 Å similar to that found in coupled binuclear copper sites.[26] The closed form results from reorganization of the Cu_A (Cu_H) and Cu_B (Cu_M) domains around a flexible connecting loop (Figure 3, loop in gray). This finding revives the debate on the ET between the two copper centers over a distance of 11 Å.[19,43–45] Based on this DβM structure, Christensen *et al.* reinvestigated the mechanism of dioxygen activation catalyzed by DβM.[42] They propose that during catalysis the two subunits could alternate between an open and a closed forms. The open form, catalytically inactive, could allow the binding of substrate, releasing of product and recycling the copper redox site, while the closed form, catalytically active, could perform the dioxygen activation and hydroxylation of substrate. This new proposal however suffers some inconsistencies. First of all, a closed form was never observed in the different published PHM structures[39,43,46–48] and a reactivity was observed *in crystalo* with a reduced PHM.[46] Secondly, as already mentioned, the DβM is not fully metalated, only one Cu_B (Cu_M) center is occupied. The catalytic domains in the open form contain unoccupied copper sites, while the closed form contains unoccupied Cu_A (Cu_H) site but occupied Cu_B (Cu_M) sites. The closed form could be the result of interactions of the unoccupied Cu_A (Cu_H) site with the copper of the occupied Cu_B (Cu_M) site. In these conditions, it is hazardous to argue that open and closed forms possess catalytic functional roles.

C. Copper/Oxygen Active Species

Comparison of the kinetic parameters for DβM and PHM with substrates of comparable reactivity indicates the same intrinsic H/D isotope effect for the C–H activation step. Additionally, similar ^{18}O isotope effects decreasing with substrate deuteration observed for these enzymes imply a chemical mechanism for substrate oxidation that is likely to be identical for both enzymes. Kinetic studies indicate that both DβM (in the presence of the fumarate dianion activator) and PHM proceed through a preferred ordered mechanism in which substrate binds to the enzyme before dioxygen interaction. Some differences have been pointed out for TβM. A distinctive feature of TβM is the very strong substrate inhibition that is dependent on the level of the co-substrates, dioxygen, and ascorbate, as well as substrate deuteration.[28] This feature has led to a model in which the substrate tyramine can bind either to the Cu(I) or Cu(II) forms of TβM, with substrate inhibition increased at very high ascorbate levels. The rate of ascorbate reduction of the Cu(II) form of TβM is also reduced at high tyramine level. Based on these results, the existence of a binding site for ascorbate, which has never been observed for DβM or PHM, has

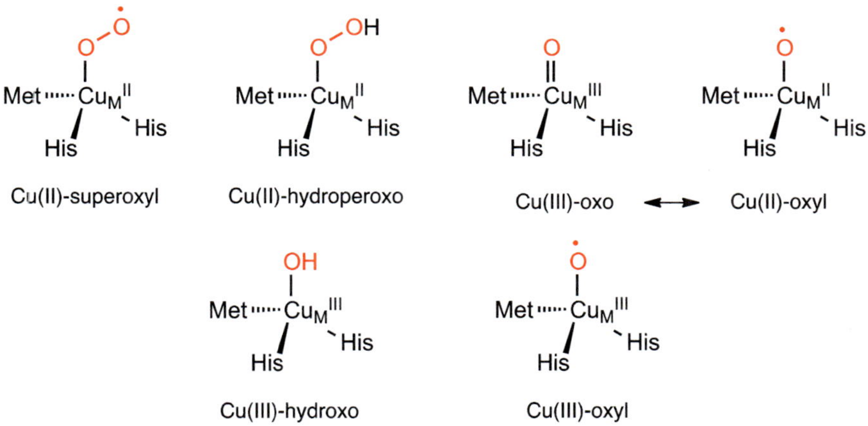

Scheme 2. Copper/oxygen adducts proposed as catalytic active species in mononuclear copper monooxygenases.

been proposed for this class of enzymes.[49] Despite this difference, all available data suggest that DβM, TβM and PHM can be regarded as interchangeable with respect to dioxygen activation mechanism and *O*-atom transfer reaction.

Regarding the copper/oxygen species involved in the substrate hydroxylation catalyzed by DβM, several active species, including copper(II)–hydroperoxo[50] and copper(II)–oxyl[51,52] have been proposed during the last two decades (Scheme 2).[34]

With PHM, a copper–dioxygen complex has been trapped by freezing protein crystals soaked in the presence of dioxygen with ascorbate and slow substrate (*N*-α-acetyl-3,5-dioiodotyrosyl-D-threonine, IYT).[46] The X-ray crystal structure of this precatalytic complex, determined to 1.85 Å resolution, showed that dioxygen binds to the Cu_M ion in the enzyme with an end-on geometry (Figure 4a).

D. Mechanism for Substrate Hydroxylation

As indicated by the kinetic isotope studies, the reaction catalyzed by DβM, TβM and PHM require a H-atom abstraction (HAA) from the substrate. In early studies on mechanism-based DβM inhibitors such as 3-phenylpropenes, it has been shown that the substrate functionalization

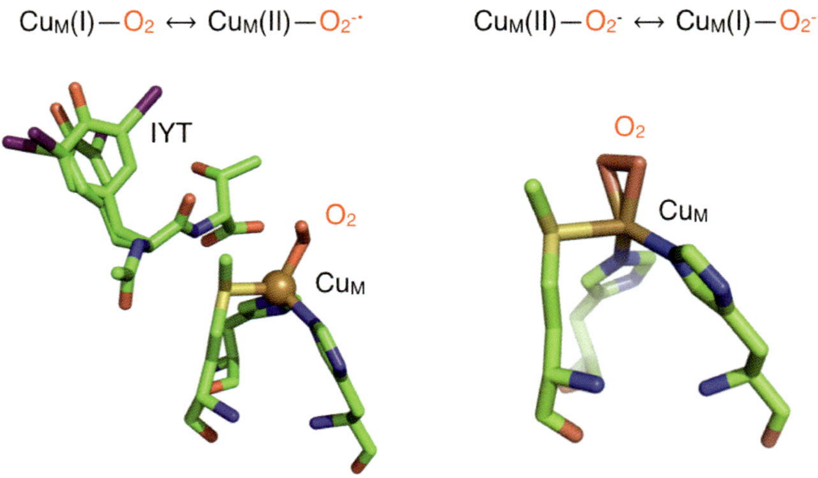

(a) (b)

Figure 4. Structure of the Cu_M center of PHMcc with: (a) with the *N*-α-acetyl-3,5-dioiodotyrosyl-D-threonine (IYT) substrate showing the binding of dioxygen in an end-on geometry (PDB code 1SDW); (b) showing the binding of hydrogen peroxide in a side-on geometry (PDB code 4E4Z).

involved a H-atom abstraction that leads to the formation of a free radical intermediate.[53] With the conformationally-restricted substrate 2-aminoindane, a two-step radical mechanism with retention of configuration at the hydroxylated carbon have been proposed.[54] With PHM substrates such as 4-phenyl-3-butenoic acids the occurrence of a delocalized free radical have also been demonstrated.[55,56]

1. *Copper(II)–Superoxyl as Active Species*

Identification of the oxygen species catalyzing H-atom abstraction has proven to be far more elusive and has been the subject of several hypothesis.[19,57] Given the structure of the precatalytic complex in PHM, it is likely that dioxygen is directly involved in the ET and HAA steps of the PHM reaction.[46] These findings suggest a mechanism in which a diamagnetic copper(II)-superoxyl species formed initially at very low level, abstracts a H-atom from the substrate to generate Cu(II)OOH and a substrate-free radical as intermediates. Subsequent participation of the second

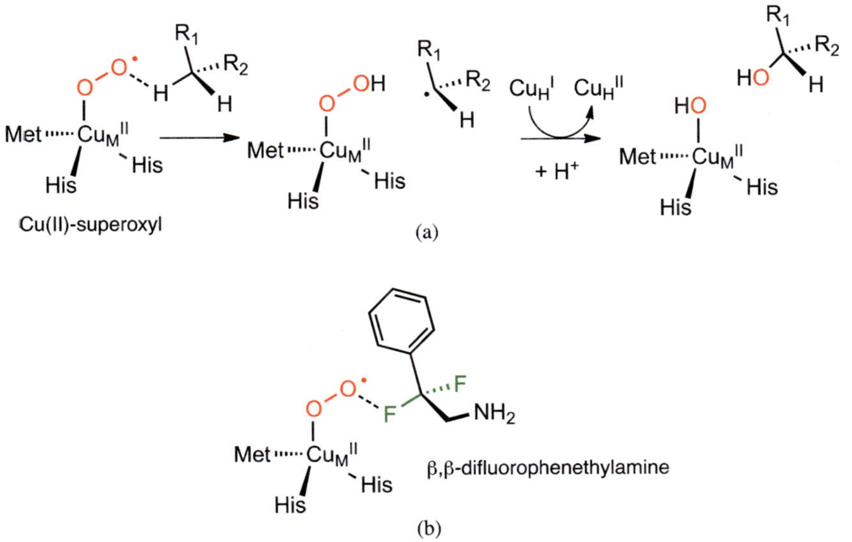

Scheme 3. (a) Two-step radical mechanism with retention of configuration proposed for the C–H bond abstraction catalyzed by mononuclear copper monooxygenases; (b) Cu(II)–$O_2^{\bullet}$ in interaction with β,β-difluorophenethylamine.

copper site (Cu_H) per subunit completes the reaction cycle, generating the hydroxylated product (Scheme 3a).[58]

Evans *et al.* reported on two experimental probes of the activated oxygen species in DβM.[59] First, the capacity of a substrate analog that cannot be functionalized (β,β-difluorophenethylamine) to induce reoxidation of the pre-reduced copper sites of DβM upon mixing with dioxygen under rapid freeze-quench conditions was examined (Scheme 3b). This experiment failed to give rise to an EPR-detectable copper species, in contrast to a substrate with an active C–H bond. This indicates either that (i) the re-oxidation of the enzyme-bound copper sites in the presence of dioxygen is tightly associated with the C–H activation or that (ii) a diamagnetic Cu(II)-$O_2^{\bullet}$ species has been formed. In the event of an active site that is open and fully solvent-accessible, as seen for the homologous PHM, the accumulation of a reduced oxygen species in DβM prior to C–H cleavage would be expected to produce an uncoupling of dioxygen and substrate consumption by analogy to cytochrome P-450. For this reason, the correlation between dioxygen and substrate consumption was examined in DβM using both end point and initial rate experimental protocols. With substrates that differ by more than three orders of magnitude in rate, a strong correlation between dioxygen uptake and product formation was observed. This led to the conclusion that no accumulation of an activated form of dioxygen occurs before H-atom abstraction in the DβM and copper(II)–superoxyl species was proposed to be responsible for the H-atom abstraction of the dopamine substrate.[59]

The formation of an end-on copper(II)–superoxyl at Cu_M (Cu_B) center that performs a H-atom abstraction on the substrate to generate a substrate radical is now well accepted. However, details of the hydroxylation mechanism and timing of long-range electron transfer (LRET) are not clear. Several mechanisms have been proposed on the base of density functional theory (DFT) calculations.[59–63] Most likely, the LRET occurs through the water-filled cavity located at the solvent interface linking the two copper binding domains.[44,60] More advanced DFT-calculations of the full reaction coordinate have been published by the Solomon's group. In this mechanism,

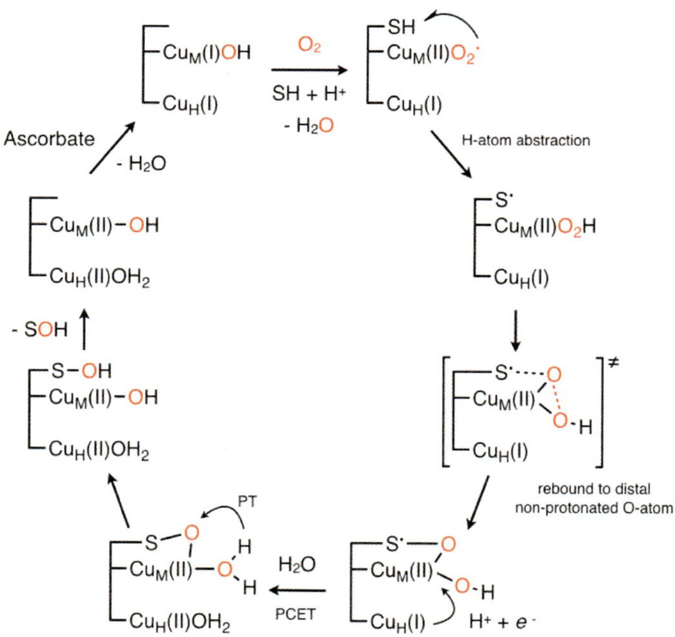

Scheme 4. Solomon's mechanism involving a substrate radical rebounds at the proximal non-protonated O-atom of the $Cu_M(II)OOH$.[64]

Solomon proposes that the substrate radical rebounds at the proximal non protonated O-atom of the Cu_M(II)OOH.[64] This process leads to the O–O bond cleavage of Cu_M(II)OOH with the formation of a new radical that is reduced through a proton-coupled electron transfer (PCET) from the Cu_H(I) center. This LRET can only occur at a late step with a large driving force to prevent the deleterious formation of hydroxy radicals.

2. *Copper(II)–Hydroperoxo as Active Species*

The observation of pH-dependent isotope effect for DβM provided evidence for the involvement of a single proton in the chemical conversion process.[50] This led to the hypothesis of a copper(II)-hydroperoxo as the reactive oxygen-centered intermediate. While reaction of oxidized DβM with hydrogen peroxide led to irreversible enzyme inactivation,[65] oxidized PHM reacts with the dansyl-tyrosine-valine-glycine-COOH (dansyl-YVG) tripeptide substrate and hydrogen peroxide to form the α-hydroxyglycine product.[66] The peroxide shunt is catalytic (K_M = 2.78 mM and $k_{cat} = 5\ s^{-1}$ for H_2O_2 *versus* 9.2 s^{-1} for ascorbate/dioxygen) and does not require the presence of additional reductant. Under anaerobic conditions with labeled $H_2{}^{18}O_2$, 100% of the hydroxyglycine compound was labeled with ^{18}O, while in $^{18}O_2$ atmosphere and with $H_2{}^{16}O_2$ only 60 % of the hydroxyglycine compound was labeled. This suggests that the reaction proceeds *via* an intermediate that can react directly with dioxygen and thus scramble the label. In addition, the peroxide shunt was eliminated by a mutation at the Cu_H center.

The structure of the oxidized form of PHM complexed with hydrogen peroxide (Figure 4a) revived the debate on the mechanism and active species.[48] This structure revealed that: (i) the (hydro)peroxide binds to Cu_M in a slightly asymmetric side-on mode with the Cu–O distances of 2.0 and 2.1 Å and (ii) the O–O distance of the copper-bound ligand of 1.5 Å is characteristic of peroxo or hydroperoxo species. On the other hand, DFT-calculations based on the first coordination sphere of the Cu_M center indicated that energies of the side-on $Cu_M(II)\text{-}O_2{}^{2-}$ species and its isomeric end-on structure Cu_M(I)–O_2 are similar, suggesting that both these intermediates are significantly populated within the protein environment and therefore they may interconvert during the catalytic cycle.

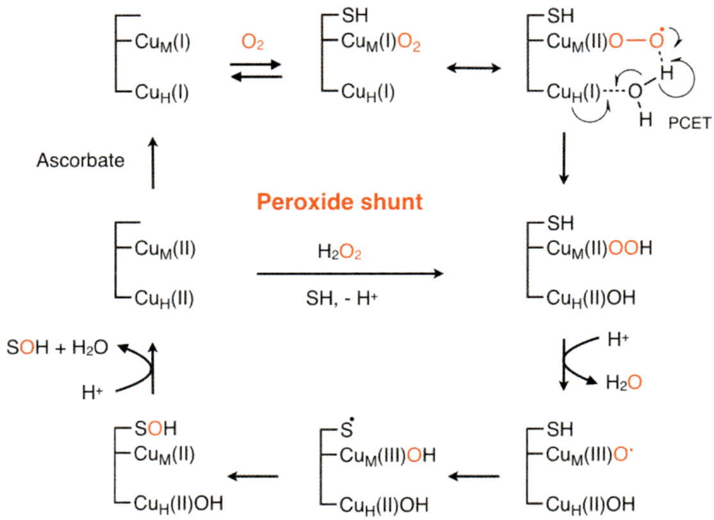

Scheme 5. Amzel's mechanism proposed for the O_2/ascorbate and hydrogen peroxide PHM-catalyzed hydroxylation.[48]

Based on this finding, Amzel *et al.* reconsidered the mechanism for the PHM-catalyzed hydroxylation (Scheme 5).[48] They proposed the formation of a copper(II)–hydroperoxide at the Cu_M center by (i) reduction of the copper(II)-superoxide through a proton-coupled long range electron transfer (PC-LRET) from Cu_H(I) or (ii) by reaction of the fully oxidized enzyme with hydrogen peroxide. The crucial step is the heterolytic O–O bond cleavage that leads to a formal Cu_M(III)–oxyl that could be responsible for the H-atom abstraction to substrate.

E. Conclusion

In the non-coupled binuclear copper containing monooxygenases, the large distance between the two copper centers (11 Å) is an essential structural feature allowing a peculiar reactivity thanks to two ET at proper stages of the catalysis. While it seems to be clear that the first 1-ET occurs at an early stage to lead to the formation of a copper(II)–superoxyl species, the timing of the second ET from the Cu_H (Cu_A) center is still a matter of an important debate. Indeed, a LRET at a late stage allows time to the copper(II)–superoxyl species to abstract the

H-atom from the substrate. The LRET occurs to restore the copper active site and liberate the oxidized substrate (Scheme 4). A LRET at an early stage would lead to the formation of a copper(II)-hydroperoxo species. In this situation, to explain the H-atom abstraction to substrate, it is needed to postulate the occurrence of the high valent Cu_M(III)–oxyl species coming from to the heterolytic O–O bond cleavage of the copper(II)–hydroperoxo species (Scheme 5). The discovery of the lytic polysaccharide monooxygenase (LPMO) which is a pure mononuclear copper-containing monooxygenase should allow to decipher the mechanism of the dioxygen activation at mononuclear copper center.

III. Lytic Polysaccharide Monooxygenase (LPMO)

A. Discovery of an Intriguing Family of Enzymes Involved in the Oxidative Degradation of Recalcitrant Polysaccharides

1. *Discovery of New Players in the Enzymatic Breakdown of Recalcitrant Polysaccharides*

Some fungi and bacteria are organisms that are well known to degrade recalcitrant polysaccharides such as plant biomass (especially lignocellulose) and chitin.[67,68] Enzymatic degradation of recalcitrant polysaccharides has traditionally been thought to occur through the synergistic action of hydrolytic enzymes that have complementary activities.[69] This hydrolytic enzymatic machinery has been extensively studied and was thought until recently to be only composed of three families of hydrolytic enzymes: endoglucanases, exoglucanases, and β-glucosidases, as classified in the CAZy database (Carbohydrate-Active enZymes Database, http://www.cazy.org).[70]

It has however been observed that an additional non-hydrolytic factor could render the cellulose less recalcitrant to enzymatic hydrolysis.[71] The exact nature of this partner has long remained unknown. Some members of the glycoside hydrolase family 61 (GH61) and carbohydrate-binding module family 33 (CBM33) have been recently reported to boost the conversion of recalcitrant cellulose or chitin, respectively.[72,73] Their action in synergy with cellulases/chitinases results in an increased depolymerization yield. Unexpectedly, GH61 and CBM33 were found to be copper-dependent enzymes that use reducing equivalents given by the cellobiose dehydrogenase (CDH), or small molecules such as ascorbic acid, gallic acid or glutathione to catalyze the oxidative cleavage of recalcitrant polysaccharide chains.[72,73] They introduce chain breaks in the biopolymer, which facilitates the action of the hydrolytic processing enzymes. GH61 and CBM33 are indeed lytic polysaccharide monooxygenases (LPMOs). In consequence, these enzymes were reclassified as AA9 (auxiliary activities) for GH61 and AA10 for CBM33 in the CAZy database.[74] LPMOs are now divided into six subfamilies since the recent discovery of four additional subfamilies, AA11,[75] AA13,[76] AA14,[77] and AA15.[78]

2. *Oxidative Cleavage of Recalcitrant Polysaccharides by LPMOs*

From a molecular point of view, LPMOs catalyze the oxidative cleavage of polysaccharide chains of cellulose or chitin. After action of a bacterial LPMO on chitin, oxidation products were identified using mass spectrometry in addition to isotope labeling techniques with $^{18}O_2$ or $H_2{}^{18}O$. These products are chitooligosaccharides having their reducing end oxidized to the corresponding carboxylic acid.[72] Interestingly, cellulose was found to be oxidized either at position C-1 (aldonic acid product) or at position C-4 (ketoaldose product) or even at both positions by some LPMOs. The identification of such reaction products is consistent with the hydroxylation of a glycosidic C–H bond (at position C-1 and/or C-4) of the polysaccharidic chain. The hemiketal intermediate formed upon hydroxylation is not stable and further decomposes. After elimination and bond cleavage, the corresponding lactone (then hydrolyzed to an aldonic acid) or ketoaldose products are released (Scheme 6).

3. *Recalcitrant Polysaccharide Breakdown in Biomass Degradation and Beyond*

The first LPMOs to be characterized were from fungi and bacteria that are well-known to participate in biomass degradation (cellulose, chitin).[72,73,79,80] Saccharides produced by the synergistic action of LPMOs and glycoside hydrolases are used by these organisms as their source of carbon. Soon after their discovery, a special attention was drawn to their potential use as biocatalysts for second-generation cellulosic biofuel production. They are

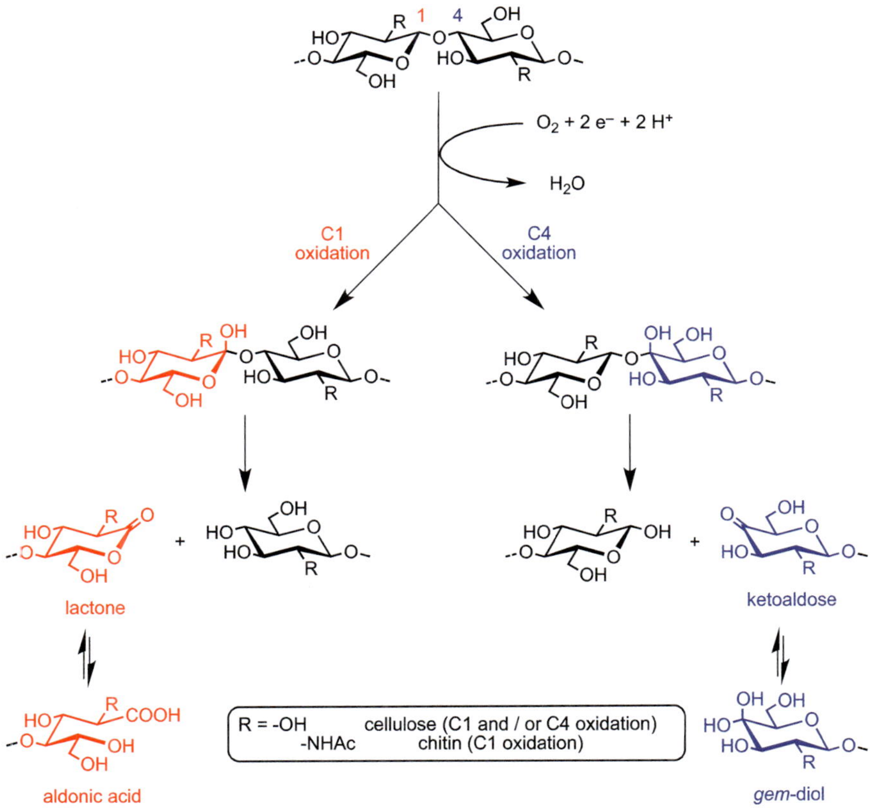

Scheme 6. Reactions catalyzed by lytic polysaccharide monooxygenases.

now included in enzymes cocktails (with cellulases) for industrial production of bioethanol from cellulosic feedstocks.

Other biological functions have also been found for LPMOs. Some of them are virulence factors for bacteria and viruses. For instance, *Paenibacillus larvae* is the causing agent of American Foulbrood of honey bees by infecting their larvae. Pathogens present in the midgut of honey bee larvae need to cross the peritrophic matrix before they can infect their host. The peritrophic matrix that normally protects the insect is mainly composed of chitin. Therefore, *Paenibacillus larvae* infection requires the secretion of chitin-degrading enzymes. In addition to some chitinases, it was recently shown that the major protein secreted by this insect pathogen is a LPMO.[81] This LPMO participates in the degradation of the chitin peritrophic matrix along with chitinases. Knocking-out the gene encoding for this LPMO impairs the peritrophic matrix degradation, resulting in a loss of virulence of *Paenibacillus larvae* due to a reduced chitinolytic activity. Similarly, some insect viruses produce spindles as virulence factor. Spindles are microcrystals of the fusolin protein.[82] The *N*-terminal part of fusolin comprises a canonical LPMO catalytic domain. The *C*-terminal part is just an extension that mediates spindolin assembly through the formation of domain-swapped fusolin dimers and additional inter-dimer disulfide bonds, which explains why spindles are so stable. Fusolin when found in spindles is not an active LPMO since each active site is capped by the *C*-terminal extension of the other monomer of the dimer. De-assembly of spindles occurs in the midgut of larvae by proteolytic degradation (host serine protease) and/or alkaline conditions.[82,83] The released LPMO catalytic domain was shown to bind to chitin and is suggested to participate in the peritrophic matrix degradation.[83]

Some human pathogens also secrete LPMOs as virulence/colonization factors.[84] GbpA from *Vibrio cholerae* is a modular colonization factor, the *N*-terminal domain of which is a LPMO catalytic domain.[85] GbpA selectively binds to chitooligosaccharides (including chitin), which might reflect the ability of *Vibrio cholerae* to colonize the chitin exoskeleton of zooplankton.[85] *Vibrio cholerae* is also the human pathogen causing cholera after colonization of the intestine. Accordingly, GbpA is also able to bind mucins, heavily glycosylated (containing *N*-acetylglucosamine) membrane proteins present at the surface of the intestine.[85,86] GbpA was later found to be a catalytically active LPMO with chitin as substrate, suggesting an active role of GbpA in the virulence of *Vibrio cholerae*.[87] Similarly, the LPMO secreted by the pathogenic bacteria *Listeria monocytogenes* is a virulence factor. This LPMO was also shown to be active on chitin but its exact biological function remains elusive.[88]

B. Architecture and Properties of the Copper Active Site of LPMOs

1. *Three-Dimensional Structure of LPMOs*

Vaaje-Kolstad *et al.* determined the first crystal structure of a LPMO: the AA10 LPMO from *Serratia marcescens*, also known as CBP21.[89] The structure is described as a "budded" fibronectin type III fold featuring two β-sheets forming a core β-sandwich tailored with a 65-amino acid "bud" (between β-strands 1 and 2) consisting of helical inserts. It also features a planar surface including conserved mostly hydrophilic residues involved in substrate (chitin) binding and the copper active site that was not identified in the early studies. Structural studies on AA9, and later on AA11 and AA13-15 families revealed that LPMOs adopt this common overall immunoglobulin/fibronectin type III-like fold with some variations (Figure 5).[73,75–78] Although the active site surface of AA11 LPMOs is

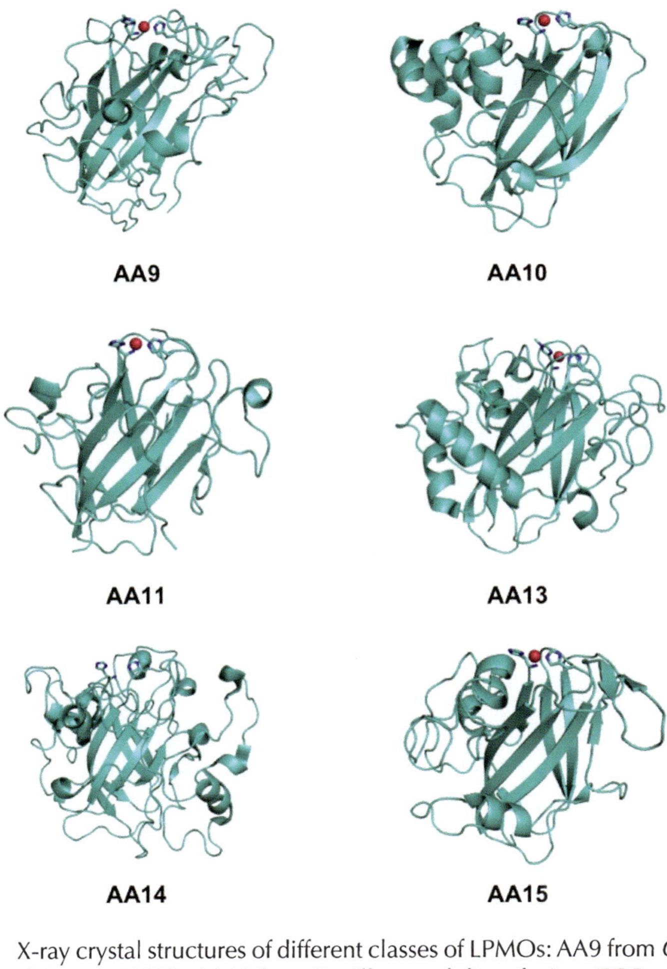

Figure 5. X-ray crystal structures of different classes of LPMOs: AA9 from *Collariella virescens* (PDB code 5NLT), AA10 from *Bacillus amyloliquefaciens* (PDB code 5IJU), AA11 from *Aspergillus oryzae* (PDB code 4MAI), AA13 from *Aspergillus oryzae* (PDB code 4OPB), AA14 from *Pycnoporus coccineus* (PDB code 5NO7), and AA15 from *Thermobia domestica* (PDB code 5MSZ).

more convex, it also includes hydrophilic residues involved in chitin binding.[75] On the other hand, several aromatic residues (mostly tyrosines) that are reminiscent of family 1 cellulose carbohydrate-binding module (CBM) are found on the flat surface of AA9 LPMOs, which indicates that these residues participate in substrate (cellulose) binding.[79] Interestingly, the active site surface of AA13, AA14 LPMOs are not as planar as that of AA9 and AA10 LPMOs, which is likely due to different substrate specificity (starch and xylan for AA13 and AA14 LPMOs, respectively).[76,77] It is also noteworthy that some LPMOs that have been characterized so far are not only constituted of a catalytic domain solely but are modular and feature a *C*-terminal CBM connected to the *N*-terminal catalytic domain by a linker region.[90–95] A more complex architecture has been observed for the AA10 LPMO from *Vibrio cholerae* which possesses three domains (including a CBM) in its *C*-terminal region in addition to the *N*-terminal catalytic domain.[85]

A more in-depth structural analysis can be found in two recent reviews focusing on the structural diversity of LPMOs.[96,97]

2. *Unique Coordination at the Copper Active Site of LPMOs*

Early crystal structures of LPMOs were ambiguous concerning the presence of an active metal site ion. Some structures did not include any metal site. Some found a divalent cation site and modeled it as Mg^{2+}, Ni^{2+} or Zn^{2+}. This led to the identification of two histidine residues as metal ligands and was corroborated by a site directed mutagenesis study.[79] The nature of the metal was not clearly known when LPMOs were first established to perform the oxidative cleavage of polysaccharides.[72] A combination of biochemical, biophysical, and structural studies led to the correct identification of the copper active site in AA9 LPMOs and shortly after in AA10 LPMOs. The topology of their active site is unique and the first coordination sphere is similar for all LPMOs: surprisingly, the copper(II) is coordinated by both the side chain and the main-chain amino group of the *N*-terminal histidine, and by another histidine side chain. This unusual *N*-terminal histidine coordination feature, called "histidine brace" is strictly conserved for LPMOs (Figures 6–11) and has only been reported for a few other copper metalloproteins: the copper transporter CopC,[98,99] the particulate methane monooxygenase,[15,100,101] and by structural homology the ammonia monooxygenase.[102] The "histidine brace" copper coordination requires a mature protein with *N*-terminal histidine. This is only possible for LPMOs, that are natively produced with a signal peptide, after its cleavage by a protease upon secretion. Remarkably, the *N*-terminal histidine residue of fungal LPMOs carry an unusual post-translational modification: its side chain ε-nitrogen atom is methylated. Interestingly, fungal LPMOs overexpressed in *Pichia pastoris* or *Escherichia coli* are not methylated but still remain active.[75,90,103] Petrovi *et al.* recently showed that the *N*-terminal methylation of the AA9 LPMO from *Thermoascus aurantiacus* affect neither its substrate preference nor its oxidation regioselectivity. However, the non-methylated enzyme is more prone to self-oxidative inactivation (by hydrogen peroxide), suggesting a protective role of the *N*-terminal methylation.[104]

The copper active site of AA9 LPMOs is classified as a T2 copper site with a Jahn–Teller distorted octahedral coordination geometry. In addition to the histidine ligands, the coordination sphere is completed by two water molecules in the remaining equatorial position and in axial position. The other axial position is occupied by the hydroxyl group of a conserved tyrosine, even though the Cu–O distance (over 2.7 Å) is somehow high for a coordination bond (Figure 6). The importance of this tyrosine residue is not yet completely understood since the phenylalanine mutant retains partial activity.[79] Copper(II) binds tightly to the histidine brace active site and its dissociation constant was estimated by ITC to be less than 1 nM.[73]

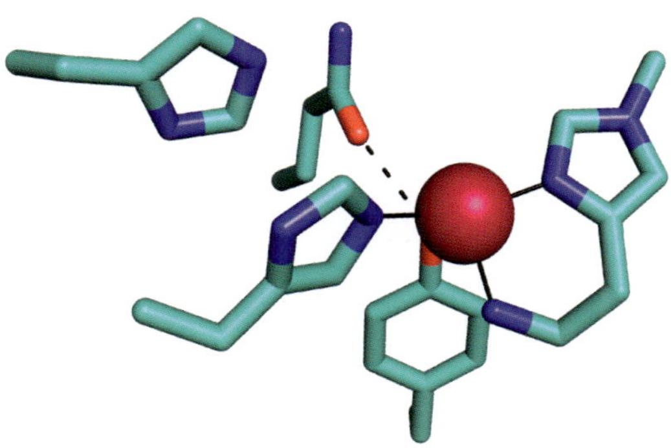

Figure 6. Active site of the AA9 LPMO from *Collariella virescens* (PDB code 5NLT) in the reduced Cu(I) state.

The redox potential of the copper(II) center of AA9 LPMOs ranges between 150 and 330 mV *versus* SHE.[73,105] The second sphere of coordination includes several conserved residues that are essential for catalysis as probed by site directed mutagenesis.[79] They include a glutamine and a histidine residue that are involved in a hydrogen bond network with water molecules. The role of these residues was recently found to be crucial for catalysis as probed by site-directed mutagenesis. Span *et al.* suggest that both residues are involved in the stabilization of copper-oxygen species over the course of the catalytic cycle. Additionally, the histidine residue was proposed to transfer a proton to a copper(II)–superoxo species during catalysis.[106]

For bacterial AA10 LPMOs active on chitin, the active site copper(II) coordination geometry is trigonal bipyramidal with histidine and water ligands. The *N*-terminal histidine side chain is not methylated for bacterial LPMOs. The conserved axial tyrosine residue of AA9 LPMOs is replaced by a conserved phenylalanine (Figure 7a). Accordingly, their corresponding EPR spectrum is different to that of AA9 LPMOs with a more rhombic character, which indicates a different copper coordination geometry.[107] A recent study suggests heterogeneity in the histidine brace copper coordination. The EPR spectrum of AA10 LPMOs might not reflect a rhombicity of the active site but a mixture of two different copper(II) species in

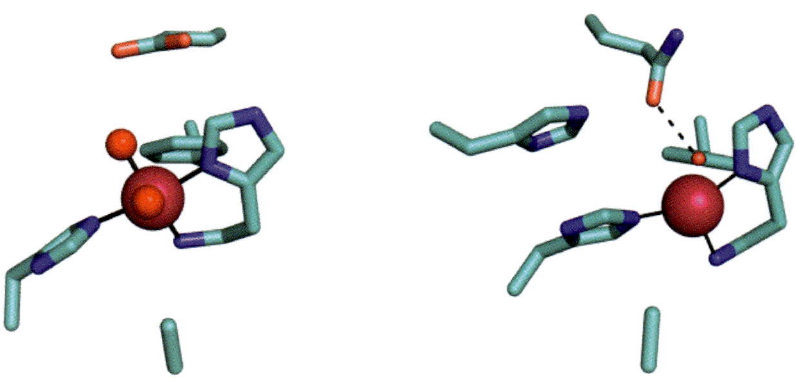
(a) (b)

Figure 7. Active site of: (a) the chitin-active AA10 LPMO from *Bacillus amyloliquefaciens* (PDB code 5IJU) in its Cu(II) oxidized state, and (b) the cellulose-active AA10 LPMO from *Streptomyces cœlicolor* (PDB code 4OY6) in its reduced Cu(I) state.

solution. It is proposed that the second species is a two-nitrogen coordinated copper(II) where the amino terminus coordination is absent and replaced by a water molecule or a chloride anion.[108] Upon reduction to copper(I), the coordination geometry of AA10 LPMOs becomes T-shaped with only nitrogen atom ligands (two histidine site chains and the amino terminus) as it is often observed in crystal structures due to easy X-ray photoreduction during data collection.[107,109] Copper(II) does not bind to AA10 LPMOs active site as tightly as for AA9 LPMOs. Indeed, dissociation constants in the low to mid-nanomolar range were determined for a few AA10 LPMOs (6 to 80 nM depending on the pH, and 55 ± 8 nM for AA10 LPMOs from *Bacillus amyloliquefaciens* and *Serratia marcescens*, respectively).[107,110,111] The redox potential of the copper(II) center of AA10 LPMOs ranges between 250 and 370 mV *versus* SHE,[107,110,112] which is slightly higher than for AA9 LPMOs. In the copper(II) state of AA10 LPMOs, a conserved alanine is located near the axial position occupied by a water molecule. This conserved residue is not found in AA9 LPMOs. It is suggested that the alanine side chain prevents the binding of other ligands at the axial position due to steric effects.[107,113] In chitin-active AA10 LPMOs, a conserved glutamate is found in place of the conserved active site glutamine in AA9 LPMOs and participate to a network of hydrogen bonds with water molecules. Interestingly, a small subset of AA10 LPMOs are active on cellulose. Their active site is either similar to that of chitin-active AA10 LPMOs or differs slightly with a tyrosine as axial residue, a glutamine and a histidine (residues observed in the active site of AA9 LPMOs) in the second sphere of coordination (Figure 7b).[112] All cellulose-active AA10 LPMOs conserve the alanine residue close to the axial water molecule ligand but a shift in its position was observed in some cases.[112] Interestingly, the EPR spectrum of cellulose-active AA10 LPMOs is closer to that obtained for cellulose-active AA9 LPMOs than for chitin-active AA10 LPMOs.[114]

Fewer members of the last AA11 and AA13-15 families have been characterized so far. AA11, AA13, and AA14 LPMOs are all fungal enzymes but are not active on the same substrates. AA11 are the first reported fungal LPMOs to catalyze the oxidative cleavage of chitin.[75] The active site of *Aspergillus oryzae* AA11 LPMO features some characteristics of both AA9 and AA10 LPMOs. Accordingly, its EPR characteristics are in between that of AA9 and AA10 LPMOs. The first sphere of coordination exhibits a non-methylated histidine brace, although fungal AA11 LPMOs are predicted to bear a post-translational methylated *N*-terminal histidine. As with some AA9 LPMOs, this is due to the overexpression of *Aspergillus oryzae* AA11 LPMO in *Escherichia coli* that does not allow this methylation. As for cellulose-active AA9 and some AA10 LPMOs, a conserved tyrosine residue is located in axial position of the copper active site of AA11 LPMOs with a Cu–O distance of 3.2 Å for *Aspergillus oryzae* AA11 LPMO. A copper dissociation constant K_d of 790 ± 150 pM was determined for this enzyme, which indicates a tight binding of copper(II) to the active site as for AA9 LPMOs. Additional conserved residues in the second sphere of coordination include an alanine residue close to the equatorial copper-coordinated water molecule, and a glutamate as found in chitin-active bacterial AA10 LPMOs (Figure 8).[75]

Unlike all other LPMOs, fungal AA13 LPMOs from *Neurospora crassa* and *Aspergillus nidulans* catalyze the oxidative cleavage of α-1,4 glycosidic bonds of starch.[76,115] Their *N*-terminal histidine is methylated as with other fungal LPMOs. An axial tyrosine residue completes the first sphere of coordination with a Cu–O distance of 2.5 Å that is significantly shorter to that found in other LPMOs harboring this conserved residue

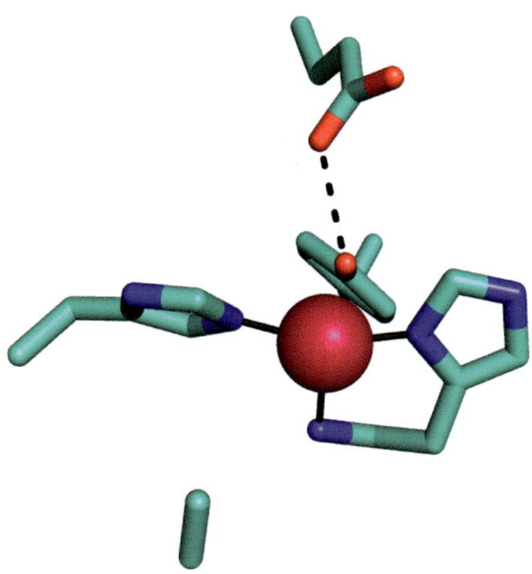

Figure 8. Active site of the AA11 LPMO from *Aspergillus oryzae* (PDB code 4MAI) in its reduced Cu(I) state.

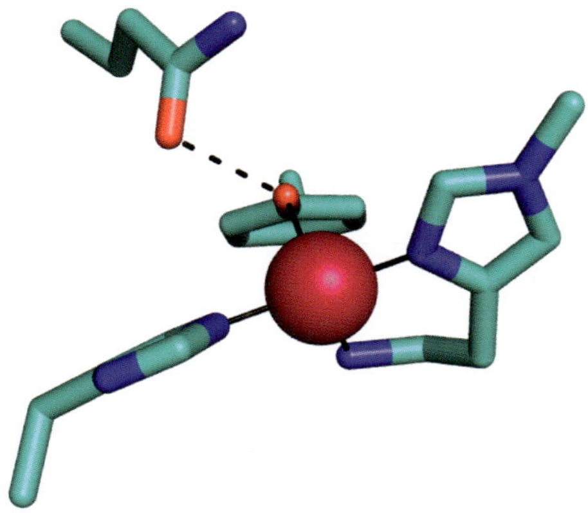

Figure 9. Active site of the AA13 LPMO from *Aspergillus oryzae* (PDB code 4OPB) in its reduced Cu(I) state.

(Figure 9). Their EPR spectrum also differs from that of other LPMOs. Notably, a super-hyperfine coupling to nitrogen ligands is well resolved in the case of *Aspergillus nidulans* AA13 LPMO.[76] Copper(II) binding affinity is similar to that of bacterial AA10 LPMOs with a K_d value of 26 ± 5 nM for the AA13 LPMO from *Aspergillus nidulans*. As observed in the crystal structure of *Aspergillus oryzae* AA13 LPMO, the conserved alanine residue found in AA10 and AA11 LPMOs is replaced by a glycine. A conserved glutamine in the second sphere of coordination is located in a similar position to that of AA9 and AA10 LPMOs that are active on cellulose.[76]

The characterization of two fungal LPMOs from *Pycnoporus coccineus* (*Pc*AA14A and *Pc*AA14B) active on cellulose-bound xylan led to the discovery of the AA14 LPMO family.[77] Similarly to other LPMO families, AA14 LPMOs strictly conserve the *N*-terminal "histidine brace" motif. These fungal LPMOs are also predicted to be methylated. However, the two characterized enzymes are not, which is due to their overexpression in *Pichia pastoris*.[77] The crystal structure of *Pc*AA14B reveals the absence of a copper(II)-bound ion, even though EPR studies have clearly shown copper(II) binding in solution (EPR parameters close to that of AA9 LPMOs). Other active site residues include a conserved tyrosine in axial position (similarly to AA9, cellulose-active AA10, AA11 and AA13 LPMOs), and a glutamine and histidine residues reminiscent of that in the AA9 LPMOs (Figure 10).

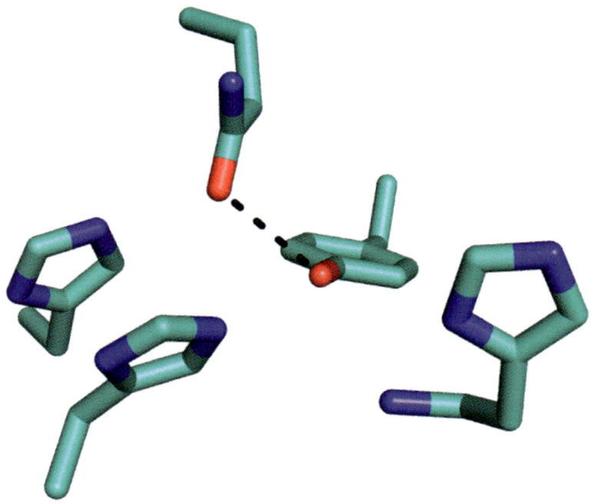

Figure 10. Active site of the AA14 LPMO from *Pycnoporus coccineus* (PDB code 5NO7) in its *apo* form.

The last family to be characterized is the AA15.[78] Interestingly, bioinformatics studies identified AA15 LPMOs in insects, crustaceans, molluscs, chelicerates, algae, and oomycetes, whereas most of the LPMOs so far were found in fungi, bacteria, and viruses. Two AA15 LPMOs from the insect *Thermobia domestica* (*Td*AA15A and *Td*AA15B) have been biochemically characterized and shown to be active on chitin, and cellulose and chitin (for *Td*AA15B and *Td*AA15A, respectively). Similarly to bacterial LPMOs, their *N*-terminal histidine is not methylated, as shown by proteomic analyses. Copper(II) EPR signal of *Td*AA15A is typical for LPMOs with some distortion from the axial symmetry. The crystal structure of *Td*AA15A reveals a tyrosine residue in axial position of the active site copper ion with a Cu–O distance of 2.6 Å, whereas the sequence of *Td*AA15A displays a phenylalanine in this position. This is likely correlated with substrate preference (chitin *versus* cellulose) as observed in the AA10 family. An alanine residue is found close to the active site copper ion of *Td*AA15A. However, this residue does not seem to be as conserved as in the AA10 family since other AA15 LPMOs (including *Td*AA15B) feature a threonine instead. A conserved second-sphere of coordination glutamate residue (in both predicted chitin and cellulose active AA15) hydrogen bonds with the axial tyrosine (Figure 11).

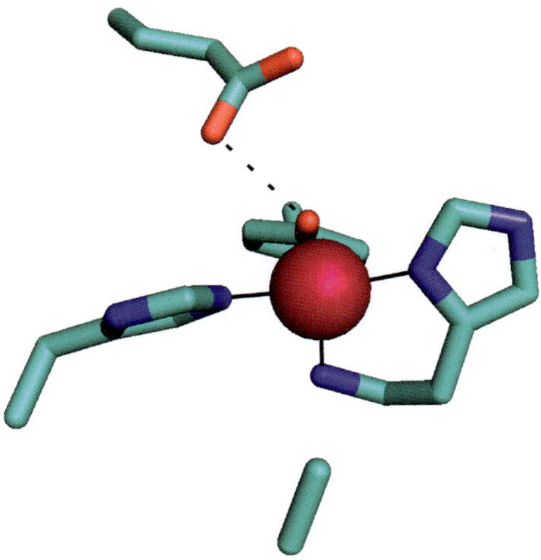

Figure 11. Active site of the AA15 LPMO from *Thermobia domestica* (PDB code 5NO7) in its reduced Cu(I) form.

3. *Regioselectivity of Oxidation*

So far, most AA9 LPMOs have been found to use cellulose as substrate. Nevertheless, some members were also shown to oxidize short cellooligosaccharides,[90,116–119] hemicelluloses such as xyloglucan,[92,117,119–123] and glucomannan.[117,119] As aforementioned, AA9 LPMOs catalyze the oxidative cleavage of cellulose either at position C-1 or C-4. Differences of regioselectivity for cellulose oxidation are observed for members of the AA9 LPMO subfamily: oxidation at position C-1 only, C-4 only, or both. This regioselectivity depends on substrate binding on the flat surface of the enzyme harboring its active site. Aromatic residues (mostly tyrosine) on this surface are responsible for cellulose binding. Different patterns of surface tyrosines (number and position on the surface) might account for cellulose orientation relatively to the active site on this surface, which in turn governs the regioselectivity of oxidation.[96,97,113,124,125]

Vu *et al.* identified additional structural determinants that might explain the regioselectivity of cellulose oxidation among the AA9 LPMO family.[126] In the absence of enzyme-substrate complex structures, several subclasses were proposed according to a phylogenetic analysis of about

500 sequences corresponding to potential AA9 LPMOs. This analysis suggests that three major subclasses, accounting for about 90% of the AA9 LPMOs, can be differentiated by their regioselectivity of cellulose oxidation: C-1 for the subclass named PMO1, C-4 for the subclass PMO2, C-1 and C-4 for the subclass PMO3, and C-1 for PMO3*, a subgroup of PMO3. The regioselectivity of oxidation was advanced for each subclass according to that of the few subclass members that were previously biochemically characterized. Sequence and structural alignments allowed to pinpoint structural features that might govern the regioselectivity of oxidation. PMO3s exhibit an inserted sequence that forms an extra helix on the active site flat surface. This insert includes a conserved tyrosine residue involved in carbohydrate recognition and binding. Truncation of this insert totally abolishes the production of ketoaldoses (from oxidation at position C-4). Interestingly, the PMO3* subgroup is unable to oxidize cellulose at position C-4 while harboring this extra insert. This might be partly due to the lack of the conserved tyrosine residue as well as other aromatic amino acids involved in substrate binding. Finally, the PMO2 subclass exhibits an extra sequence that forms a short helix near the active site surface. Additionally, two conserved residues (lysine and serine) found close to the active site for this subclass might also account for the regioselectivity of oxidation at position C-4. Further characterization of additional AA9 LPMOs led to some disagreement with this sequence-based classification. Indeed, the AA9 LPMO H from *Podospora anserina* is predicted to oxidize cellulose at position C-4 only (PMO2 subclass) but oxidation products arising from C-1 and C-4 oxidation were identified.[90] This indicates that additional structural features might participate to the regioselectivity of oxidation.

The structure of an enzyme-substrate complex has long been missing to understand the molecular and structural basis of substrate binding and regioselectivity of oxidation. Difficulties due to the insolubility of the cellulose substrate were recently overcome.[118] Frandsen *et al.* took advantage of the ability of the AA9 LPMO from *Lentinus similis* to oxidize soluble short cellooligosaccharides into ketoaldoses (oxidation at position C-4 only). The corresponding crystal structures revealed interactions between the cellooligosaccharides and non-conserved polar residues of the active site surface (Figure 12). These residues are also found in other cellooligosaccharide-active AA9 LPMOs but not in AA9 LPMOs that do

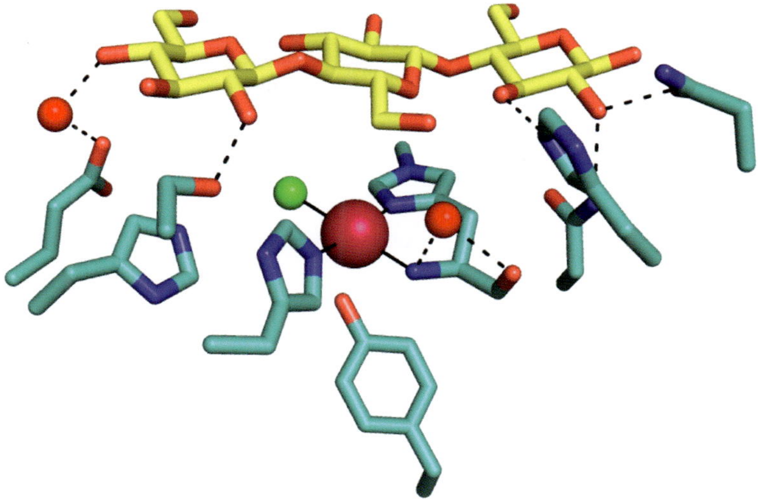

Figure 12. X-ray crystal structure of the complex between the AA9 LPMO from *Lentinus similis* and its cellotriose substrate (PDB code 5NKW).

not oxidize soluble substrates.[90,127] Surprisingly, in the substrate-bound crystal structure, a chloride ion is coordinated to the copper(II) active site ion in the equatorial position. This could be reminiscent of a copper-oxygen species involved in the C–H bond activation. Therefore, Frandsen *et al.* propose that the slightly shorter distance of 3.7 Å *versus* 3.9 Å between the chloride and the H atom in position C-4 or C-1, respectively, accounts for the regioselectivity of oxidation of this enzyme.[118]

Cellulose-active bacterial AA10 LPMOs are able to oxidize their substrate either at position C-1 or at both positions C-1 and C-4. Surprisingly, no cellulose-active AA10 LPMO oxidizing at position C-4 only has been discovered so far. Chitin-active AA10 and AA11 LPMOs as well as starch-active AA13 LPMOs are all pure C-1 oxidizers.[72,75,76] No further study to determine the molecular basis for the regioselectivity of oxidation for the aforementioned enzymes has been undergone yet.

C. Redox Partners

As mentioned above, LPMOs require a source of electrons (formally two electrons) to catalyze the oxidative cleavage of recalcitrant polysaccharides.

In this part, the current knowledge on systems able to transfer electrons to LPMOs and initiate the oxidative polysaccharide degradation is summarized.

1. *Small Molecule Reductants*

Dioxygen activation at the copper center of LPMOs for the oxidative breakdown of recalcitrant polysaccharides can be initiated by small molecule reductants with appropriate redox potentials such as ascorbate, thiols (cysteine or glutathione), diphenols, gallic acid, or pyrogallol.[72,73,80] Although these small molecule reductants are effective electron donors for *in vitro* studies, it remains unclear if they are biologically relevant for LPMO catalysis in terms of reducing efficiency and co-occurrence with LPMOs in biological systems.

Lignins which are part of the plant cell walls exhibit a boosting effect on the activity of fungal AA9 LPMOs towards cellulose. Although no clear evidence of electron transfer to the LPMO active site copper ion was shown, this observation suggests that lignins may act as electron donors for LPMOs. Recently, Westereng *et al.* demonstrated that soluble low-molecular weight lignins (LMWL) or insoluble high-molecular weight lignins (HMWL) could only poorly transfer electrons to LPMOs.[128] However, a synergistic effect was observed when both LMWL and HMWL were used as a source of electron for LPMOs. This effect is not achieved when replacing LMWL by small molecule reductants. Westereng *et al.* propose that LMWL mediate long range electron transfers between HMWL and LPMO's copper center.[128]

Recently, Cannella *et al.* demonstrated that photosynthetic pigments such as chlorophyll (directly from thylakoids) and derivatives (water soluble chlorophyllin) are able to reduce AA9 LPMOs upon light activation.[129] Indeed, light excitation of the pigment generates its excited state that is a strong reductant. Therefore, the authors propose that it efficiently transfers an electron to the copper ion in LPMO, which initiates polysaccharide oxidative cleavage upon dioxygen activation. To be fully catalytic, a sacrificial reductant such as ascorbate or lignin is used in this system to reduce the oxidized pigment back to its reduced form (ground state). Interestingly, the use of photo-excited pigments as electron donors for LPMOs leads to a 100-fold increased catalytic activity in comparison with standard reductants. Möllers *et al.* further demonstrated that cellulose oxidation is not performed by reactive oxygen species that are commonly formed in this light-driven system.[130] Accordingly, polysaccharide oxidative breakdown is most likely catalyzed by electron transfer from the photoexcited pigment to the LPMO as it was initially proposed. The biological relevance of light-driven cellulose breakdown by LPMOs remains unclear. However, the unprecedented efficiency of this system may be of applied interest in biotechnology such as for the production of second generation biofuels.

2. *Redox Enzymes*

An early study on the basidiomycete fungus *Trametes versicolor* has shown that cellobiose dehydrogenase (CDH) knock-out strains could not grow on cellulose as sole source of carbon, suggesting that CDH is essential for cellulose degradation and utilization by this fungus.[131] A further transcriptome and secretome analysis of the basidiomycete *Phanerochaete chrysosporium* revealed that the secreted enzymes AA9 LPMOs and CDH are co-expressed upon growth on cellulose and are also both upregulated by xylan.[132,133] Co-transcription and/or co-expression of secreted AA9 LPMOs and CDH were similarly described for other fungi.[134,135] These studies clearly link CDH to cellulose degradation even though its exact biological function was not clearly established. CDH was originally shown to oxidize the reducing end of cellobiose and other cellooligosaccharides to the corresponding lactones,[136] but the biological relevance of this activity was not clearly understood. CDH was eventually demonstrated to be the biological partner of fungal LPMOs for electron transfer and to trigger recalcitrant polysaccharide oxidative breakdown.[137,138] CDHs are secreted two-domain hemoflavoenzymes from the glucose-methanol-choline (GMC) oxidoreductase superfamily and classified as AA3_1 in the CAZy database. Tan *et al.* evidenced by rapid kinetics experiments that electrons are first transferred from FAD of the flavin dehydrogenase-binding domain to heme b of the cytochrome domain upon cellobiose oxidation. Reduction of the Cu(II) center of LPMO is subsequently achieved by electron transfer from the cytochrome domain since no direct electron transfer from the flavin-binding domain to LPMO is observed.[139]

Surprisingly, some fungi producing LPMOs lack a gene encoding CDH, suggesting that alternative redox partners might also be secreted by these organisms. Garajova *et al.* explored other members of the AA3 family as alternative redox partners to CDH.[105] Indeed, flavo-oxidases from the AA3_2 subfamily are also secreted by fungi such as *Pycnoporus cinnabarinus* and *Ustilago maydis* when grown under lignocellulolytic conditions. Several flavo-oxidases such as aryl-alcohol quinone oxidoreductase and glucose dehydrogenase (but not aryl-alcohol oxidase) were found to efficiently transfer electrons to AA9 LPMOs and initiate the oxidative cleavage of cellulose. Contrary to CDHs, members of the AA3_2 subfamily lack the cytochrome domain. However, electron transfer from aryl-alcohol quinone oxidoreductase and glucose dehydrogenase to LPMOs is thermodynamically favored given their redox potentials.[105] Similarly, Várnai *et al.* recently showed that another class of oxidoreductases, the pyrroloquinoline-quinone-dependent pyranose dehydrogenases (members of the AA12 family), are also able to trigger LPMO catalysis.[122] This report expands the scope of potential redox partners for LPMOs.

For bacterial AA10 LPMOs, no associated reductase has been identified to date. However, a recent systems biology study identified two candidates Cbp2D and Cbp2E as potential redox partners of a bacterial AA10 LPMO from *Cellvibrio japonicus* for cellulose degradation.[140] Interestingly, Cbp2D and Cpb2E feature predicted redox active domains in addition to their *N*-terminal cellulose-binding CBM2 domain: two cytochrome c domains for Cbp2D and a putative ubiquinone binding YceI domain. Electron transfers between these predicted extracellular proteins and bacterial AA10 LPMOs still remain to be demonstrated. Although no bacterial CDH equivalent as redox partner for AA10 LPMOs has been unambiguously discovered so far, Loose *et al.* recently reported that the fungal CDH from *Myriococcum thermophilum* can activate chitin- or cellulose-active bacterial LPMOs for polysaccharide oxidative breakdown.[141]

3. *Synergy between Small Molecule Reductants and Enzymatic Redox Partners*

As mentioned before, small molecule reductants are often used *in vitro* as electron donors to assess LPMOs activity. However, in a biological context, only a limited supply of reducing agent is likely to be available. Therefore, their enzymatic production might be required to ensure proper electron supply for LPMO activity. For instance, LPMOs were reported to be activated by catechols produced by oxidation of lignin-derived polyphenols by different polyphenol oxidases.[142,143] Additionally, the recycling of reducing agents is required to maintain the enzymatic LPMO activity. As described above, such a recycling is achieved for lignin derivatives for which soluble low molecular weight lignin acts as mediators for electron transfer to LPMOs from insoluble high molecular weight lignins.[128] Kracher *et al.* recently reported on the ability of some hydroquinones to transfer electrons to LPMOs. Additionally, quinone by-products were shown to be efficiently reduced back to hydroquinones by members of the glucose-methanol-choline (GMC) oxidoreductase superfamily such as glucose dehydrogenase (GDH), glucose oxidase (GOx), or pyranose dehydrogenase (PDH).[144] GMC oxidoreductases are widely found in genomes and secretomes of biomass-degrading fungi. This suggests that diphenol recycling by GMC oxidoreductases might also serve as an alternative source of electrons to CDH for LPMOs.

D. Mechanism

1. *Dioxygen Activation at the Copper Center of LPMOs*

LPMOs exhibit functional similarities (i.e. the catalysis of a C–H bond hydroxylation) with other copper-monooxygenases including dopamine β-monooxygenase (DβM),[50] tyramine β-monooxygenase (TβM),[28] and peptidylglycine α-hydroxylating monooxygenase (PHM).[29] Shortly after their discovery, a catalytic mechanism was proposed for LPMOs based on previous studies on DβM, TβM, and PHM.[138] In this mechanism (Scheme 7), the active site Cu(II) is first reduced to Cu(I). This reduction allows dioxygen to bind, forming a Cu(II)–O–O$^•$ intermediate. This copper–superoxo species was initially believed to be responsible for the hydrogen abstraction, generating a radical on position C-1 or C-4 of the substrate. The copper–hydroperoxo intermediate (Cu(II)–OOH) would subsequently be reduced to a copper–oxyl species (Cu(II)–O$^•$) that would further recombine with the substrate radical. Hydroxylation of the substrate at position C-1 or C-4 destabilizes the glycosidic bond and promotes the elimination reaction, leading to either a lactone or a ketoaldose, for a C-1 or C-4 oxidation, respectively (Scheme 6).

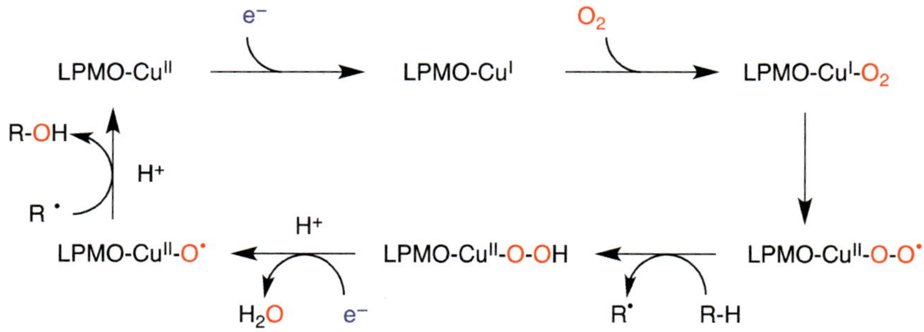

Scheme 7. First proposed LPMO mechanism with a Cu(II)–superoxo intermediate as active species.[138]

The first steps of the mechanism are commonly admitted in the field of mononuclear copper monooxygenases. Accordingly, the formation of the copper–superoxo species has been evidenced for *Ta*AA9 by EPR and stopped flow absorption spectroscopies.[145] It has been shown that the reaction with the reduced Cu(I) LPMO and dioxygen results in the one-electron reduction of dioxygen to superoxide. The transient copper–superoxo species is not stable in the absence of substrate and quickly decomposes to the superoxide anion and the oxidized Cu(II) state of the enzyme. DFT calculations corroborate the prediction that the copper–superoxo intermediate adopt an end-on geometry.[145,146] This was recently confirmed by X-ray crystallography for an AA9 LPMO from *Neurospora crassa*.[147] Soaking of a crystal of the Cu(II) state of the enzyme with ascorbate led to a crystal structure that exhibits electron density (near the copper center) attributed to a superoxo. Interestingly, the modeled copper–superoxo species adopts an end-on geometry as previously predicted. Contrary to what was initially proposed, additional DFT calculations support that oxidation of polysaccharides proceed *via* an oxygen-rebound mechanism from the copper–oxyl species (Scheme 8) rather than the copper–superoxo species (Scheme 7).[146]

2. *Hydrogen Peroxide Activation at the Copper Center of LPMOs*

Recently, Bissaro *et al.* reported that LPMOs could actually be peroxygenases instead of monooxygenases, i.e. use hydrogen peroxide as a co-substrate in

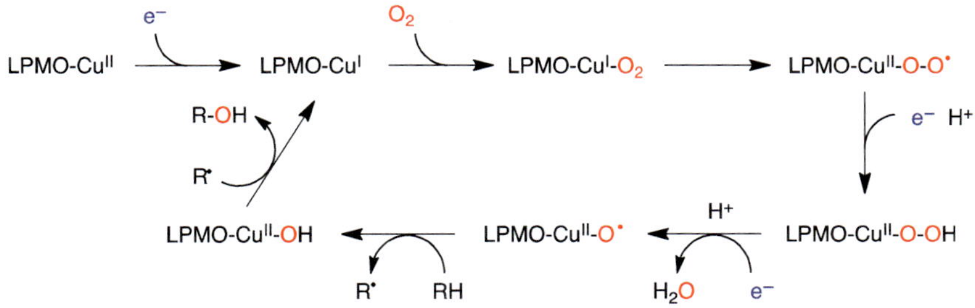

Scheme 8. Proposed LPMO mechanism with a Cu(II)–oxyl intermediate as active species.[145,146]

place of dioxygen.[148] In this study, initial results indicate that hydrogen peroxide in the presence of reductant is able to promote LPMO catalysis (AA9 and AA10 families), yielding much higher concentrations of oxidized products (over 20 times) than with dioxygen as co-substrate. The effect of hydrogen peroxide is dose-dependent but higher concentration lead to LPMO inactivation through protein oxidative damage. Bissaro *et al.* show that hydrogen peroxide consumption by LPMOs correlates with the production of oxidized products. Isotopic labeling experiments with $H_2{}^{18}O_2$ lead to the incorporation of ^{18}O in released soluble oxidized oligosaccharidic products, even in the presence of a 10-fold excess of dioxygen. Notably, catalytic amounts of reducing agents are used to prime the reaction with hydrogen peroxide whereas reaction with dioxygen as co-substrate requires stoichiometric amounts (two equivalents per turnover reaction). Moreover, the reaction without any added hydrogen peroxide is inhibited by a peroxide in a dose-dependent manner. This suggests that the reaction under usual aerobic conditions could actually be initiated by the production of hydrogen peroxide through uncoupling of the enzyme as previously reported, and that consequently that dioxygen is not a co-substrate for LPMOs. In a follow-up study, Kuusk *et al.* provided a detailed kinetic analysis of chitin degradation by the AA10 LPMO from *Serratia marcescens* (*Sm*AA10) in the presence of hydrogen peroxide.[149] They notably report a k_{cat} (5.6 s^{-1}) that is two orders of magnitude higher than the apparent rate constant (0.02 s^{-1})[72] previously reported for the same enzyme with dioxygen. Given an estimated K_m value of 2.8 μM, the catalytic efficiency of *Sm*AA10 (k_{cat}/K_m) is about 10^6 M^{-1} s^{-1}. This value is about three orders of magnitude higher than the apparent inactivation constant of the enzyme by hydrogen peroxide ($k_i \sim 10^3\ M^{-1}\ s^{-1}$). Altogether, these compelling data show that hydrogen peroxide acts as a co-substrate for LPMOs. Additionally, Bissaro *et al.* argue that this reaction with hydrogen peroxide is not comparable with the "peroxide shunt" as seen for other monooxygenases such as cytochromes P450.[148] Indeed, "peroxide shunt" reactions are usually associated with less turnover numbers and less stability of the enzyme than during usual catalysis with dioxygen as co-substrate, which is not the case for LPMOs. Accordingly, hydrogen peroxide as a co-substrate was later found to significantly improve the saccharification efficiency of a commercial cellulase cocktail that includes LPMOs. For instance, the anaerobic controlled supply of hydrogen peroxide in the presence of sub-stoichiometric amounts of reductant led to substrate (cellulose or lignocellulose) conversion to glucose that is almost two orders of magnitude better than standard aerobic conditions.[150]

In their mechanism, Bissaro *et al.* propose a catalytic role of the reductant to prime the catalysis (Scheme 9). Reduction of the Cu(II) to Cu(I) is followed by reaction with hydrogen peroxide leading to a copper(II)–oxyl species either *via* an heterolytic or an homolytic cleavage of the O–O bond. The homolytic cleavage of the O–O bond produces an intermediate hydroxyl radical that further abstracts a hydrogen atom from the Cu(II)–OH to yield the copper–oxyl species. Similarly to the mechanism with dioxygen as co-substrate, the active copper–oxyl species hydroxylates the substrate *via* an oxygen-rebound mechanism. Another putative pathway includes the generation of a hydroxyl free radical

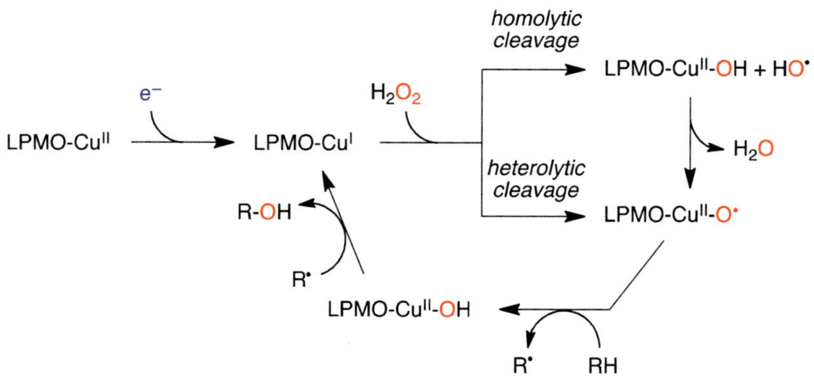

Scheme 9. Proposed LPMO mechanism with hydrogen peroxide as co-substrate.[148]

(Fenton-type chemistry) within the active site of the LPMO as an active species for substrate hydrogen atom abstraction. This pathway was considered less likely given the high regioselectivity (C-1 and / or C-4) displayed by LPMOs.[148]

Even though LPMOs seem to be using hydrogen peroxide as co-substrate, a recent report by Hangasky *et al.* provides additional clues in favor of dioxygen as the natural co-substrate for AA9 LPMOs. This study focused on an AA9 LPMO from *Myceliophthora thermophila* (*Mt*AA9E) that is active on both insoluble cellulose and soluble cello-oligosaccharides with a C-4 regioselectivity. As previously observed for other LPMOs, it is shown that *Mt*AA9E can use both dioxygen and hydrogen peroxide as co-substrate for polysaccharide oxidation. Using single and multiple turnover reactions on both soluble and insoluble substrates, Hangasky *et al.* confirmed that reaction with hydrogen peroxide is faster than with dioxygen.[151] However, it leads to more self-inactivation of the enzyme through oxidative damage and the regioselectivity of oxidation is altered. Indeed, although *Mt*AA9E almost exclusively oxidizes polysaccharide in C-4 position, the reaction with hydrogen peroxide leads to a mixture of C-1 and C-4 oxidation products as well as additional oxidized products (whose site of oxidation could not be determined, but that is neither C-1 nor C-4), consistent with Fenton-like chemistry. Additionally, the type of substrate and its concentration is shown to clearly affect the outcome of the reaction. Indeed, *Mt*AA9E has weaker affinity for insoluble substrates. This results in a fraction of the enzyme remaining unbound in solution that can generate the more reactive hydrogen peroxide co-substrate by uncoupled reaction with dioxygen. On the other hand, *Mt*AA9E has higher affinity for soluble cellohexaose, and thus can be fully saturated, which yields in these conditions to purely monooxygenase chemistry. Additionally, the nature of the reducing agent seems to be crucial as CDH limits the uncoupling reactions (by comparison with ascorbate) thanks to timely delivered electron governing the formation of the active copper(II)-oxyl species. Finally, Hangasky *et al.* propose that the combination of low $k_{cat}/K_M(O_2)$ and high $K_M(O_2)$ (365 and 230 μM in the absence and the presence of 1 mM cellohexaose, respectively) prevents the enzyme from self-inactivation by minimizing uncoupled reactions.[151]

3. *Cu(II)–Oxyl and Cu(III)–Hydroxo as the Active Species*

Although the debate is still open on whether dioxygen or hydrogen peroxide is the natural co-substrate for LPMOs, recent studies tend to all agree that the copper(II)–oxyl intermediate is the reactive intermediate leading to the hydroxylation of a strong substrate C–H bond (~100 kcal mol^{-1}).[152]

As previously mentioned, Kim *et al.* were the first to investigate LPMOs mechanism with theoretical calculations (DFT).[146] Based on an active site model of *Ta*AA9, Kim *et al.* predict an end-on configuration for the Cu(II)–superoxo intermediate. However, the activation barrier for hydrogen atom abstraction by this species is quite high (34.9 kcal mol^{-1}), ruling out the mechanism of substrate hydroxylation *via* the Cu(II)–superoxo intermediate. Instead, it is proposed that the more powerful Cu(II)–oxyl species is the active species catalyzing the C–H bond hydroxylation. Accordingly, the calculated activation barrier is more favorable (15.4 kcal mol^{-1}). Additional reports confirmed that the Cu(II)–superoxo is not the active species responsible for substrate hydrogen atom abstraction. Hedegård *et al.* calculated bond dissociation energies for different intermediates and suggest that the Cu(II)–oxyl, the Cu(III)–oxyl and the Cu(III)–hydroxide could potentially be the active species for hydrogen atom abstraction unlike the Cu(II)–superoxo.[153] Similarly, theoretical calculations (DFT) performed by Bertini *et al.* led to the identification of the Cu(II)–oxyl as the intermediate responsible for hydrogen atom abstraction from the substrate.[154] Nevertheless, their calculated activation barrier is significantly lower than that of Kim *et al.* (6.2 kcal mol^{-1} *versus* 15.4 kcal mol^{-1}).[146] Interestingly, Bertini *et al.* highlight the important role of conserved residues such as the second sphere of coordination glutamine and histidine that were previously found to be crucial for catalysis as probed by site directed mutagenesis.[106] These residues are involved in a large network of hydrogen bonds within the active site as observed in their optimized structures, similarly to X-ray crystal structures.[118] In the optimized structure of the reduced enzyme, these two residues form hydrogen bonds with a water molecule that was previously bound to the copper in the oxidized form of the enzyme. Throughout the catalytic cycle, these two residues hold this water molecule within the active site, which was found to be important for substrate hydroxylation by assisting first Cu(II)–OH bond cleavage and R–OH formation during the oxygen rebound step. Additionally, Bertini *et al.* show that dioxygen binding to Cu(I) is more favorable (4.0 kcal mol^{-1}) in the presence of substrate and thus propose a mutually binding of the two substrates (dioxygen and polysaccharide).[154] Experimentally, copper reduction was shown to increase LPMO's affinity for its substrate.[155] Hangasky *et al.* performed a steady-state kinetics of dioxygen-dependent oxidation of cellohexaose (Glc_6) oxidation by *Mt*AA9E.[156] $k_{cat}/K_M(O_2)$ and $k_{cat}/K_M(Glc_6)$ were shown to both depend on the concentration of the other substrate. Similarly, the reduced enzyme was found to have a better affinity for Glc6 than the oxidized form (K_D 175 and 994 μM for reduced and oxidized *Mt*AA9E, respectively). Hangasky *et al.* thus propose a random-sequential kinetic mechanism in which the enzyme is first reduced, and then binds the substrate (either polysaccharide or dioxygen). The sequential binding of the second substrate forms the Cu(II)–superoxo ternary complex that proceeds to substrate hydroxylation after subsequent evolution to the reactive Cu(II)–oxyl species.[156]

Wang *et al.* investigated the hydrogen peroxide-dependent activation mechanism of LPMOs through a combination of DFT calculations, classical molecular dynamics, and quantum mechanical/molecular mechanical (QM/MM) calculations,[157] with the structure of *Ls*AA9 in complex with its oligosaccharide substrate.[118] It is shown that hydrogen peroxide binds within the cavity formed by the enzyme and its substrate, but does not coordinate to the active site Cu(I) (with a calculated Cu–O distance of 2.77 Å). Hydrogen peroxide is hold in place within the active site thanks to an extended network of hydrogen bonds with the substrate and second sphere of coordination residues such as the conserved histidine and glutamine as pointed by other groups.[106,154] Activation of hydrogen peroxide then occurs by homolytic cleavage of the O–O bond yielding a Cu(II)–OH along with a hydroxyl free radical (Scheme 10). Several hydrogen bond interactions perfectly position this "caged" hydroxyl radical within the active site and preclude this reactive species from reacting with the substrate C–H bond. Instead, the hydroxyl radical abstracts a hydrogen atom from the Cu(II)–hydroxo intermediate to form the reactive Cu(II)–oxyl species. Substrate hydroxylation proceeds *via* hydrogen abstraction followed by oxygen-rebound. Additionally, theoretical calculations are

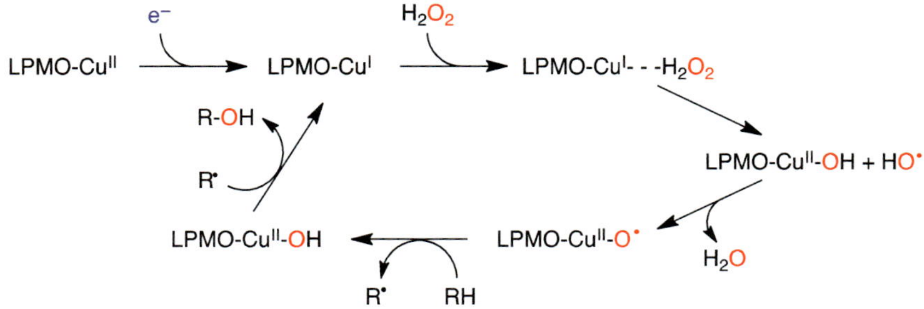

Scheme 10. Proposed mechanism with hydrogen peroxide as co-substrate *via* a caged hydroxyl free radical intermediate leading to the Cu(II)–oxyl active species.[157]

consistent with the C-4 regioselectivity observed for *Ls*AA9. Accordingly, Wang *et al.* report a low activation barrier for hydrogen atom abstraction from C-4 position (5.5 kcal mol^{-1}) whereas that for C-1 position is much higher (21.9 kcal mol^{-1}), which is due to a long distance between the oxygen atom of the Cu(II)-oxyl and the hydrogen in position C-1 (2.89 Å) along with an unfavorable geometry.

Hedegård *et al.* recently reported a comprehensive study investigating the mechanism of LPMOs by computational chemistry (Scheme 11).[158] Similarly to previous reports,[146,154] activation barriers calculations for hydrogen atom abstraction definitely rule out substrate activation by the Cu(II)–superoxo species (activation barrier of 37.3 kcal mol^{-1}). Similarly to previous theoretical and experimental reports,[106,118,154,157] the second sphere of coordination conserved histidine is crucial for catalysis. Indeed, the Cu(II)–superoxo intermediate is readily protonated by this conserved residue, and then undergo a reduction to the corresponding Cu(II)–hydroperoxo species. Likewise, the same residue acts as a general acid to protonate the Cu(II)–hydroperoxo generating after reduction and O–O bond cleavage (water elimination) the Cu(II)–oxyl intermediate. The Cu(III)–hydroxo intermediate is accessible by simple protonation of the oxyl species. Hedegård *et al.* also investigated the reaction of the reduced form of the enzyme with H_2O_2 as co-substrate. Hydrogen peroxide does not bind to the active site Cu(I) but instead reacts *via* an outer-sphere mechanism (Scheme 11). Contrary to what was proposed by Wang *et al.*, the intermediate formation of a Cu(II)–hydroxo along with a "caged" hydroxyl radical is not the most favorable pathway. It is instead proposed

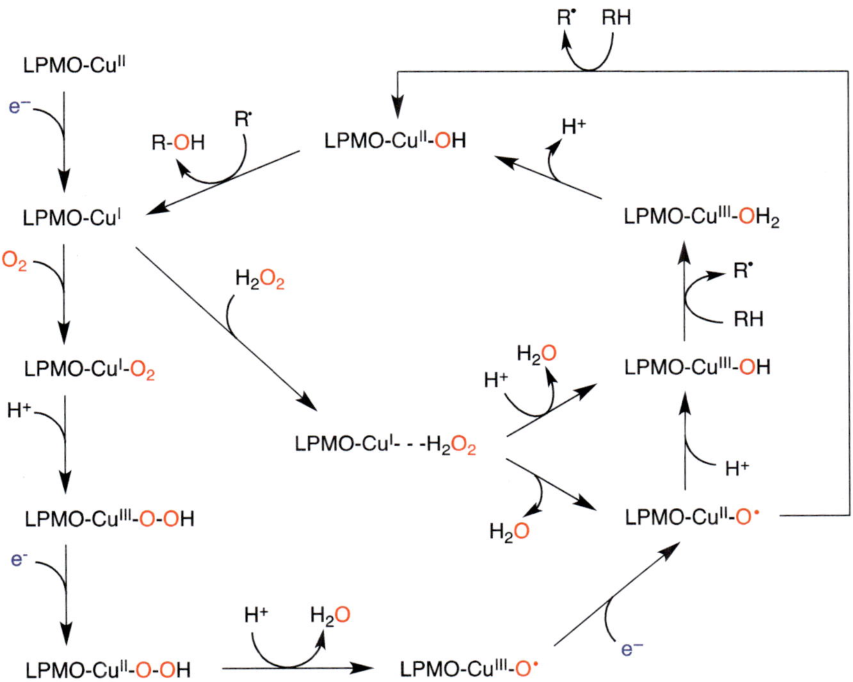

Scheme 11. Comprehensive LPMO mechanism with either dioxygen or hydrogen peroxide as co-substrates.[158]

that Cu(II)–oxyl or Cu(III)–hydroxo species are directly formed depending on the protonation state of the histidine residue. Formation of these reactive species from hydrogen peroxide and the reduced enzyme confirms that hydrogen peroxide is a co-substrate for LPMOs. Calculated activation barriers for the hydrogen atom abstraction from the substrate by the Cu(II)–oxyl and Cu(III)–hydroxo are 16.5 and 22.5 kcal mol^{-1}, respectively, suggesting that both intermediates can be the reactive species responsible for substrate hydroxylation. This proceeds *via* hydrogen atom abstraction and oxygen-rebound as previously proposed by other groups.

A putative Cu(III)–OH intermediate, a tautomeric form of the Cu(II)–$O^•$ after deprotonation of the ligated *N*-terminal amine, was also postulated as a potential reactive species for hydrogen atom abstraction from the substrate. Such a highly oxidizing species was only identified for bioinspired small molecule copper complexes.[159] However, the recent resolution of an enzyme-substrate structure (for the AA9 LPMO from *Lentinus similis*) revealed that the two protons of the ligated *N*-terminal amine group are not equivalent (also confirmed by HYSCORE) since one of them is involved in a hydrogen bond network mediated by a water molecule.[118] Accordingly, the deprotonation of the *N*-terminal amine could be facilitated by this hydrogen bond network. Such deprotonation is possible as it was reported in a neutron crystal structure of the AA10 LPMO from *Jonesia denitrificans* (*Jd*AA10A).[160] However, even though Hedegård *et al.* showed that the Cu(III)–OH would be a strong enough oxidant for substrate C–H bond abstraction,[153,158] "histidine brace" deprotonation does not lead to increased reactivity as supported by theoretical calculations.[153] Nonetheless, direct protonation of the Cu(II)–oxyl species could generate the corresponding reactive Cu(III)–hydroxo intermediate (with a protonated "histidine brace").

Further insights into LPMOs mechanism can be found in recent reviews on the subject.[161–164] Nevertheless, most if not all the current theoretical mechanistic studies have been performed on AA9 LPMOs, the only class for which the structure of an enzyme-substrate complex is available. Conserved second sphere of coordination residues identified as playing key roles during catalysis are not the same in other LPMO classes (for instance, in chitin active AA10 LPMOs, the conserved histidine is not found and the glutamine is replaced by a glutamate). This raises interesting mechanistic questions that remain unanswered so far. Future studies in this direction combining experimental and theoretical, as well as biomimetic/bioinspired small molecule complexes are likely to advance our knowledge on this intriguing enzyme family.

IV. Model Complexes and Cu/O_2 Intermediates

As previously mentioned, mononuclear copper-containing monooxygenases achieve reductive dioxygen activation to perform C–H bond hydroxylation reactions. Although, several Cu/O_2 intermediates have been postulated as key species in the catalytic cycle of single copper-containing monooxygenases, few have been directly evidenced in the enzymatic systems. Synthetic model complexes have therefore been used to better understand the nature and properties of the mononuclear copper–oxygen intermediates possibly involved.[165–170] Synthetic Cu/O_2 adducts have been prepared to reflect the great variety of species arising from various redox states of the copper ion and various redox states and coordination geometries of the O_2-derived ligand (Scheme 2) Indeed, Cu(II) or Cu(III) ions can be ligated to superoxide, peroxide in either end-on (η^1) or side-on (η^2) coordination modes.[167,168]

In most cases, Cu/O_2 intermediate species are highly unstable and their isolation requires both the synthesis of appropriate ligand systems to provide enough stability and avoid dimerization as well as the use of low-temperature conditions. Several copper-dioxygen species have been sufficiently stabilized for structural characterization using X-ray diffraction analysis. Alternatively, Cu K-edge EXAFS analysis can be performed both in solution and in the solid state, providing important structural parameters such as Cu-ligand distances. In addition to structural data, the understanding of the structure/reactivity properties of such intermediates relies on the use of a combination of approaches ranging from spectroscopic techniques to reactivity studies including theoretical calculations. Ultra violet-visible (UV-vis.) spectroscopy is a powerful tool to monitor the formation of Cu/O_2 species that often display strong oxygen-to-copper charge-transfer transitions.[167] Resonance Raman (rR) spectroscopy has been widely used to analyze the Cu–O and O–O vibrational modes, the nature of which can be confirmed by shifts obtained upon ^{18}O labeling. When applicable, O–O vibrations and their isotope sensitivity provide key information on the nature of the O_2-derived ligand.[26]

In the following section, the main properties of mononuclear Cu/O_2 adducts will be exposed in line with their reactivity and their possible implication in the catalytic cycle of the biological systems.

A. Copper(II)–Superoxo Intermediates

1. *Preparation of Synthetic Copper(II)–Superoxo Species*

Oxygenation of Cu(I) complexes leads to the formation of a copper(II)–superoxo species (Scheme 12) and thermodynamics of O_2-binding on Cu(I) has been well documented.[171] Copper(II)–superoxyl intermediates have generally been synthesized by mixing the Cu(I) precursors with O_2 at very low temperatures (–50 to –80°C) in non-coordinating solvents

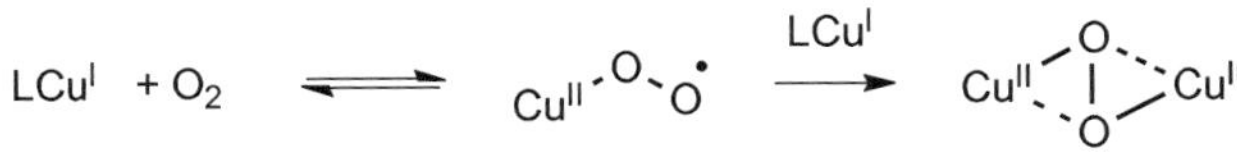

Scheme 12. Formation of Cu(II)–superoxo intermediate and dimerization reaction.

such as acetone, tetrahydrofuran (THF) or dichloromethane (DCM). Binding of dioxygen on the Cu(I) precursors is a reversible process and cycles between O_2 exposure and O_2 removal (e.g. by bubbling with Ar) can often be performed.[172]

In solution, the Cu(II)–superoxo species tends to react with a second equivalent of the Cu(I) precursor providing well-characterized bridged peroxo dicopper intermediates that are not covered in this revue (Scheme 12).[167,171] Therefore, to stabilize synthetic copper(II)–superoxyl mononuclear species, supporting ligands having bulky substituents have generally been used to prevent dimerization reactions.[165] Strongly electro-donating ligands (as anionic ligands) have also been widely used to stabilize the intermediates by lowering their reactivity. The main ligands that have been used to prepare synthetic superoxide intermediates are presented in Scheme 13.

2. *Structural, Spectroscopic and Electronic Properties*

Only two copper(II)–superoxo intermediates have been structurally characterized so far, displaying the end-on (η^1) and the side-on (η^2) coordination modes respectively.[173,174] The structures are presented in Figure 13. In the other cases, structures of the copper(II)–superoxyl intermediates were predicted using the structures of copper(I) and copper(II) precursors and in many cases thanks to theoretical calculations.

a. *Side-on Copper–Superoxo Intermediates*

By using trispyrazolylborate-derived ligand (**Tp$^{\mathbf{tBu,iPr}}$**), the group of Kitajima has isolated and crystallized the first copper(II)–superoxo intermediate (Figure 13).[173] In this case the superoxide is bound in a symmetric side-on coordination mode displaying Cu–O and O–O distances of

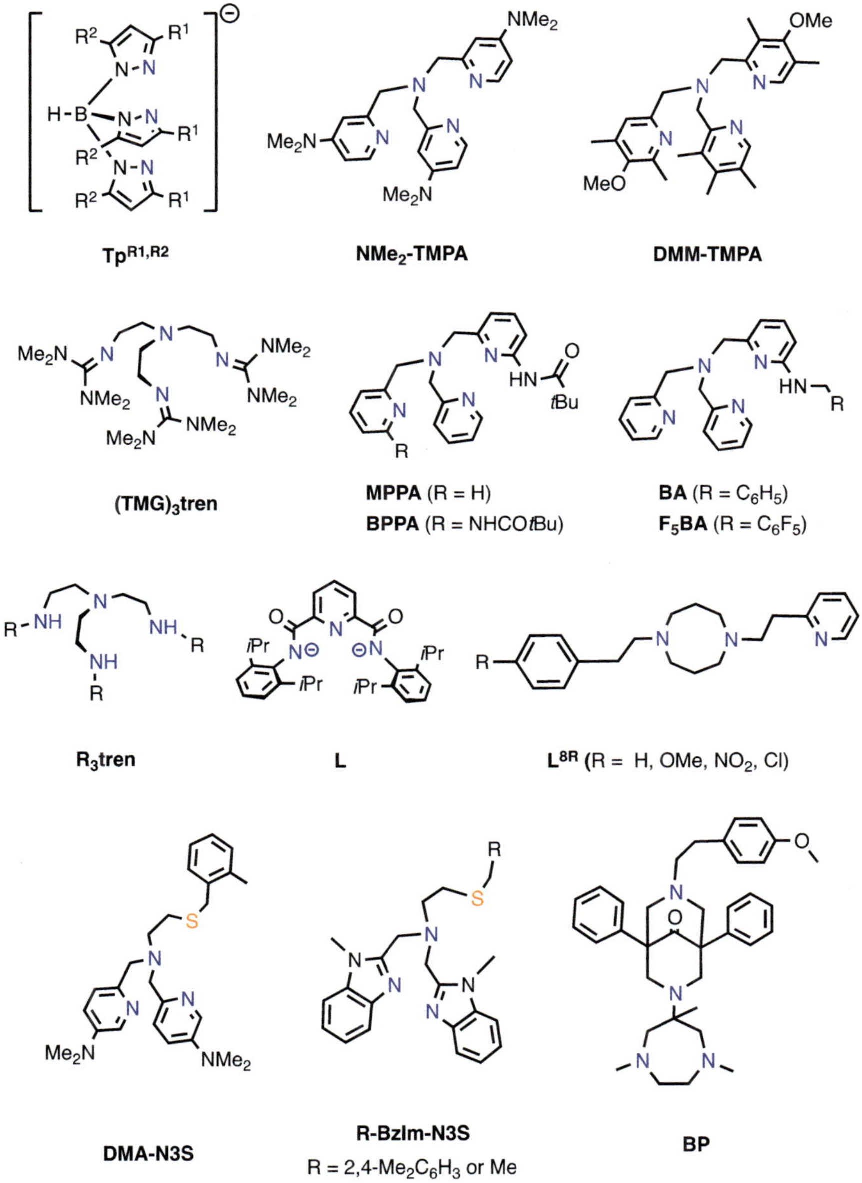

Scheme 13. Examples of ligands used to synthesize copper(II)–superoxyl species.

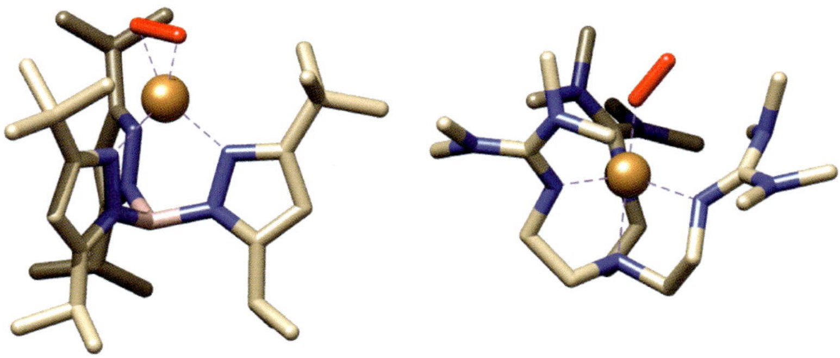

Figure 13. Crystal structure of (left) side-on superoxo intermediate $[Cu(O_2)$ $(\mathbf{Tp}^{\mathbf{tBu,iPr}})]$ and (right) end-on superoxo intermediate $[Cu(O_2)[\mathbf{(TMG)_3tren}]^+$. H-atoms and the counter anion were omitted for clarity.

1.84 and 1.22 Å, respectively. An O–O vibrational frequency of 1112 cm^{-1} is observed, which shifts to 1060 cm^{-1} in the ^{18}O isotope labeled sample. The intermediate does not display intense absorption bands in UV-vis region that is dominated by weak d–d transitions.[175] DFT-calculations provided a detailed electronic structure description of the intermediate that has a diamagnetic singlet ground-state and an excited triplet state lying 1500 cm^{-1} higher in energy.

Using **BzIm-N3S** type ligands and introducing a thioether function in order to reproduce the methionine ligation found around the Cu_M/Cu_B center in DβM or PHM, the group of Castillo have isolated EPR-silent intermediates that present charge-transfer (CT) transitions centered around 335–340 nm (ε = 1500–2000 $mol^{-1}Lcm^{-1}$) and d–d transitions around 600 nm.[176] Although no rR data have been reported, DFT-calculations support the assignment to side-on copper(II)–superoxyl intermediates with triplet ground-states arising from ferromagnetic coupling between the two unpaired electrons of the Cu(II) and the superoxide radical.

Using bidentate and anionic β-diketiminate or anilido ligands, the group of Tolman has isolated several Cu/O_2 intermediates displaying (η^2) coordination mode of the O_2-moiety.[177,178] However, due to the strong donating ability of the ligand, the intermediates are best described as copper(III)–peroxide intermediates.[179,180]

b. *End-on Copper–Superoxyl Intermediates*

Using tripodal tetradentate ligands derived from tris(2-aminoethyl)amine (**R_3tren**), Schindler and coworkers succeeded in trapping several copper(II)–superoxo intermediates.[181,182] The species could be stabilized and crystallized by introducing electrodonating tetramethylguanidine substituents in the **(TMG)$_3$tren** ligand.[174] The [**(TMG)$_3$tren**Cu(O_2)]$^+$ complex displays a five-coordinated bipyramide trigonal geometry and the superoxide is found in end-on (η^1) coordination mode with O–O distance of 1.28 Å and Cu–O–O angle of 123° close to the 110° value found in the oxy-PHMcc.[46]

Using tetradentate ligands derived from **tren**, tris(2-pyridylmethyl) amine **(TMPA)** or bispidine (**BP**) several η^1 copper-superoxide intermediates have been isolated and well-characterized (although not crystallized).[172,183–186] A tetradentate ligand displaying a thioether moiety (**DMA-N3S** on Scheme 13) has also been used.[187] All these intermediates are predicted to have five-coordinated copper ions in trigonal bipyramidal geometries. They all have characteristic spectral features that include an intense CT transition at *ca*. 400–430 nm ($\varepsilon = 3000$–$8000\ mol^{-1}Lcm^{-1}$) and two weaker transitions around 550–600 and 700–750 nm (Figure 14).

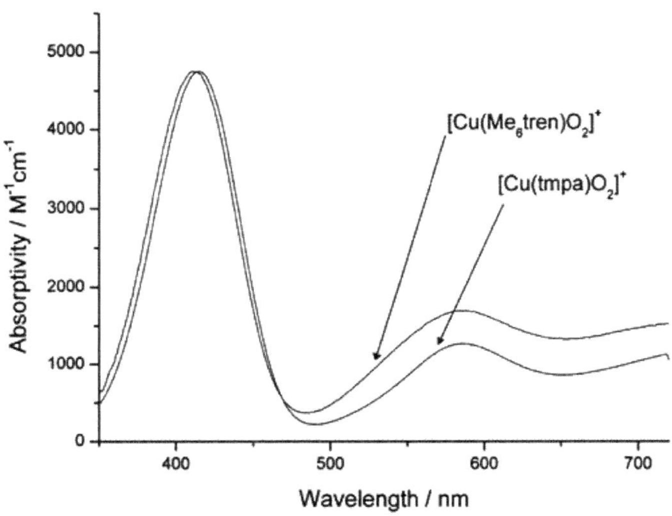

Figure 14. Calculated spectra for two Cu(II)–superoxide adducts. Reprinted with permission from Ref. 181. Copyright 2016 American Chemical Society.

Raman spectra provided information on the O–O frequencies that are found in the 1060–1130 cm^{-1} range and are shifted upon ^{18}O substitution ($\Delta[^{18}O_2] \approx 60\ cm^{-1}$). All these intermediates have triplet ground-states.[188]

The group of Tolman has reported a copper(II)–superoxide complex supported by a sterically hindered pyridinedicarboxamide-type ligand (**L** in Scheme 13).[189] The intermediate displays a CT band at 627 nm ($\varepsilon \approx 1700\ mol^{-1}Lcm^{-1}$) and an O-isotope sensitive rR shift at 1104 cm^{-1} ($\Delta[^{18}O_2] \approx 60\ cm^{-1}$) confirming its assignment as copper-superoxide complex. An end-on superoxide structure is proposed based on DFT-calculations and the geometry around the copper ion is proposed to be square-planar.

To synthesize superoxide species with tetrahedral geometry as found in the enzymatic system,[46] the group of Itoh has prepared a set of ligands based on a 1,5-diazacyclooctane motif bearing a 2-(pyridyn-2-yl)ethyl pendant arm ($\mathbf{L}^{8R}$ in Scheme 13).[190,191] The spectral and electronic properties of such intermediates are very similar to the ones obtained with tetradentate ligands.

3. *Reactivity*

Since the first isolation of copper(II)–superoxyl intermediates, efforts were developed towards the understanding of their reactivity and in particular their ability to activate C–H bonds.

The intermediates based on **tren** or **TMPA** ligands, ([((**TMG**)$_3$**tren**)Cu(O_2)]$^+$, [(**NMe$_2$-TMPA**)Cu(O_2)]$^+$ and [(**DMM-TMPA**)Cu(O_2)]$^+$) were shown to react with phenolic substrates or H-atom donors such as TEMPO-H (2,2,6,6-tetramethylpiperidine-*N*-hydroxide) to yield the phenoxyl radicals or TEMPO• together with the corresponding copper(II)–hydroperoxo intermediates that could not be detected.[183,185,192] Detailed kinetic data supported a H-atom transfer (HAT) as the rate-limiting step in phenol O–H bond activation by copper–superoxide intermediates rather than a PCET mechanism.[185] Subsequent phenol oxygenation products or aliphatic C–H bond oxidation of the ligand (hydroxylation of a methyl group of **TMG** substituent) were detected. However, although these Cu(II)–OO• complexes are able to perform O–H bond activation of phenols, the "true" oxidants in oxygenation reactions are proposed to be species derived from them, such as a high-valent copper–oxo intermediates.[165]

Using the tetradentate [bis(pyrid-2-ylmethyl){[6-(pivalamido)pyrid-2-yl]methyl}amine] ligand (**MPPA**, Scheme 13), in which the pivalamido group provides stabilization of the copper(II)–superoxo intermediate thanks to H-bonding, Karlin's group have provided the first example of intermolecular C–H bond activation by a cupric superoxo complex.[184] Indeed, [(**MPPA**)Cu(O_2)]$^+$ species is able to perform oxidation of relatively weak C–H bonds of 1-benzyl-1,4-dihydronicotinamide (BNAH) and of 1,3-dimethyl-2,3-dihydrobenzimidazole (BzImH). With the same type of ligands (**BA** and **F$_5$BA**, Scheme 13), Karlin's group showed how the H-bonding strength in the ligand secondary coordination sphere tunes the stability and the reactivity of the superoxide species.[193]

The introduction of thioether moiety and sulfur ligation led to more electrophilic copper–superoxide intermediates that were then able to oxidize C–H bond of *N*-methyl-9,10-dihydroacridine (AcrH2) at very low temperature (−135°C) although the analogs based on N_4-containing ligands (**NMe$_2$-TMPA** or **DMM-TMPA**) did not react at that temperature.[187] The substitution of a $N_{pyridine}$ ligand by a $S_{thioether}$ ligand was found to raise the redox potential of the Cu(II)/Cu(I) couple by *ca.* 200 mV, which correlates with the increased electrophilicity of the intermediate. A similar tendency was described in the case of the η^2-bound superoxide intermediate isolated using benzimidazole-based N_3S ligands.[176] The copper–superoxide intermediates were found able to perform C–H bond activation of dihydroanthracene (DHA) and THF.

Tetrahedral copper(II)–superoxyl intermediates based on $\mathbf{L}^{8R}$ type ligands exhibited C–H bond hydroxylation on the benzylic position of the ligand providing copper(II)–alcoholate complexes (Scheme 14).[190] The

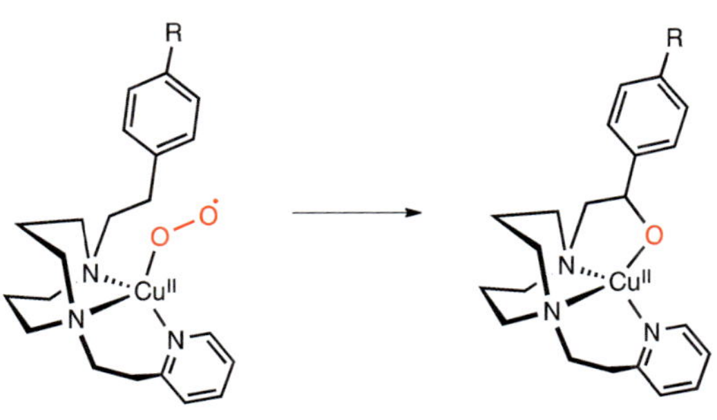

Scheme 14. Intramolecular hydroxylation of the ligand by [($\mathbf{L}^{8R}$)Cu(O_2)]$^+$ species.

addition of TEMPO-H, and thus the generation of the corresponding copper(II)–hydroperoxide intermediate (as observed in the case of tetradentate ligands)[183,192] did not lead to hydroxylation products, supporting a superoxo-mediated oxygenation reactivity.[191] Kinetic and DFT studies have suggested that the ligand hydroxylation reaction involves H-atom abstraction and oxygen rebound mechanism.[194]

When the dianionic ligand derived from pyridinecarboxamide moiety was used, the electron-rich copper–superoxide intermediate displayed nucleophilic character.[195] The intermediate efficiently reacted with acyl chloride substrates (giving carboxylic acids), with aldehyde to perform deformylation reactions as well as Baeyer–Villiger oxidations. Such nucleophilic reactivity has also been reported for some iron–superoxo intermediates.[196]

With a copper(I) complex derived from the calix[6]arene-tren ligand, Reinaud's group observed the formation of a Cu(II)–superoxyl species upon the reaction with dioxygen at low temperature.[63,197] In acetonitrile at –40°C, the copper(I) complex is rapidly oxidized into copper(II) with the concomitant release of superoxide. In contrast, in dichloromethane (DCM) at –90°C, without any molecule susceptible to displace the superoxide, the transient Cu(II)–superoxyl species undergoes an intramolecular redox process leading to the insertion of an O-atom into the ligand. While the precise site of the O-atom insertion into the ligand has not been identified, isotope labeling experiments have confirmed that the O-atom

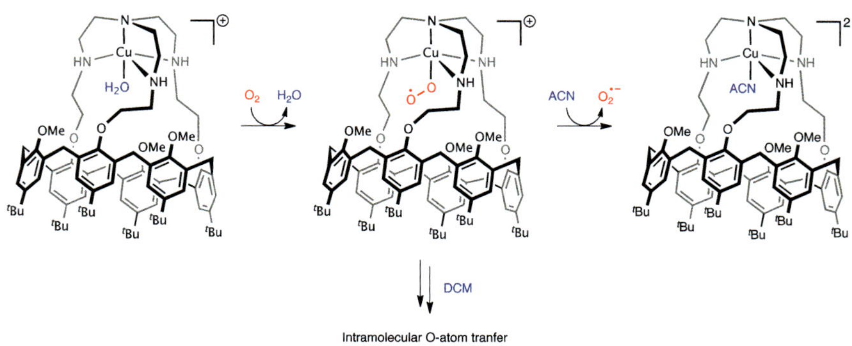

Scheme 15. Reactivity of the {calix[6]arene-trenCu(I)}$^+$ with dioxygen in coordinating (ACN) and non-coordinating (DCM) solvents.

comes from dioxygen and not from a residual water molecule. Electrospray ioniszation mass (ESI-MS) analyses and spectroscopic data are consistent with oxygenation of the OCH_2 moieties linking the **tren** cap to the calix[6]arene core.

In summary, most of the above-mentioned complexes display the ability to perform a variety of transformations (HAT, ET, hydroxylation) that mimic the electrophilic chemistry of copper oxygenases. The observed effects of geometry (tetrahedral *versus* bipyramide trigonal) and ligand variation (S *versus* N-donors) on their reactivity make them relevant active species for the chemistry performed by DβH and PHM enzymes. Their involvement in the catalytic cycle of LPMO enzymes has still to be proven since the geometry and electronic properties of the metal center differ significantly from that of DβH and PHM enzymes.

B. Copper(II)–Hydroperoxo Intermediates

1. *Preparation of Synthetic Copper(II)–Hydroperoxo Species*

In few cases, copper(II)–hydroperoxo transient intermediates have been generated by reaction of Cu(I) complexes with O_2 and the initial formation of a copper(II)–superoxyl species.[186,192,198] In most cases, mononuclear copper(II)–hydroperoxo complexes have been prepared at low temperature (–80°C) by addition of H_2O_2 to a solution of copper(II) complexes in a presence of a base, such as triethylamine, that promotes H_2O_2 deprotonation.[165,167,168] This last procedure (called peroxide shunt) was mainly employed for the preparation of synthetic copper(II)–hydroperoxo complexes, due to difficulties in controlling copper(II)–superoxyl reactivity or reduction (Scheme 16).[199]

Ligands used for the preparation of synthetic copper(II) hydroperoxides are mostly derived from the tripodal tetradentate **TMPA** ligand using different substituents on one or several pyridine groups.[186,200–206] Particularly, one or more H-bonding donor groups have been introduced to stabilize the hydroperoxo by interaction with the proximal O-atom. Tridentate ligands have also been used.[200] Finally, several ligands such as (i) a **N3S** ligand bearing a thioether function;[207] (ii) **Me6-tren**[208] or a **tren**-based cryptate;[209] (iii) **N4Py**;[210] (iv) ligands introducing imidazole

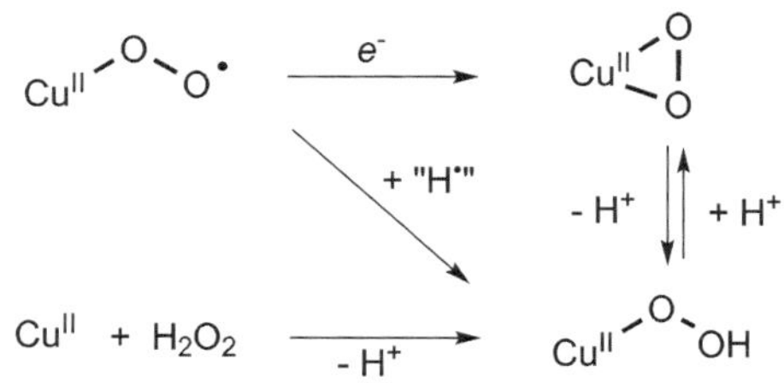

Scheme 16. Possible pathways for the formation of copper(II)–hydroperoxo intermediates.

functions[211,212] and (v) shiff-base derived ligands[213] have been used to prepare copper(II)–hydroperoxo complexes. Example of ligands employed to synthesize mononuclear copper(II)–hydroperoxo complexes are displayed in Scheme 17.

2. *Structural, Spectroscopic and Electronic Properties*

Masuda and coworkers have obtained the crystal structure of a stable copper(II)–hydroperoxo complex using **BPPA** ligand.[214] The structure of this intermediate is presented in Figure 15. The complex displays a five-coordinated axially compressed trigonal bipyramide geometry with the hydroperoxide moiety bound in axial position in end-on (η^1) coordination mode. Bulky hydrophobic *tert*-butyl groups as well as hydrogen bonding involving the amide substituents probably contribute to the great stabilization of this hydroperoxyl species that is stable for a month at room temperature. The Cu–O distance is 1.888 Å, the O–O distance is 1.46 Å and Cu–O–O angle is 114.5°.

The structural data for other Cu(II)OOH intermediates were deducted from spectroscopic and theoretical analysis. Most of the described complexes harbor distorted bipyramide trigonal or square-pyramidal geometries. However, using a trispyrazolylborate-derived ligand (**Tp**^{iPr2}), Chen *et al.* have reported copper(II)–hydroperoxo or copper(II)–alkylperoxo intermediates displaying distorted tetrahedral geometries.[215]

Copper(II)–hydroperoxo species generally display intense peroxo-to-copper LMCT absorption bands at 350-390 nm (ε = 1000–5000 mol^{-1}Lcm^{-1}).

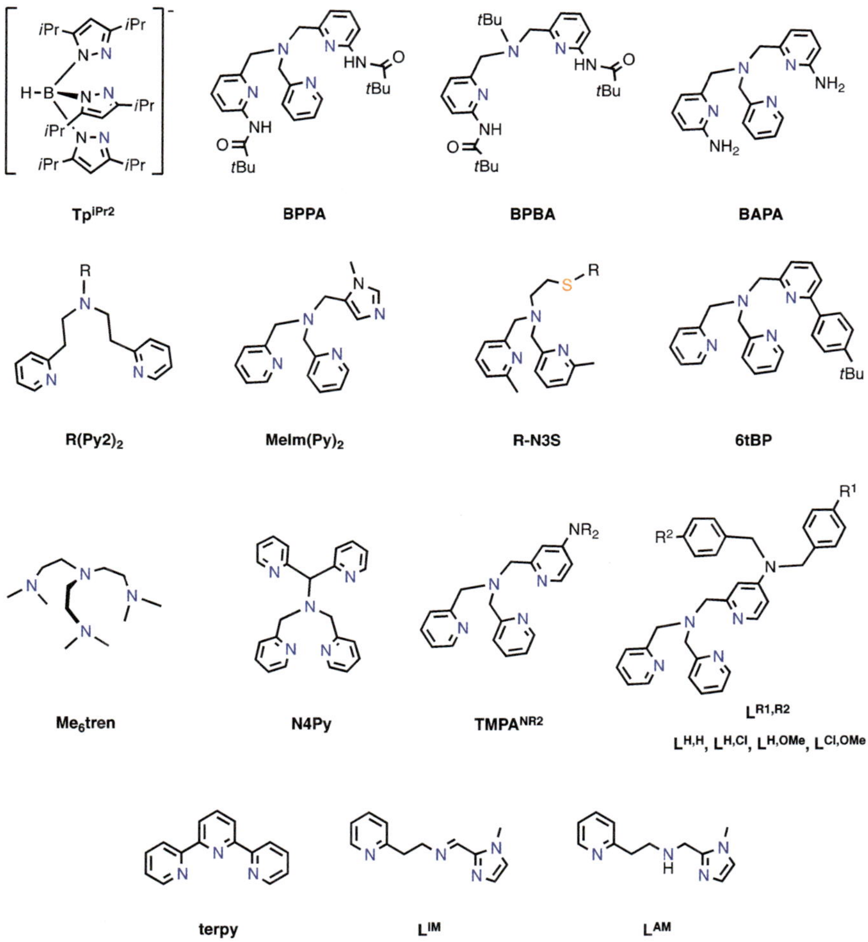

Scheme 17. Examples of ligands used to synthesize mononuclear copper(II)–hydroperoxo complexes.

Vibrational data reveal ν_{O-O} stretches (mainly obtained using rR) at 820–860 cm^{-1} that shift to lower energy by *ca.* 45 cm^{-1} upon ^{18}O substitution.

3. *Reactivity*

The stability of copper(II)-hydroperoxide intermediates were evaluated as a function of the H-bonding interactions provided by the ligands.[199,216]

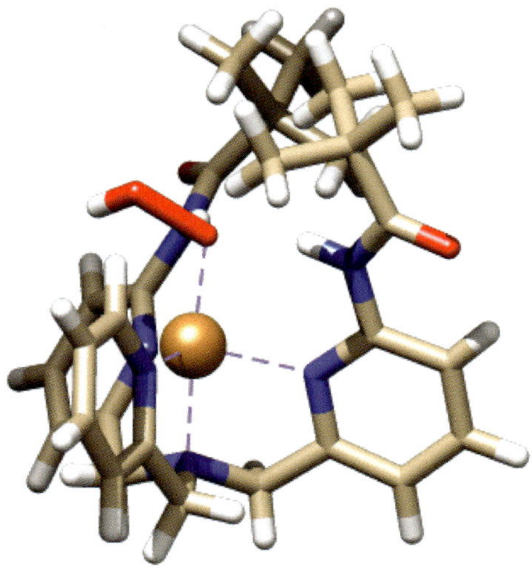

Figure 15. Crystal structure of $[Cu(OOH)(\mathbf{BPPA})]^{+}$. Counter anion was omitted for clarity.

Ligands bearing substituent providing H-bonding to the two different O-atoms of the hydroperoxide moieties were prepared. Hydrogen-bonding interactions involving the coordinated oxygen of hydroperoxide (proximal oxygen, O_p) were shown to enhance the stability of the Cu(II)OOH intermediates.[204,214] On contrary, H-bonding involving the non-coordinated O_d-atom (distal oxygen) from the hydroperoxide moieties resulted in decreased stabilization (activation) of the intermediates (Scheme 18).[200]

Interestingly, using biotin–streptavidin (Sav) technology developed by Ward,[217,218] artificial copper proteins between Sav and $\mathbf{R(Py2)}_2$ type copper(II) complex ($[Cu^{II}(\mathbf{dpea})(H_2O)_2]^{2+}$) conjugated to biotin ($[Cu^{II}(\text{biot-et-}\mathbf{dpea})(H_2O)_2]^{2+}$) have been prepared in order to stabilize a Cu(II)OOH species.[219] The crystal structure of the hydrogen peroxide adduct ($[Cu^{II}(\text{biot-et-}\mathbf{dpea})(H_2O)(O_2H]^{+}\subset$SavWT) has been solved. It shows that the Sav host provides a local environment that stabilizes the Cu(II)OOH thanks to a H-bonding network around a hydroperoxo ligand involving both O-atoms. Indeed, the **dpea** copper(II) complex is anchored within Sav in such a way that the asparagine residue N49 forms an H-bond with the O_d-atom of the hydroperoxido ligand. A second H-bond is formed between a water molecule and the O_p-atom of the hydroperoxido

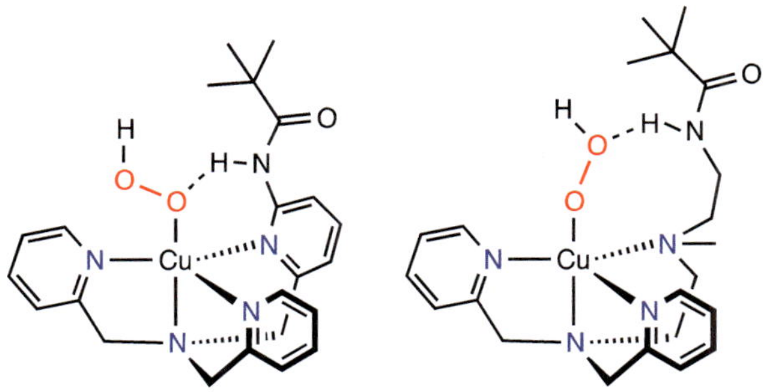

Scheme 18. Possible structures of Cu(II)OOH intermediates displaying hydrogen-bonding interactions with either proximal (right) or distal (left) oxygen of the hydroperoxide moiety.

ligand. This water molecule (O_w) is maintained in the close environment of the hydroperoxo ligand thanks to the serine residue S112 and the carboxy O-atom of the alanine residue A86 that both form a H-bonding network around the water molecule.

Selected deletions of individual H-bonds to the Cu(II)OOH species corroborate the Masuda's proposal (*vide supra*).[200] The Sav N49A variant that disrupts the H-bond to the proximal O-atom of the hydroperoxido ligand maintained the stability of the hydroperoxido ligand and demonstrated that stability is achieved with a single hydrogen bond to the proximal O-atom of the hydroperoxido ligand. The Sav S112A variant that disrupts the network of H-bonding to the distal O-atom of the hydroperoxido ligand produced a reactive variant that oxidized an external substrate such as 4-chlorobenzylamine.

The effect of S *versus* N-donors on the stability of Cu(II)OOH species was also evaluated. Kodera *et al.* have prepared **N3S** ligands (in particular a PhS-thioether containing ligand) mimicking the methionine residue in DβM and PHM. The Cu(II)OOH intermediates were found stabilized by sulfur ligation and it was proposed that polarizable sulfur donor could enhance the covalent character of the Cu–O bond leading to greater stabilization of the intermediates.[207]

In general Cu(II)OOH intermediates have been described as sluggish oxidants towards external substrates. Catalytic oxygenation reactions performed by copper(II) complexes in the presence of H_2O_2 have been

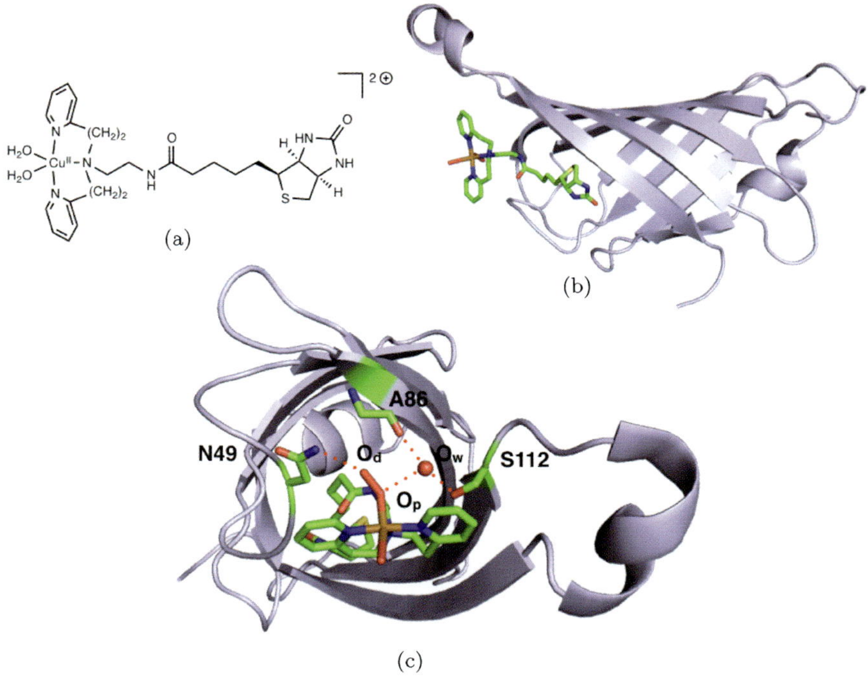

Scheme 19. (a) Representation of biotinylated copper(II) **dpea** complex $[Cu^{II}$(biot-et-**dpea**)$(H_2O)_2]^{2+}$; (b) the $[Cu^{II}$(biot-et-**dpea**)$(H_2O)(HO_2)]^{+}\subset$SavWT (PDB: 6ANX) and (c) left view of (b) showing the local environment around the Cu(II)OOH that stabilizes it due to a H-bonding network (PDB code 5VKX).

reported,[213,220,221] but proof of the direct involvement of the hydroperoxo intermediates in these oxidation reactions are rare. Using Me_6tren or pentadentate **N4Py** ligands, Nam and coworkers have reported that copper(II)-hydroperoxide intermediates are inactive towards electrophilic and nucleophilic substrates.[208,210]

Using tridentate **BPBA** ligand, Masuda and coworkers have characterized a Cu(II)OOH displaying a square-planar geometry around the copper(II) ion.[222] Direct reaction between the Cu(II)OOH intermediate and sulfide substrates was examined. This intermediate was shown to perform oxygenation of dimethylsulfide and quantitative production of sulfoxide was obtained. The reactivity was enhanced as compared to that of Cu(II)OOH displaying trigonal bipyramidal geometries. On the basis of the Masuda proposal,[222] Yamaguchi *et al.* investigated the reactivity of the square-planar $[Cu(\mathbf{terpy})]^{2+}$ complex encapsulated into Na-Y zeolite ($[Cu\text{-}(\mathbf{terpy})]^{2+}\subset Y$).[223] The reaction of $[Cu(\mathbf{terpy})]^{2+}\subset Y$ with H_2O_2 led to a thermally stable Cu(II)OOH species characterized by spectroscopic methods (EPR, UV-vis. and rR). In the presence of an excess of H_2O_2, $[Cu(\mathbf{terpy})]^{2+}$ free complex and $[Cu(\mathbf{terpy})]^{2+}\subset Y$ complex exhibited both high catalytic selectivity for the oxidation of thioanisole into methylphenylsulfoxide (TON = 22). In the same conditions, $[Cu(\mathbf{terpy})]^{2+}\subset Y$ complex performed stoichiometric transformations of phenyethylamine into 2-amino-1-phenylethanol, and benzene into phenol.

Using tridentate $\mathbf{L}^{AM}$ and $\mathbf{L}^{IM}$ ligands, Simaan and coworkers reported two copper complexes as first functional catalytic models for LPMO (Figure 16).[224] Both complexes feature structural and spectroscopic properties similar to those of the enzyme: (i) X-ray diffraction

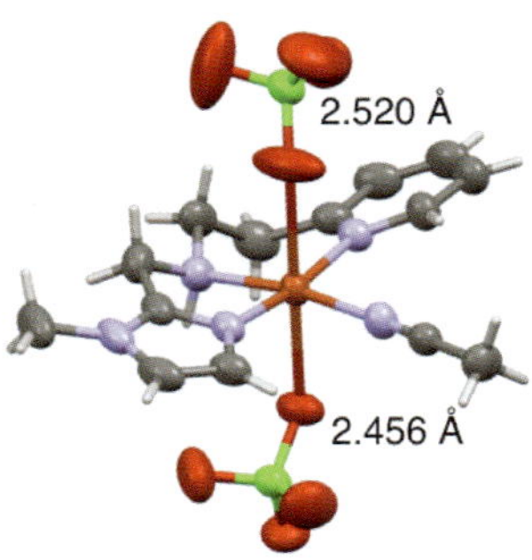

$[(\mathbf{L}^{AM})Cu^{II}(ACN)](ClO_4)_2$

(a)

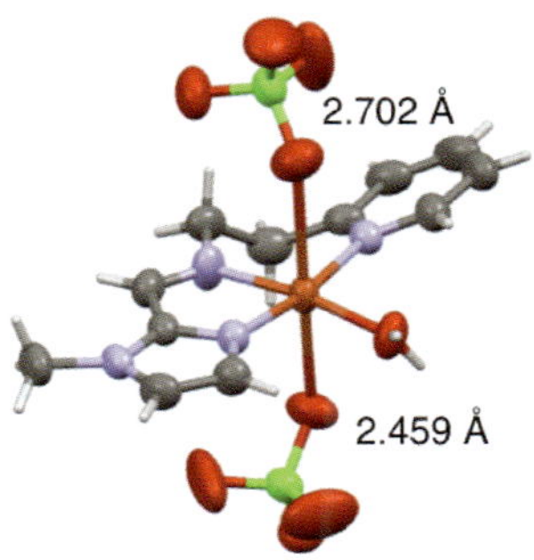

$[(\mathbf{L}^{IM})Cu^{II}(H_2O)](ClO_4)_2$

(b)

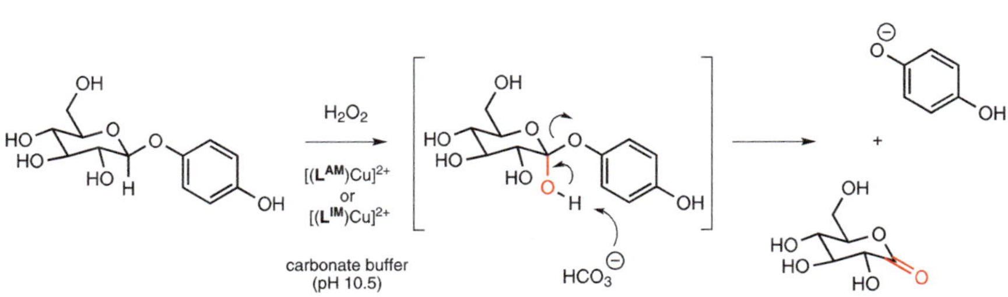

(c)

Figure 16. Structures of copper complexes (a) $[(\mathbf{L}^{AM})Cu^{II}(ACN)](ClO_4)_2$; (b) $[(\mathbf{L}^{iM})Cu^{II}(H_2O)](ClO_4)_2$ and (c) Oxidative cleavage of *p*-nitrophenyl-β-D- glucopyranoside.

analysis showed that the copper(II) ions are in pseudooctahedral geometry and are coordinated in the equatorial plane by the three N-donors with Cu–N distances within the range of distances found in the enzymatic systems and (ii) EPR parameters are in good agreement with those obtained for LPMOs.[72,161,225] In addition, they catalyze oxidative cleavage of the model substrate *p*-nitrophenyl-β-D-glucopyranoside. A particularly stable copper(II) hydroperoxide intermediate was detected in the reaction conditions. Direct involvement of this intermediate in the H-atom abstraction of the substrate or O-atom transfer was however not established.

Karlin and coworkers have used a series of **TMPA**-derived ligands, in which substituents were attached at the 6-position of one pyridine moiety. Using *p-tert*-butylphenyl substituents on **TMPA** ligand (**6tBP**), the Cu(II) OOH intermediate was generated at low temperature and upon warming, aromatic ligand hydroxylation was observed leading to a phenolate-Cu(II) final product.[202] Aromatic hydroxylation was not observed when the bis-*μ*-oxo-dicopper(III) intermediate was prepared using the same ligand. Direct involvement of Cu(II)OOH species was not confirmed and high-valent copper–oxo species derived from O–O bond cleavage were rather proposed as active species. Using *N,N*-dialkylamino substituents (**TMPA**^{NR2}, R = Me or CH_2Ph), oxidative ligand *N*-dealkylation was observed (Scheme 20).[203,226] Observed kinetic isotope effects upon deuteration (KIE = 2) are rather low for an H-atom abstraction mechanism. To obtain insights into the mechanism of this process, additional **TMPA**^{NR2} ligands were used (**L**^{R1,R2} in Scheme 17).[206] The corresponding Cu(II)OOH intermediates were generated at low temperature. Based on experimental data and DFT-calculations, a mechanism involving direct

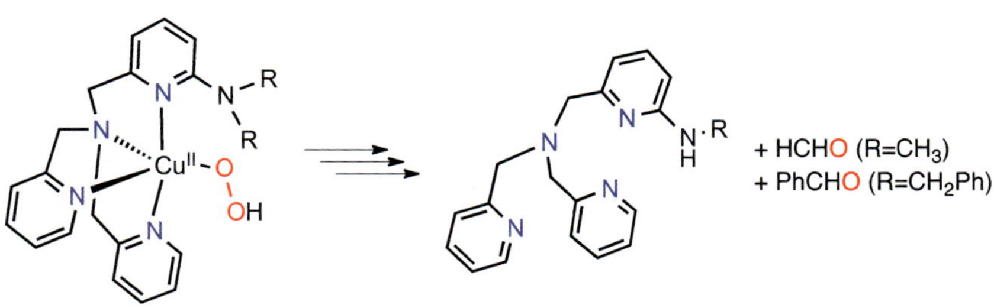

Scheme 20. *N*-dealkylation reaction observed using **TMPA**^{NR2} ligands after warming-up a solution of Cu(II)OOH intermediate.[203,226]

H-atom abstraction by the Cu(II)OOH intermediate was ruled out. It was concluded that the chemistry rather proceeds through Cu–O homolytic cleavage followed by reduction of the copper center to Cu(I). This copper(I) species would further undergo reaction with H_2O_2 to yield a Cu(I)OOH intermediate responsible for Fenton-type reactions and for the generation of hydroxyl radicals that are thought to be responsible for the observed reactivity. Interestingly, reduction of Cu(II) to Cu(I) has been reported in several cases upon reaction of copper(II) complexes with hydrogen peroxide.[227,228] Cu(I)OOH species were also postulated as intermediates in oxidation reactions although their direct involvement was not established.[228–230]

C. High-Valent Mononuclear Copper–Oxo or Hydroxo Species

1. *Copper(II)–Oxyl/Copper(III)–Oxo species, $[CuO]^+$*

Mononuclear copper–oxo species have been postulated as important reactive intermediates in mononuclear copper-containing biological systems and in their synthetic low molecular weight models. However, no copper–oxo intermediate has ever been stabilized or isolated and no direct evidence of the formation of such species in either biological and biomimetic systems has been obtained.[231] The reactivity of such $[CuO]^+$ core has rather been evaluated using indirect methods.

One way to generate such species is by reaction of Cu(I) precursors with O-atom transfer reagent such as PhIO. In 1989, Réglier *et al.* have reported that the reaction of Cu(I) complexes using **RPy2** ligands (R = CH_2Py or CH_2CH_2Py) with PhIO led to hydroxylation of a pyridine ligand in *ortho* position.[232] Possible formation of a high-valent copper–oxo species was discussed. Later, the group of Karlin examined the reaction of $[(\mathbf{(TMG)_3tren})Cu]^+$ complex with PhIO in ACN and observed hydroxylation of a methyl substituent on the ligand and the almost quantitative formation of final copper(II)–alkoxo complex (Scheme 21).[192] Interestingly, using $\mathbf{TMPA}^{NR2}$ (R = Ch_2Ph in Scheme 17), the reaction between Cu(I) precursor and PhIO was followed by low-temperature ESI-MS spectrometry allowing the detection of the cupryl intermediate (*e.g.* $[(\mathbf{TMPA}^{NR2})CuO]^+$) at the early stage of the reaction before the predominant formation of the final alkoxo complex.[203]

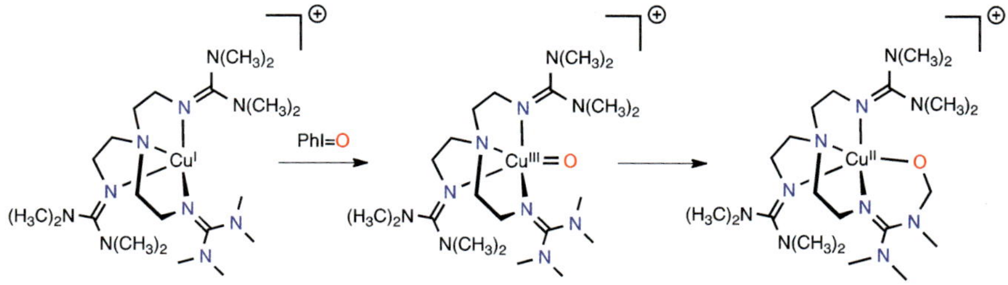

Scheme 21. Proposed reaction of [(**TMG**$_3$**tren**)Cu]$^{1+}$ complex with PhIO, *via* the formation of cupryl intermediates.

One means for generating high-valent copper–oxo species is to get inspiration from non-heme iron enzymes that generate analogous $[FeO]^{2+}$ species. In particular, 2-oxoglutarate dependent dioxygenases are non-heme iron(II) containing enzymes that react with dioxygen and generate high-valent Fe(IV)=O species upon oxidative decarboxylation of the 2-oxoglutarate cofactor.[233] Tolman and coworkers have therefore used oxidative decarboxylation of Cu(I)-2–oxoacid complexes to generate high-valent copper–oxo species.[234] Pyridine-imine ligands with arene substituents appropriately positioned for intramolecular reaction with transient copper–oxygen species were used (Scheme 22). Quantum calculations predicted that the observed intramolecular arene hydroxylation is performed by $[CuO]^+$ intermediates that are described as $[Cu^{II}–O^{\bullet}]^+$ species with triplet ground states.[235]

Finally, the only compelling evidences for the reactivity of such high-valent copper–oxo species came from gas-phase reactions. The group of Schwarz has performed ESI-MS of copper(II) nitrate solution in the presence of phenanthroline (**phen**) leading to the detection of the [(**phen**)CuO]$^+$ cation.[236] No reaction of the mass-selected [(**phen**)CuO]$^+$ ion with methane or ethane was detected. However, when reacted with propane, two new ionic products were detected, [(**phen**)CuO]$^+$ and [(**phen**)Cu]$^+$ probably arising from H-atom abstraction and O-atom transfer reactions respectively. Several years later, the same group has generated gaseous $[CuO]^+$ species and reported their reactivity towards methane.[237] These species were found efficient for the strong C–H bond activation of methane at room temperature both by H-atom abstraction (40%) and O-atom transfer (60%), the latter probably arising from

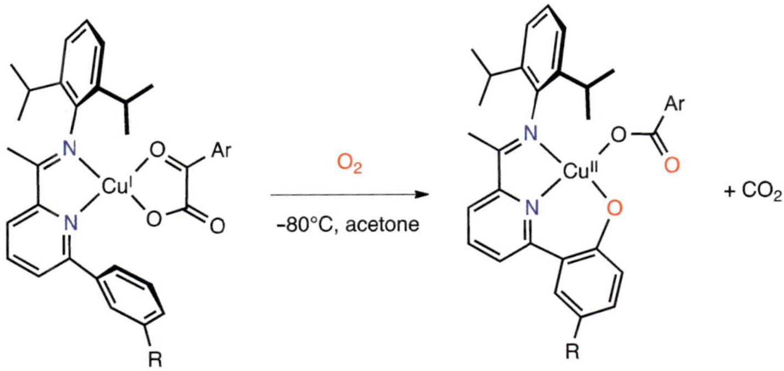

Scheme 22. Reactivity of Cu(I)-2–oxoacid complexes with dioxygen. R = H or OMe; Ar = Ph or Mes (mesityl).

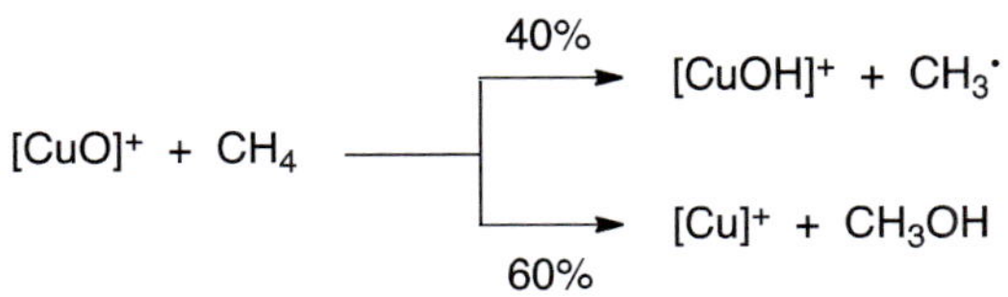

Scheme 23. Activation of methane by gaseous $[CuO]^+$ species.

rebound mechanism (Scheme 23). High-level quantum chemical calculations were conducted and revealed that the electronic state of $[CuO]^+$ could be best described as $[Cu^{II}–O\bullet]^+$ species with a triplet (S = 1) ground-state and the importance of the oxygen-centered radical in the reactivity towards methane was demonstrated.

2. *Copper(III)–Hydroxo Species, [CuOH]$^+$*

Tolman and coworkers have prepared Cu(III)–OH species by one-electron oxidation of the Cu(II)–OH precursors using electron donating-ligands that efficiently stabilize Cu(III) redox state (Scheme 24). Using *N,N'*-bis(2,6-diisopropylphenyl)-2,6-pyridinedicarboxamide ligand, the Cu(III)–OH species was generated at low temperature upon chemical oxidation of the starting Cu(II)–OH complex.[238] This oxidized species was characterized by an intense ligand-to-metal charge transfer transition centered at 540 nm (ε = 12,400 mol.$^{-1}$L.cm^{-1}). X-ray absorption spectroscopy (XAS) and DFT studies confirmed the presence of a Cu(III)–OH moiety.

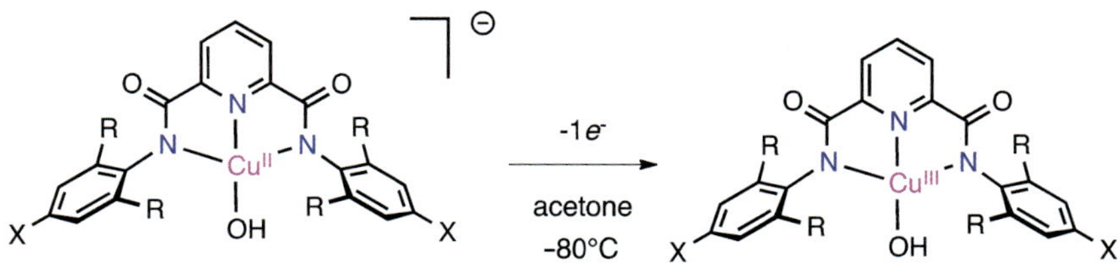

Scheme 24. Generation of Cu(III)–OH from 1-electron chemical oxidation of the Cu(II) precursors (R = *i*Pr and X = H; R = *i*Pr and X = NO_2; R = Me and X = H).

The intermediate was found able to activate C–H bonds of dihydroanthracene that is converted into anthracene and LCu(II)–OH_2 complex was concomitantly formed. Further studies revealed the ability of the intermediate to activate strong C–H bonds with high rates including that of cyclohexane (bond dissociation energy, BDE = 99 kcal.mol^{-1}).[159] Detailed kinetic studies revealed large kinetic isotope effects (k_D/k_H) consistent with a HAT of C–H bonds as rate-determining steps. Several ligand sets displaying different electron-donating or electron-withdrawing capabilities were prepared in order to evaluate the electronic effects on the property and reactivity of the corresponding Cu(III)–OH cores.[239,240]

Overall, the reactivity of $[CuOH]^{2+}$ species in H-atom abstraction reactions was found significantly higher than that of most reported metal–oxo or metal–hydroxo complexes.[159] O–H BDEs of *ca.* 90 kcal.mol^{-1} were determined and are among the highest compared to the other metal–oxo/hydroxo complexes. The high basicity of the hydroxide moieties (pKa 19–20 in THF) consistent with the dianionic strongly electron donating ligand, was found an essential factor to contribute to the high O–H BDE. On the basis of these studies it is therefore possible to consider such $[CuOH]^{2+}$ cores in the mechanism of mononuclear copper-containing enzymes and biomimetic systems. In the case of LPMO, Tolman *et al.* have proposed that such Cu(III)–OH species could be formed upon protonation of $[CuO]^+$ species *via* tautomerization involving the primary amine of the histidine-brace.[231] Deprotonation of the primary amine could then greatly contribute to the stabilization of the Cu(III)–OH moiety and provide the necessary electronic donation in order to get appropriate basicity and reactivity of the core, as observed in the model complexes (Scheme 25). Deprotonation of NH_2-terminal function in LPMO is still debated.[164]

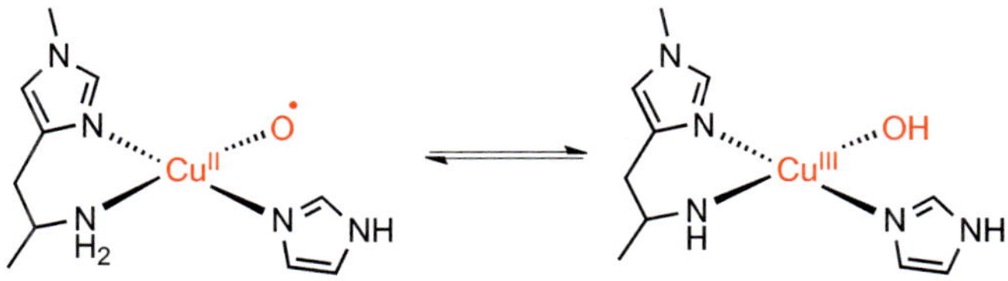

Scheme 25. Possible deprotonation of the histidine-brace and tautomerization of $[CuO]^+$ and $[CuOH]^{2+}$ species in the active site of LPMO.

3. *Copper(III)–Oxyl Species, $[CuO]^{2+}$*

Looking for the copper/oxygen intermediate responsible for the hydroxylation of substrate in PHM, Amzel and coworkers proposed that this intermediate could be a copper(III)–oxyl species resulting from the heterolytic O–O bond cleavage of a Cu(II)OOH entity (Scheme 5).[48] This highly reactive copper(III)–oxyl species has been also more recently proposed as active species in LPMO catalysis (Scheme 11).[158] With model compounds, only one paper reports on such high-valent metal–oxo species. Chen and coworkers have prepared a series of copper(II) complexes based on the iminodiacetic acid (**IDA**) ligand on which they studied the effect of various pyridines substituted on the *para* position.[241] Catalytic activity tests on the azo dye RR195 degradation in the presence of H_2O_2 showed that the use of 4-aminopyridine (**4-ampy**) resulted in an enhanced catalytic activity of the Cu^{II}(**IDA**) complex. The high-valent copper–oxo species [•OCu^{III}(**IDA**)(**4-ampy**)$_2$] were detected by *in situ* high definition ESI-MS spectrometry. In addition, DFT-calculations showed that the [•OCu^{III}(**IDA**)(**4-ampy**)$_2$] species has a stable structure in the S = 3/2 ground state. Considering this, they propose that electron-donating **4-ampy** provides a "push" effect that allow the heterolytic cleavage of the O–O bond of the Cu(II)OOH intermediate leading the copper(III)–oxyl species (Scheme 26).

V. Conclusions and Perspectives

The accumulation of biochemical and structural data, the development of bioinspired/biomimetic copper complexes as well as the help of theoretical calculations allowed us a better understanding of the functioning of the mononuclear copper-containing monooxygenases. While, the formation in

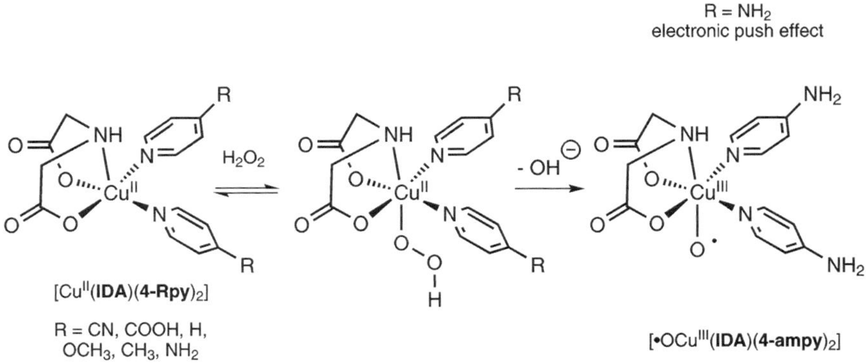

Scheme 26. Hydrogen peroxide activation by [Cu^{II}(**IDA**)(**4-Rpy**)$_2$] complexes.

the first step of a copper(II)–superoxyl species now seems well admitted, the timing of the second ET is less obvious and is always the subject of discussion. This ET is the *chef d'orchestre* of the reaction catalyzed by these enzymes. A late ET will maintain a predominant role for the copper(II)–superoxyl species while an early ET that will reduce the copper(II)–superoxyl species, will give the predominant role to a copper(II)–hydroperoxo species. If in the case of PHM, DβH and TβH, the ET timing is imposed by the second copper center (Cu_H and Cu_A), in the case of LPMOs this role is played by the reductant, which is still unknown.

To date, a lot of copper complexes that activate dioxygen have been studied. Most of them reproduce very well the coordination of oxygen species (dioxygen, superoxide and peroxide) to mononuclear copper center, but few are able to activate an exogenous substrate in a catalytic manner. One of the reasons of this unsuccessful approach could be the difficulty to control the electron transfer steps to the copper/oxygen core. In our opinion, the challenge for the future will be to develop new systems integrating a catalytic core with an ET core in the manner of the non-coupled binuclear copper containing monooxygenases, PHM, DβH and TβH.

VI. References and Citations

1. Anastas, P.; Eghbali, N. *Chem. Soc. Rev.* **2010**, *39*, 301.
2. Derouane, E. G. *CATTECH* **2001**, *5*, 214.
3. Arakawa, H.; Aresta, M.; Armor, J. N.; Barteau, M. A.; Beckman, E. J.; Bell, A. T.; Bercaw, J. E.; Creutz, C.; Dinjus, E.; Dixon, D. A.; Domen, K.; DuBois, D. L.;

Eckert, J.; Fujita, E.; Gibson, D. H.; Goddard, W. A.; Goodman, D. W.; Keller, J.; Kubas, G. J.; Kung, H. H.; Lyons, J. E.; Manzer, L. E.; Marks, T. J.; Morokuma, K.; Nicholas, K. M.; Periana, R.; Que, L., Jr; Rostrup-Nielson, J.; Sachtler, W. M.; Schmidt, L. D.; Sen, A.; Somorjai, G. A.; Stair, P. C.; Stults, B. R.; Tumas, W. *Chem. Rev.* **2001**, *101*, 953.

4. Yee, G. M.; Tolman, W. B. In *Metal Ions in Life Sciences*; Springer International Publishing: Cham, 2014; Vol. 15, pp. 131–204.
5. Schmid, R. D.; Urlacher, V. In *Modern biooxidation: enzymes, reactions and applications*; Wiley, 2007.
6. Pazmino, D. E. T.; Winkler, M.; Glieder, A.; Fraaije, M. W. *J. Biotechnol.* **2010**, *146*, 9.
7. Que, L., Jr; Tolman, W. B. *Nature* **2008**, *455*, 333.
8. Matoba, Y.; Kumagai, T.; Yamamoto, A.; Yoshitsu, H.; Sugiyama, M. *J. Biol. Chem.* **2006**, *281*, 8981.
9. Halfen, J. A.; Mahapatra, S.; Wilkinson, E. C.; Kaderli, S.; Young, V. G.; Que, L., Jr; Zuberbühler, A. D.; Tolman, W. B. *Science* **1996**, *271*, 1397.
10. Halvagar, M. R.; Solntsev, P. V.; Lim, H.; Hedman, B.; Hodgson, K. O.; Solomon, E. I.; Cramer, C. J.; Tolman, W. B. *J. Am. Chem. Soc.* **2014**, *136*, 7269.
11. Isaac, J. A.; Gennarini, F.; López, I.; Thibon-Pourret, A.; David, R.; Gellon, G.; Gennaro, B.; Philouze, C.; Meyer, F.; Demeshko, S.; Le Mest, Y.; Réglier, M.; Jamet, H.; Le Poul, N.; Belle, C. *Inorg. Chem.* **2016**, *55*, 8263.
12. Gennarini, F.; Kochem, A.; Isaac, J.; Mansour, A.-T.; López, I.; Le Mest, Y.; Thibon-Pourret, A.; Faure, B.; Jamet, H.; Le Poul, N.; Belle, C.; Simaan, A. J.; Réglier, M. *Inorg. Chim. Acta.* **2018**, *481*, 113.
13. Gennarini, F.; David, R.; López, I.; Le Mest, Y.; Réglier, M.; Belle, C.; Thibon-Pourret, A.; Jamet, H.; Le Poul, N. *Inorg. Chem.* **2017**, *56*, 7707.
14. Kochem, A.; Gennarini, F.; Yemloul, M.; Orio, M.; Le Poul, N.; Rivière, E.; Giorgi, M.; Faure, B.; Le Mest, Y.; Réglier, M.; Simaan, A. J. *ChemPlusChem* **2017**, *82*, 615.
15. Cao, L.; Caldararu, O.; Rosenzweig, A. C.; Ryde, U. *Angew. Chem. Int. Ed. Engl.* **2018**, *57*, 162.
16. Shiota, Y.; Yoshizawa, K. *Inorg. Chem.* **2009**, *48*, 838.
17. Culpepper, M. A.; Rosenzweig, A. C. *Crit. Rev. Biochem. Mol. Biol.* **2012**, *47*, 483.
18. Sirajuddin, S.; Rosenzweig, A. C. *Biochemistry* **2015**, *54*, 2283.
19. Osborne, R. L.; Klinman, Insights into the proposed copper–oxygen intermediates that regulate the mechanism of reactions catalyzed by dopamine β-monooxygenase, peptidylglycine α-hydroxylating monooxygenase, and tyramine β-monooxygenase, in *Copper-Oxygen Chemistry*, K. D. Karlin, S. Itoh, Eds.; John Wiley & Sons Inc., Hoboken, NJ, 2011; pp. 1–22.
20. Réglier, M.; Belle, C. In *Encyclopedia of Metalloproteins*; Kretsinger, R. H., Uversky, V. N., Permyakov, E. A., Eds.; Encyclopedia of Metalloproteins; Springer New York: New York, NY, 2013; pp. 723–729.
21. Chambers, K. J.; Tonkin, L. A.; Chang, E.; Shelton, D. N.; Linskens, M. H.; Funk, W. D. *Gene* **1998**, *218*, 111.
22. Xin, X.; Mains, R. E.; Eipper, B. A. *J. Biol. Chem.* **2004**, *279*, 48159.

23. Dhawan, S.; Duong, L. T.; Ornberg, R. L.; Fleming, P. J. *J. Biol. Chem.* **1987**, *262*, 1869.
24. Gray, E. E.; Small, S. N.; McGuirl, M. M. *Protein Expr. Purif.* **2006**, *47*, 162.
25. Murthy, A. S.; Mains, R. E.; Eipper, B. A. *J. Biol. Chem.* **1986**, *261*, 1815.
26. Solomon, E. I.; Heppner, D. E.; Johnston, E. M.; Ginsbach, J. W.; Cirera, J.; Qayyum, M.; Kieber-Emmons, M. T.; Kjaergaard, C. H.; Hadt, R. G.; Tian, L. *Chem. Rev.* **2014**, *114*, 3659.
27. Stewart, L. C.; Klinman, J. P. *Annu. Rev. Biochem.* **1988**, *57*, 551.
28. Hess, C. R.; McGuirl, M. M.; Klinman, J. P. *J. Biol. Chem.* **2008**, *283*, 3042.
29. Kulathila, R.; Merkler, K. A.; Merkler, D. J. *Nat. Prod. Rep.* **1999**, *16*, 145.
30. Eipper, B. A.; Perkins, S. N.; Husten, E. J.; Johnson, R. C.; Keutmann, H. T.; Mains, R. E. *J. Biol. Chem.* **1991**, *266*, 7827.
31. Chufan, E. E.; De, M.; Eipper, B. A.; Mains, R. E.; Amzel, L. M. *Structure* **2009**, *17*, 965.
32. Mueller, G. P.; Driscoll, W. J. Post-translational modifications of proteins; In *Methods in Molecular Biology*; Humana Press: Totowa, NJ, 2008; Vol. 446, pp. 67–84.
33. Southan, C.; Kruse, L. I. *FEBS Lett.* **1989**, *255*, 116.
34. Klinman, J. P. *Chem. Rev.* **1996**, *96*, 2541.
35. Hasnain, S. S.; Diakun, G. P.; Knowles, P. F.; Binsted, N.; Garner, C. D.; Blackburn, N. J. *Biochem. J.* **1984**, *221*, 545.
36. Blackburn, N. J.; Hasnain, S. S.; Pettingill, T. M.; Strange, R. W. *J. Biol. Chem.* **1991**, *266*, 23120.
37. Walker, G. A.; Kon, H.; Lovenberg, W. *Biochim. Biophys. Acta* **1977**, *482*, 309.
38. Ljones, T.; Flatmark, T.; Skotland, T.; Petersson, L.; Bäckström, D.; Ehrenberg, A. *FEBS Lett.* **1978**, *92*, 81.
39. Prigge, S. T.; Kolhekar, A. S.; Eipper, B. A.; Mains, R. E.; Amzel, L. M. *Science* **1997**, *278*, 1300.
40. Chufan, E. E.; Prigge, S. T.; Siebert, X.; Eipper, B. A.; Mains, R. E.; Amzel, L. M. *J. Am. Chem. Soc.* **2010**, *132*, 15565.
41. Kapoor, A.; Shandilya, M.; Kundu, S. *Plos One* **2011**, *6*, e26509.
42. Vendelboe, T. V.; Harris, P.; Zhao, Y.; Walter, T. S.; Harlos, K.; Omari, El, K.; Christensen, H. E. M. *Sci. Adv.* **2016**, *2*, e1500980.
43. Prigge, S. T.; Kolhekar, A. S.; Eipper, B. A.; Mains, R. E.; Amzel, L. M. *Nat. Struct. Biol.* **1999**, *6*, 976.
44. La Lande, de, A.; Mart, S.; Parisel, O.; Moliner, V. *J. Am. Chem. Soc.* **2007**, *129*, 11700.
45. Zhu, H.; Sommerhalter, M.; Nguy, A. K. L.; Klinman, J. P. *J. Am. Chem. Soc.* **2015**, *137*, 5720.
46. Prigge, S. T.; Eipper, B. A.; Mains, R. E.; Amzel, L. M. *Science* **2004**, *304*, 864.
47. Siebert, X.; Eipper, B. A.; Mains, R. E.; Prigge, S. T.; Blackburn, N. J.; Amzel, L. M. *Biophys. J.* **2005**, *89*, 3312.
48. Rudzka, K.; Moreno, D. M.; Eipper, B. A.; Mains, R. E.; Estrin, D. A.; Amzel, L. M. *J. Biol. Inorg. Chem.* **2013**, *18*, 223.
49. Hess, C. R.; Klinman, J. P.; Blackburn, N. J. *J. Biol. Inorg. Chem.* **2010**, *15*, 1195.

50. Miller, S. M.; Klinman, J. P. *Biochemistry* **1985**, *24*, 2114.
51. Tian, G.; Berry, J. A.; Klinman, J. P. *Biochemistry* **1994**, *33*, 226.
52. Tian, G.; Berry, J.; Klinman, J. P. *Biochemistry* **1994**, *33*, 14650.
53. Fitzpatrick, P. F.; Flory, D. R.; Villafranca, J. J. *Biochemistry* **1985**, *24*, 2108.
54. Mitrochkine, A. A.; Eydoux, F.; Gil, G.; Réglier, M. *Eur. J. Org. Chem.* **1998**, *1998*, 1171.
55. Driscoll, W. J.; König, S.; Fales, H. M.; Pannell, L. K.; Eipper, B. A.; Mueller, G. P. *Biochemistry* **2000**, *39*, 8007.
56. Langella, E.; Pierre, S.; Ghattas, W.; Giorgi, M.; Réglier, M.; Saviano, M.; Esposito, L.; Hardré, R. *ChemMedChem* **2010**, *5*, 1568.
57. Blain, I.; Slama, P.; Giorgi, M.; Tron, T.; Réglier, M. *J. Biotechnol.* **2002**, *90*, 95.
58. Evans, J. P.; Blackburn, N. J.; Klinman, J. P. *Biochemistry* **2006**, *45*, 15419.
59. Evans, J. P.; Ahn, K.; Klinman, J. P. *J. Biol. Chem.* **2003**, *278*, 49691.
60. Prigge, S. T.; Mains, R. E.; Eipper, B. A.; Amzel, L. M. *Cell. Mol. Life Sci.* **2000**, *57*, 1236.
61. Chen, P.; Solomon, E. I. *J. Am. Chem. Soc.* **2004**, *126*, 4991.
62. La Lande, de, A.; Parisel, O.; Gérard, H.; Moliner, V.; Reinaud, O. *Chem. Eur. J.* **2008**, *14*, 6465.
63. Meliá, C.; Ferrer, S.; Rezáč, J.; Parisel, O.; Reinaud, O.; Moliner, V.; La Lande, de, A. *Chem. Eur. J.* **2013**, *19*, 17328.
64. Cowley, R. E.; Tian, L.; Solomon, E. I. *Proc. Natl. Acad. Sci. USA* **2016**, *113*, 12035.
65. Slama, P.; Jabre, F.; Tron, T.; Réglier, M. *FEBS Lett.* **2001**, *491*, 55.
66. Bauman, A. T.; Yukl, E. T.; Alkevich, K.; McCormack, A.; Blackburn, N. J. *J. Biol. Chem.* **2006**, *281*, 4190.
67. Cragg, S. M.; Beckham, G. T.; Bruce, N. C.; Bugg, T. D.; Distel, D. L.; Dupree, P.; Etxabe, A. G.; Goodell, B. S.; Jellison, J.; McGeehan, J. E.; McQueen-Mason, S. J.; Schnorr, K.; Walton, P. H.; Watts, J. E.; Zimmer, M. *Curr. Opin. Chem. Biol.* **2015**, *29*, 108.
68. Gupta, V. K.; Kubicek, C. P.; Berrin, J.-G.; Wilson, D. W.; Couturier, M.; Berlin, A.; Filho, E. X. F.; Ezeji, T. *Trends. Biochem. Sci.* **2016**, *41*, 633.
69. Henrissat, B.; Driguez, H.; Viet, C.; Schülein, M. *Nat. Biotechnol.* **1985**, *3*, 722.
70. Cantarel, B. L.; Coutinho, P. M.; Rancurel, C.; Bernard, T.; Lombard, V.; Henrissat, B. *Nucleic Acids Res.* **2009**, *37* (Database), D233.
71. Reese, E. T.; Siu, R. G. H.; Levinson, H. S. *J. Bacteriol.* **1950**, *59*, 485.
72. Vaaje-Kolstad, G.; Westereng, B.; Horn, S. J.; Liu, Z.; Zhai, H.; Sørlie, M.; Eijsink, V. G. H. *Science* **2010**, *330*, 219.
73. Quinlan, R. J.; Sweeney, M. D.; Leggio, Lo, L.; Otten, H.; Poulsen, J.-C. N.; Johansen, K. S.; Krogh, K. B. R. M.; Jørgensen, C. I.; Tovborg, M.; Anthonsen, A.; Tryfona, T.; Walter, C. P.; Dupree, P.; Xu, F.; Davies, G. J.; Walton, P. H. *Proc. Natl. Acad. Sci. USA* **2011**, *108*, 15079.
74. Levasseur, A.; Drula, E.; Lombard, V.; Coutinho, P. M.; Henrissat, B. *Biotechnol. Biofuels* **2013**, *6*, 41.
75. Hemsworth, G. R.; Henrissat, B.; Davies, G. J.; Walton, P. H. *Nat. Chem. Biol.* **2014**, *10*, 122.

76. Leggio, Lo, L.; Simmons, T. J.; Poulsen, J.-C. N.; Frandsen, K. E. H.; Hemsworth, G. R.; Stringer, M. A.; Freiesleben, von, P.; Tovborg, M.; Johansen, K. S.; De Maria, L.; Harris, P. V.; Soong, C.-L.; Dupree, P.; Tryfona, T.; Lenfant, N.; Henrissat, B.; Davies, G. J.; Walton, P. H. *Nat. Commun.* **2015**, *6*, 5961.
77. Couturier, M.; Ladevèze, S.; Sulzenbacher, G.; Ciano, L.; Fanuel, M.; Moreau, C.; Villares, A.; Cathala, B.; Chaspoul, F.; Frandsen, K. E.; Labourel, A.; Herpoël-Gimbert, I.; Grisel, S.; Haon, M.; Lenfant, N.; Rogniaux, H.; Ropartz, D.; Davies, G. J.; Rosso, M.-N.; Walton, P. H.; Henrissat, B.; Berrin, J.-G. *Nat. Chem. Biol.* **2018**, *14*, 306.
78. Sabbadin, F.; Hemsworth, G. R.; Ciano, L.; Henrissat, B.; Dupree, P.; Tryfona, T.; Marques, R. D. S.; Sweeney, S. T.; Besser, K.; Elias, L.; Pesante, G.; Li, Y.; Dowle, A. A.; Bates, R.; Gomez, L. D.; Simister, R.; Davies, G. J.; Walton, P. H.; Bruce, N. C.; McQueen-Mason, S. J. *Nat. Commun.* **2018**, *9*, 756.
79. Harris, P. V.; Welner, D.; McFarland, K. C.; Re, E.; Navarro Poulsen, J.-C.; Brown, K.; Salbo, R.; Ding, H.; Vlasenko, E.; Merino, S.; Xu, F.; Cherry, J.; Larsen, S.; Leggio, Lo, L. *Biochemistry* **2010**, *49*, 3305.
80. Westereng, B.; Ishida, T.; Vaaje-Kolstad, G.; Wu, M.; Eijsink, V. G. H.; Igarashi, K.; Samejima, M.; Ståhlberg, J.; Horn, S. J.; Sandgren, M. *Plos One* **2011**, *6*, e27807.
81. Garcia-Gonzalez, E.; Poppinga, L.; Fünfhaus, A.; Hertlein, G.; Hedtke, K.; Jakubowska, A.; Genersch, E. *PLoS Pathog.* **2014**, *10*, e1004284.
82. Chiu, E.; Hijnen, M.; Bunker, R. D.; Boudes, M.; Rajendran, C.; Aizel, K.; Oliéric, V.; Schulze-Briese, C.; Mitsuhashi, W.; Young, V.; Ward, V. K.; Bergoin, M.; Metcalf, P.; Coulibaly, F. *Proc. Natl. Acad. Sci. USA* **2015**, *112*, 3973.
83. Takemoto, Y.; Mitsuhashi, W.; Murakami, R.; Konishi, H.; Miyamoto, K. *J. Virol.* **2008**, *82*, 12406.
84. Frederiksen, R. F.; Paspaliari, D. K.; Larsen, T.; Storgaard, B. G.; Larsen, M. H.; Ingmer, H.; Palcic, M. M.; Leisner, J. J. *Microbiol.* **2013**, *159*, 833.
85. Wong, E.; Vaaje-Kolstad, G.; Ghosh, A.; Hurtado-Guerrero, R.; Konarev, P. V.; Ibrahim, A. F. M.; Svergun, D. I.; Eijsink, V. G. H.; Chatterjee, N. S.; van Aalten, D. M. F. *PLoS Pathog.* **2012**, *8*, e1002373.
86. Bhowmick, R.; Ghosal, A.; Das, B.; Koley, H.; Saha, D. R.; Ganguly, S.; Nandy, R. K.; Bhadra, R. K.; Chatterjee, N. S. *Infect. Immun.* **2008**, *76*, 4968.
87. Loose, J. S. M.; Forsberg, Z.; Fraaije, M. W.; Eijsink, V. G. H.; Vaaje-Kolstad, G. *FEBS Lett.* **2014**, *588*, 3435.
88. Paspaliari, D. K.; Loose, J. S. M.; Larsen, M. H.; Vaaje-Kolstad, G. *FASEB J.* **2015**, *282*, 921.
89. Vaaje-Kolstad, G.; Horn, S. J.; van Aalten, D. M. F.; Synstad, B.; Eijsink, V. G. H. *J. Biol. Chem.* **2005**, *280*, 28492.
90. Bennati-Granier, C.; Garajova, S.; Champion, C.; Grisel, S.; Haon, M.; Zhou, S.; Fanuel, M.; Ropartz, D.; Rogniaux, H.; Gimbert, I.; Record, E.; Berrin, J.-G. *Biotechnol. Biofuels* **2015**, *8*, 90.
91. Crouch, L. I.; Labourel, A.; Walton, P. H.; Davies, G. J.; Gilbert, H. J. *J. Biol. Chem.* **2016**, *291*, 7439.

92. Nekiunaite, L.; Petrović, D. M.; Westereng, B.; Vaaje-Kolstad, G.; Hachem, M. A.; Várnai, A.; Eijsink, V. G. H. *FEBS Lett.* **2016**, *590*, 3346.
93. Hansson, H.; Karkehabadi, S.; Mikkelsen, N.; Douglas, N. R.; Kim, S.; Lam, A.; Kaper, T.; Kelemen, B.; Meier, K. K.; Jones, S. M.; Solomon, E. I.; Sandgren, M. *J. Biol. Chem.* **2017**, *292*, 19099.
94. Courtade, G.; Forsberg, Z.; Heggset, E. B.; Eijsink, V. G. H.; Aachmann, F. L. *J. Biol. Chem.* **2018**, *293*, 13006.
95. Mutahir, Z.; Mekasha, S.; Loose, J. S. M.; Abbas, F.; Vaaje-Kolstad, G.; Eijsink, V. G. H.; Forsberg, Z. *FEBS Lett.* **2018**, *592*, 2562.
96. Frandsen, K. E. H.; Leggio, Lo, L. *IUCrJ* **2016**, *3*, 448.
97. Vaaje-Kolstad, G.; Forsberg, Z.; Loose, J. S.; Bissaro, B.; Eijsink, V. G. *Curr. Opin. Struc. Biol.* **2017**, *44*, 67.
98. Zhang, L.; Koay, M.; Maher, M. J.; Xiao, Z.; Wedd, A. G. *J. Am. Chem. Soc.* **2006**, *128*, 5834.
99. Lawton, T. J.; Kenney, G. E.; Hurley, J. D.; Rosenzweig, A. C. *Biochemistry* **2016**, *55*, 2278.
100. Lieberman, R. L.; Rosenzweig, A. C. *Nature* **2005**, *434*, 177.
101. Balasubramanian, R.; Smith, S. M.; Rawat, S.; Yatsunyk, L. A.; Stemmler, T. L.; Rosenzweig, A. C. *Nature* **2010**, *465*, 115.
102. Lawton, T. J.; Ham, J.; Sun, T.; Rosenzweig, A. C. *Proteins* **2014**, *82*, 2263.
103. Wu, M.; Beckham, G. T.; Larsson, A. M.; Ishida, T.; Kim, S.; Payne, C. M.; Himmel, M. E.; Crowley, M. F.; Horn, S. J.; Westereng, B.; Igarashi, K.; Samejima, M.; Ståhlberg, J.; Eijsink, V. G. H.; Sandgren, M. *J. Biol. Chem.* **2013**, *288*, 12828.
104. Petrovi, D. M.; Bissaro, B.; Chylenski, P.; Skaugen, M.; Sørlie, M.; Jensen, M. S.; Aachmann, F. L.; Courtade, G.; Várnai, A.; Eijsink, V. G. H. *Protein Sci.* **2018**, *27*, 1636.
105. Garajova, S.; Mathieu, Y.; Beccia, M. R.; Bennati-Granier, C.; Biaso, F.; Fanuel, M.; Ropartz, D.; Guigliarelli, B.; Record, E.; Rogniaux, H.; Henrissat, B.; Berrin, J.-G. *Sci. Rep.* **2016**, *17*, 28276.
106. Span, E. A.; Suess, D. L. M.; Deller, M. C.; Britt, R. D.; Marletta, M. A. *ACS Chem. Biol.* **2017**, *12*, 1095.
107. Hemsworth, G. R.; Taylor, E. J.; Kim, R. Q.; Gregory, R. C.; Lewis, S. J.; Turkenburg, J. P.; Parkin, A.; Davies, G. J.; Walton, P. H. *J. Am. Chem. Soc.* **2013**, *135*, 6069.
108. Chaplin, A. K.; Wilson, M. T.; Hough, M. A.; Svistunenko, D. A.; Hemsworth, G. R.; Walton, P. H.; Vijgenboom, E.; Worrall, J. A. R. *J. Biol. Chem.* **2016**, *291*, 12838.
109. Gudmundsson, M.; Kim, S.; Wu, M.; Ishida, T.; Momeni, M. H.; Vaaje-Kolstad, G.; Lundberg, D.; Royant, A.; Ståhlberg, J.; Eijsink, V. G. H.; Beckham, G. T.; Sandgren, M. *J. Biol. Chem.* **2014**, *289*, 18782.
110. Aachmann, F. L.; Sørlie, M.; Skjåk-Bræk, G.; Eijsink, V. G. H.; Vaaje-Kolstad, G. *Proc. Natl. Acad. Sci. USA* **2012**, *109*, 18779.
111. Gregory, R. C.; Hemsworth, G. R.; Turkenburg, J. P.; Hart, S. J.; Walton, P. H.; Davies, G. J. *Dalton Trans.* **2016**, *45*, 16904.

112. Forsberg, Z.; Mackenzie, A. K.; Sorlie, M.; Rohr, A. K.; Helland, R.; Arvai, A. S.; Vaaje-Kolstad, G.; Eijsink, V. G. H. *Proc. Natl. Acad. Sci.* **2014**, *111*, 8446.
113. Hemsworth, G. R.; Davies, G. J.; Walton, P. H. *Curr. Opin. Struc. Biol.* **2013**, *23*, 660.
114. Forsberg, Z.; Røhr, Å. K.; Mekasha, S.; Andersson, K. K.; Eijsink, V. G. H.; Vaaje-Kolstad, G.; Sørlie, M. *Biochemistry* **2014**, *53*, 1647.
115. Vu, V. V.; Beeson, W. T.; Span, E. A.; Farquhar, E. R.; Marletta, M. A. *Proc. Natl. Acad. Sci.* **2014**, *111*, 13822.
116. Isaksen, T.; Westereng, B.; Aachmann, F. L.; Agger, J. W.; Kracher, D.; Kittl, R.; Ludwig, R.; Haltrich, D.; Eijsink, V. G. H.; Horn, S. J. *J. Biol. Chem.* **2014**, *289*, 2632.
117. Fanuel, M.; Garajova, S.; Ropartz, D.; McGregor, N.; Brumer, H.; Rogniaux, H.; Berrin, J.-G. *Biotechnol. Biofuels* **2017**, *10*, 63.
118. Frandsen, K. E. H.; Simmons, T. J.; Dupree, P.; Poulsen, J.-C. N.; Hemsworth, G. R.; Ciano, L.; Johnston, E. M.; Tovborg, M.; Johansen, K. S.; Freiesleben, von, P.; Marmuse, L.; Fort, S.; Cottaz, S.; Driguez, H.; Henrissat, B.; Lenfant, N.; Tuna, F.; Baldansuren, A.; Davies, G. J.; Leggio, Lo, L.; Walton, P. H. *Nat. Chem. Biol.* **2016**, *12*, 298.
119. Simmons, T. J.; Frandsen, K. E. H.; Ciano, L.; Tryfona, T.; Lenfant, N.; Poulsen, J. C.; Wilson, L. F. L.; Tandrup, T.; Tovborg, M.; Schnorr, K.; Johansen, K. S.; Henrissat, B.; Walton, P. H.; Leggio, Lo, L.; Dupree, P. *Nat. Commun.* **2017**, *8*, 1064.
120. Agger, J. W.; Isaksen, T.; Varnai, A.; Vidal-Melgosa, S.; Willats, W. G. T.; Ludwig, R.; Horn, S. J.; Eijsink, V. G. H.; Westereng, B. *Proc. Natl. Acad. Sci.* **2014**, *111*, 6287.
121. Jagadeeswaran, G.; Gainey, L.; Prade, R.; Mort, A. J. *Appl. Microbiol. Biot.* **2016**, *100*, 4535.
122. Kojima, Y.; Várnai, A.; Ishida, T.; Sunagawa, N.; Petrovic, D. M.; Igarashi, K.; Jellison, J.; Goodell, B.; Alfredsen, G.; Westereng, B.; Eijsink, V. G. H.; Yoshida, M. *Appl. Environ. Microbiol.* **2016**, *82*, 6557.
123. Ladevèze, S.; Haon, M.; Villares, A.; Cathala, B.; Grisel, S.; Herpoël-Gimbert, I.; Henrissat, B.; Berrin, J.-G. *Biotechnol. Biofuels* **2017**, *10*, 215.
124. Li, X.; Beeson, W. T., IV; Phillips, C. M.; Marletta, M. A.; Cate, J. H. D. *Structure* **2012**, *20*, 1051.
125. Wu, M.; Beckham, G. T.; Larsson, A. M.; Ishida, T.; Kim, S.; Payne, C. M.; Himmel, M. E.; Crowley, M. F.; Horn, S. J.; Westereng, B.; Igarashi, K.; Samejima, M.; Ståhlberg, J.; Eijsink, V. G. H.; Sandgren, M. *J. Biol. Chem.* **2013**, *288*, 12828.
126. Vu, V. V.; Beeson, W. T., IV; Phillips, C. M.; Cate, J. H. D.; Marletta, M. A. *J. Am. Chem. Soc.* **2014**, *136*, 562.
127. Borisova, A. S.; Isaksen, T.; Dimarogona, M.; Kognole, A. A.; Mathiesen, G.; Várnai, A.; Røhr, Å. K.; Payne, C. M.; Sørlie, M.; Sandgren, M.; Eijsink, V. G. H. *J. Biol. Chem.* **2015**, *290*, 22955.
128. Westereng, B.; Cannella, D.; Agger, J. W.; Jørgensen, H.; Andersen, M. L.; Eijsink, V. G. H.; Felby, C. *Sci. Rep.* **2015**, *5*, 18561.

129. llers, K. B. M. O.; Frigaard, N. U.; Jensen, P. E.; Bjerrum, M. J.; Johansen, K. S.; Cannella, D.; Felby, C. *Nat. Commun.* **2016**, *7*, 1.
130. Möllers, K. B.; Mikkelsen, H.; Simonsen, T. I.; Cannella, D.; Johansen, K. S.; Bjerrum, M. J.; Felby, C. *Carbohydr. Res.* **2017**, *448*, 182.
131. Dumonceaux, T.; Bartholomew, K.; Valeanu, L.; Charles, T.; Archibald, F. *Enzyme Microb. Tech.* **2001**, *29*, 478.
132. Vanden Wymelenberg, A.; Gaskell, J.; Mozuch, M.; Sabat, G.; Ralph, J.; Skyba, O.; Mansfield, S. D.; Blanchette, R. A.; Martinez, D.; Grigoriev, I.; Kersten, P. J.; Cullen, D. *Appl. Environ. Microbiol.* **2010**, *76*, 3599.
133. Hori, C.; Igarashi, K.; Katayama, A.; Samejima, M. *FEMS Microbiol. Lett.* **2011**, *321*, 14.
134. Phillips, C. M.; Iavarone, A. T.; Marletta, M. A. *J. Proteome Res.* **2011**, *10*, 4177.
135. Yakovlev, I.; Vaaje-Kolstad, G.; Hietala, A. M.; Stefańczyk, E.; Solheim, H.; Fossdal, C. G. *Appl. Microbiol. Biot.* **2012**, *95*, 979.
136. Westermark, U.; Eriksson, K.-E. *Acta Chem. Scand. B* **1974**, *28*, 209.
137. Langston, J. A.; Shaghasi, T.; Abbate, E.; Xu, F.; Vlasenko, E.; Sweeney, M. D. *Appl. Environ. Microbiol.* **2011**, *77*, 7007.
138. Phillips, C. M.; Beeson, W. T., IV; Cate, J. H. D.; Marletta, M. A. *ACS Chem. Biol.* **2011**, *6*, 1399.
139. Tan, T.-C.; Kracher, D.; Gandini, R.; Sygmund, C.; Kittl, R.; Haltrich, D.; Hällberg, B. M.; Ludwig, R.; Divne, C. *Nat. Commun.* **2015**, *6*, 7542.
140. Gardner, J. G.; Crouch, L.; Labourel, A.; Forsberg, Z.; Bukhman, Y. V.; Vaaje-Kolstad, G.; Gilbert, H. J.; Keating, D. H. *Mol. Microbiol.* **2014**, *94*, 1121.
141. Loose, J. S. M.; Forsberg, Z.; Kracher, D.; Scheiblbrandner, S.; Ludwig, R.; Eijsink, V. G. H.; Vaaje-Kolstad, G. *Protein Sci.* **2016**, *25*, 2175.
142. Frommhagen, M.; Mutte, S. K.; Westphal, A. H.; Koetsier, M. J.; Hinz, S. W. A.; Visser, J.; Vincken, J.-P.; Weijers, D.; Berkel, W. J. H.; Gruppen, H.; Kabel, M. A. *Biotechnol. Biofuels* **2017**, *10*, 121.
143. Brenelli, L.; Squina, F. M.; Felby, C.; Cannella, D. *Biotechnol. Biofuels* **2018**, *11*, 10.
144. Kracher, D.; Scheiblbrandner, S.; Felice, A. K. G.; Breslmayr, E.; Preims, M.; Ludwicka, K.; Haltrich, D.; Eijsink, V. G. H.; Ludwig, R. *Science* **2016**, *352*, 1098.
145. Kjaergaard, C. H.; Qayyum, M. F.; Wong, S. D.; Xu, F.; Hemsworth, G. R.; Walton, D. J.; Young, N. A.; Davies, G. J.; Walton, P. H.; Johansen, K. S.; Hodgson, K. O.; Hedman, B.; Solomon, E. I. *Proc. Natl. Acad. Sci. USA* **2014**, *111*, 8797.
146. Kim, S.; Ståhlberg, J.; Sandgren, M.; Paton, R. S.; Beckham, G. T. *Proc. Natl. Acad. Sci. USA* **2014**, *111*, 149.
147. O'Dell, W. B.; Agarwal, P. K.; Meilleur, F. *Angew. Chem.* **2016**, *129*, 785.
148. Bissaro, B.; Røhr, Å. K.; Müller, G.; Chylenski, P.; Skaugen, M.; Forsberg, Z.; Horn, S. J.; Vaaje-Kolstad, G.; Eijsink, V. G. H. *Nat. Chem. Biol.* **2017**, *13*, 1123.
149. Kuusk, S.; Bissaro, B.; Kuusk, P.; Forsberg, Z.; Eijsink, V. G. H.; Sørlie, M.; Väljamäe, P. *J. Biol. Chem.* **2018**, *293*, 523.

150. Müller, G.; Chylenski, P.; Bissaro, B.; Eijsink, V. G. H.; Horn, S. J. *Biotechnol. Biofuels* **2018**, *11*, 209.
151. Hangasky, J. A.; Iavarone, A. T.; Marletta, M. A. *Proc. Natl. Acad. Sci.* **2018**, *115*, 4915.
152. Gao, J.; Thomas, D. A.; Sohn, C. H.; Beauchamp, J. L. *J. Am. Chem. Soc.* **2013**, *135*, 10684.
153. Hedegård, E. D.; Ryde, U. *JBIC J. Biol. Inorg. Chem.* **2017**, *22*, 1029.
154. Bertini, L.; Breglia, R.; Lambrughi, M.; Fantucci, P.; de Gioia, L.; Borsari, M.; Sola, M.; Bortolotti, C. A.; Bruschi, M. *Inorg. Chem.* **2018**, *57*, 86.
155. Kracher, D.; Andlar, M.; Furtmüller, P. G.; Ludwig, R. *J. Biol. Chem.* **2018**, *293*, 1676.
156. Hangasky, J. A.; Marletta, M. A. *Biochemistry* **2018**, *57*, 3191.
157. Wang, B.; Johnston, E. M.; Li, P.; Shaik, S.; Davies, G. J.; Walton, P. H.; Rovira, C. *ACS Catal.* **2018**, *8*, 1346.
158. Hedegård, E. D.; Ryde, U. *Chem. Sci.* **2018**, *9*, 3866.
159. Dhar, D.; Tolman, W. B. *J. Am. Chem. Soc.* **2015**, *137*, 1322.
160. Bacik, J.-P.; Mekasha, S.; Forsberg, Z.; Kovalevsky, A. Y.; Vaaje-Kolstad, G.; Eijsink, V. G. H.; Nix, J. C.; Coates, L.; Cuneo, M. J.; Unkefer, C. J.; Chen, J. C. H. *Biochemistry* **2017**, *56*, 2529.
161. Beeson, W. T., IV; Vu, V. V.; Span, E. A.; Phillips, C. M.; Marletta, M. A. *Annu. Rev. Biochem.* **2015**, *84*, 923.
162. Walton, P. H.; Davies, G. J. *Curr. Opin. Chem. Biol.* **2016**, *31*, 195.
163. Meier, K. K.; Jones, S. M.; Kaper, T.; Hansson, H.; Koetsier, M. J.; Karkehabadi, S.; Solomon, E. I.; Sandgren, M.; Kelemen, B. *Chem. Rev.* **2017**, *118*, 2593.
164. Ciano, L.; Davies, G. J.; Tolman, W. B.; Walton, P. H. *Nature Catalysis* **2018**, *1*, 571.
165. Itoh, S. *Acc. Chem. Res.* **2015**, *48*, 2066.
166. Cramer, C. J.; Tolman, W. B. *Acc. Chem. Res.* **2007**, *40*, 601.
167. Mirica, L.; Ottenwaelder, X.; Stack, T. D. P. *Chem. Rev.* **2004**, *104*, 1013.
168. Hatcher, L. Q.; Karlin, K. D. *J. Biol. Inorg. Chem.* **2004**, *9*, 669.
169. Liu, J. J.; Diaz, D. E.; Quist, D. A.; Karlin, K. D. *Isr. J. Chem.* **2016**, *56*, 738.
170. Elwell, C. E.; Gagnon, N. L.; Neisen, B. D.; Dhar, D.; Spaeth, A. D.; Yee, G. M.; Tolman, W. B. *Chem. Rev.* **2017**, *117*, 2059.
171. Lewis, E. A.; Tolman, W. B. *Chem. Rev.* **2004**, *104*, 1047.
172. Comba, P.; Haaf, C.; Helmle, S.; Karlin, K. D.; Pandian, S.; Waleska, A. *Inorg. Chem.* **2012**, *51*, 2841.
173. Fujisawa, K.; Tanaka, M.; Moro-oka, Y. *J. Am. Chem. Soc.* **1994**, *116*, 12079.
174. Würtele, C.; Gaoutchenova, E.; Harms, K.; Holthausen, M. C.; Sundermeyer, J.; Schindler, S. *Angew. Chem. Int. Ed. Engl.* **2006**, *45*, 3867.
175. Chen, P.; Root, D. E.; Campochiaro, C.; Fujisawa, K.; Solomon, E. I. *J. Am. Chem. Soc.* **2003**, *125*, 466.
176. Sánchez-Eguía, B. N.; Flores-Alamo, M.; Orio, M.; Castillo, I. *Chem. Comm.* **2015**, *51*, 11134.

177. Spencer, D. J. E.; Aboelella, N. W.; Reynolds, A. M.; Holland, P. L.; Tolman, W. B. *J. Am. Chem. Soc.* **2002**, *124*, 2108.
178. Spencer, D. J. E.; Reynolds, A. M.; Holland, P. L.; Jazdzewski, B. A.; Duboc-Toia, C.; Le Pape, L.; Yokota, S.; Tachi, Y.; Itoh, S.; Tolman, W. B. *Inorg. Chem.* **2002**, *41*, 6307.
179. Aboelella, N. W.; Lewis, E. A.; Reynolds, A. M.; Brennessel, W. W.; Cramer, C. J.; Tolman, W. B. *J. Am. Chem. Soc.* **2002**, *124*, 10660.
180. Aboelella, N. W.; Kryatov, S. V.; Gherman, B. F.; Brennessel, W. W.; Young, V. G.; Sarangi, R.; Rybak-Akimova, E. V.; Hodgson, K. O.; Hedman, B.; Solomon, E. I.; Cramer, C. J.; Tolman, W. B. *J. Am. Chem. Soc.* **2004**, *126*, 16896.
181. Weitzer, M.; Schindler, S.; Brehm, G.; Schneider, S.; Hörmann, E.; Jung, B.; Kaderli, S.; Zuberbühler, A. D. *Inorg. Chem.* **2003**, *42*, 1800.
182. Schatz, M.; Raab, V.; Foxon, S. P.; Brehm, G.; Schneider, S.; Reiher, M.; Holthausen, M. C.; Sundermeyer, J.; Schindler, S. *Angew. Chem. Int. Ed. Engl.* **2004**, *43*, 4360.
183. Maiti, D.; Fry, H. C.; Woertink, J. S.; Vance, M. A.; Solomon, E. I.; Karlin, K. D. *J. Am. Chem. Soc.* **2007**, *129*, 264.
184. Peterson, R. L.; Himes, R. A.; Kotani, H.; Suenobu, T.; Tian, L.; Siegler, M. A.; Solomon, E. I.; Fukuzumi, S.; Karlin, K. D. *J. Am. Chem. Soc.* **2011**, *133*, 1702.
185. Lee, J. Y.; Peterson, R. L.; Ohkubo, K.; Garcia-Bosch, I.; Himes, R. A.; Woertink, J.; Moore, C. D.; Solomon, E. I.; Fukuzumi, S.; Karlin, K. D. *J. Am. Chem. Soc.* **2014**, *136*, 9925.
186. Fujii, T.; Hirota, S.; Yamaguchi, S.; Masuda, H. *Dalton Trans.* **2008**, 164.
187. Kim, S.; Lee, J. Y.; Cowley, R. E.; Ginsbach, J. W.; Siegler, M. A.; Solomon, E. I.; Karlin, K. D. *J. Am. Chem. Soc.* **2015**, *137*, 2796.
188. Ginsbach, J. W.; Peterson, R. L.; Cowley, R. E.; Karlin, K. D.; Solomon, E. I. *Inorg. Chem.* **2013**, *52*, 12872.
189. Donoghue, P. J.; Gupta, A. K.; Boyce, D. W.; Cramer, C. J.; Tolman, W. B. *J. Am. Chem. Soc.* **2010**, *132*, 15869.
190. Kunishita, A.; Kubo, M.; Sugimoto, H.; Ogura, T.; Sato, K.; Takui, T.; Itoh, S. *J. Am. Chem. Soc.* **2009**, *131*, 2788.
191. Tano, T.; Okubo, Y.; Kunishita, A.; Kubo, M.; Sugimoto, H.; Fujieda, N.; Ogura, T.; Itoh, S. *Inorg. Chem.* **2013**, *52*, 10431.
192. Maiti, D.; Lee, D.-H.; Gaoutchenova, K.; Wuertele, C.; Holthausen, M. C.; Sarjeant, A. A. N.; Sundermeyer, J.; Schindler, S.; Karlin, K. D. *Angew. Chem. Int. Ed. Engl.* **2008**, *47*, 82.
193. Bhadra, M.; Lee, J. Y. C.; Cowley, R. E.; Kim, S.; Siegler, M. A.; Solomon, E. I.; Karlin, K. D. *J. Am. Chem. Soc.* **2018**, *140*, 9042.
194. Kunishita, A.; Ertem, M. Z.; Okubo, Y.; Tano, T.; Sugimoto, H.; Ohkubo, K.; Fujieda, N.; Fukuzumi, S.; Cramer, C. J.; Itoh, S. *Inorg. Chem.* **2012**, *51*, 9465.
195. Pirovano, P.; Magherusan, A. M.; McGlynn, C.; Ure, A.; Lynes, A.; McDonald, A. R. *Angew. Chem. Int. Ed. Engl.* **2014**, *53*, 5946.
196. Nam, W. *Acc. Chem. Res.* **2015**, *48*, 2415.

197. Izzet, G.; Zeitouny, J.; Akdas-Killig, H.; Frapart, Y.; Menage, S.; Douziech, B.; Jabin, I.; Le Mest, Y.; Reinaud, O. *J. Am. Chem. Soc.* **2008**, *130*, 9514.
198. Fujii, T.; Yamaguchi, S.; Funahashi, Y.; Ozawa, T.; Tosha, T.; Kitagawa, T.; Masuda, H. *Chem. Comm.* **2006**, 4428.
199. Yamaguchi, S.; Masuda, H. *Sci. Tech. Adv. Mater.* **2005**, *6*, 34.
200. Yamaguchi, S.; Nagatomo, S.; Kitagawa, T.; Funahashi, Y.; Ozawa, T.; Jitsukawa, K.; Masuda, H. *Inorg. Chem.* **2003**, *42*, 6968.
201. Kunishita, A.; Kubo, M.; Ishimaru, H.; Ogura, T.; Sugimoto, H.; Itoh, S. *Inorg. Chem.* **2008**, *47*, 12032.
202. Maiti, D.; Lucas, H. R.; Sarjeant, A. A. N.; Karlin, K. D. *J. Am. Chem. Soc.* **2007**, *129*, 6998.
203. Maiti, D.; Narducci Sarjeant, A. A.; Karlin, K. D. *Inorg. Chem.* **2008**, *47*, 8736.
204. Kim, S.; Saracini, C.; Siegler, M. A.; Drichko, N.; Karlin, K. D. *Inorg. Chem.* **2012**, *51*, 12603.
205. Kakuda, S.; Peterson, R. L.; Ohkubo, K.; Karlin, K. D.; Fukuzumi, S. *J. Am. Chem. Soc.* **2013**, *135*, 6513.
206. Kim, S.; Ginsbach, J. W.; Lee, J. Y.; Peterson, R. L.; Liu, J. J.; Siegler, M. A.; Sarjeant, A. A. N.; Solomon, E. I.; Karlin, K. D. *J. Am. Chem. Soc.* **2015**, *137*, 2867.
207. Kodera, M.; Hirota, S.; Kita, T.; Miura, I.; Nakayama, N.; Kawata, T.; Kano, K. *J. Am. Chem. Soc.* **2001**, *123*, 7715.
208. Choi, Y. J.; Cho, K.-B.; Kubo, M.; Ogura, T.; Karlin, K. D.; Cho, J.; Nam, W. *Dalton Trans.* **2011**, *40*, 2234.
209. Chaloner, L.; Khomutovskaya, A.; Thomas, F.; Ottenwaelder, X. *Dalton Trans.* **2016**, *45*, 11109.
210. Kamachi, T.; Lee, Y.-M.; Nishimi, T.; Cho, J.; Yoshizawa, K.; Nam, W. *J. Phys. Chem. A* **2008**, *112*, 13102.
211. Ohtsu, H.; Itoh, S.; Nagatomo, S.; Kitagawa, T.; Ogo, S.; Watanabe, Y.; Fukuzumi, S. *Inorg. Chem.* **2001**, *40*, 3200.
212. Cheruzel, L. E.; Cecil, M. R.; Edison, S. E.; Mashuta, M. S.; Baldwin, M. J.; Buchanan, R. M. *Inorg. Chem.* **2006**, *45*, 3191.
213. Biswas, S.; Dutta, A.; Debnath, M.; Dolai, M.; Das, K. K.; Ali, M. *Dalton Trans.* **2013**, *42*, 13210.
214. Wada, A.; Harata, M.; Hasegawa, K.; Jitsukawa, K.; Masuda, H.; Mukai, M.; Kitagawa, T.; Einaga, H. *Angew. Chem. Int. Ed. Engl.* **1998**, *37*, 798.
215. Chen, P.; Fujisawa, K.; Solomon, E. I. *J. Am. Chem. Soc.* **2000**, *122*, 10177.
216. Yamaguchi, S.; Wada, A.; Nagatomo, S.; Kitagawa, T.; Jitsukawa, K.; Masuda, H. *Chem. Lett.* **2004**, *33*, 1556.
217. Ward, T. R. *Acc. Chem. Res.* **2011**, *44*, 47.
218. Schwizer, F.; Okamoto, Y.; Heinisch, T.; Gu, Y.; Pellizzoni, M. M.; Lebrun, V.; Reuter, R.; Köhler, V.; Lewis, J. C.; Ward, T. R. *Chem. Rev.* **2018**, *118*, 142.
219. Mann, S. I.; Heinisch, T.; Ward, T. R.; Borovik, A. S. *J. Am. Chem. Soc.* **2017**, *139*, 17289.

220. Okuno, T.; Ohba, S.; Nishida, Y. *Polyhedron* **1997**, *16*, 3765.
221. Champloy, F.; Benali-Chérif, N.; Bruno, P.; Blain, I.; Pierrot, M.; Réglier, M.; Michalowicz, A. *Inorg. Chem.* **1998**, *37*, 3910.
222. Fujii, T.; Naito, A.; Yamaguchi, S.; Wada, A.; Funahashi, Y.; Jitsukawa, K.; Nagatomo, S.; Kitagawa, T.; Masuda, H. *Chem. Comm.* **2003**, 2700.
223. Yamaguchi, S.; Suzuki, A.; Togawa, M.; Nishibori, M.; Yahiro, H. *ACS Catal.* **2018**, *8*, 2645.
224. Concia, A. L.; Beccia, M. R.; Orio, M.; Ferre, F. T.; Scarpellini, M.; Biaso, F.; Guigliarelli, B.; Réglier, M.; Simaan, A. J. *Inorg. Chem.* **2017**, *56*, 1023.
225. Hemsworth, G. R.; Johnston, E. M.; Davies, G. J.; Walton, P. H. *Trends Biotechnol.* **2015**, *33*, 747.
226. Maiti, D.; Sarjeant, A. A. N.; Karlin, K. D. *J. Am. Chem. Soc.* **2007**, *129*, 6720.
227. Osako, T.; Nagatomo, S.; Tachi, Y.; Kitagawa, T.; Itoh, S. *Angew. Chem. Int. Ed. Engl.* **2002**, *41*, 4325.
228. Ghattas, W.; Giorgi, M.; Mekmouche, Y.; Tanaka, T.; Rockenbauer, A.; Réglier, M.; Hitomi, Y.; Simaan, A. J. *Inorg. Chem.* **2008**, *47*, 4627.
229. Oikawa, S.; Shosuke K. *Biochemistry* **1996**, *35*, 4584.
230. Park, J.-H.; Gopishetty, S.; Szewczuk, L. M.; Troxel, A. B.; Harvey, R. G.; Penning, T. M. *Chem. Res. Toxicol.* **2005**, *18*, 1026.
231. Gagnon, N.; Tolman, W. B. *Acc. Chem. Res.* **2015**, *48*, 2126.
232. Réglier, M.; Amadeï, E.; Tadayoni, R.; Waegell, B. *J. Chem. Soc., Chem. Commun.* **1989**, 447.
233. *2-Oxoglutarate-Dependent Oxygenases*; Schofield, C., Hausinger, R., Eds.; Royal Society of Chemistry: Cambridge, 2015; pp. P001–P487.
234. Hong, S.; Huber, S. M.; Gagliardi, L.; Cramer, C. C.; Tolman, W. B. *J. Am. Chem. Soc.* **2007**, *129*, 14190.
235. Huber, S. M.; Ertem, M. Z.; Aquilante, F.; Gagliardi, L.; Tolman, W. B.; Cramer, C. J. *Chem. Eur. J.* **2009**, *15*, 4886.
236. Schröder, D.; Holthausen, M. C.; Schwarz, H. *J. Phys. Chem. B* **2004**, *108*, 14407.
237. Dietl, N.; van der Linde, C.; Schlangen, M.; Beyer, M. K.; Schwarz, H. *Angew. Chem. Int. Ed. Engl.* **2011**, *50*, 4966.
238. Donoghue, P. J.; Tehranchi, J.; Cramer, C. J.; Sarangi, R.; Solomon, E. I.; Tolman, W. B. *J. Am. Chem. Soc.* **2011**, *133*, 17602.
239. Dhar, D.; Yee, G. M.; Spaeth, A. D.; Boyce, D. W.; Zhang, H.; Dereli, B.; Cramer, C. J.; Tolman, W. B. *J. Am. Chem. Soc.* **2016**, *138*, 356.
240. Dhar, D.; Yee, G. M.; Markle, T. F.; Mayer, J. M.; Tolman, W. B. *Chem. Sci.* **2017**, *8*, 1075.
241. Wang, Y.; Xia, H.; Sun, K.; Wu, S.; Lu, W.; Xu, J.; Li, N.; Pei, K.; Zhu, Z.; Chen, W. *Chem. Eng. J.* **2016**, *304*, 1000.

Index